Plant Cell Biology

The Practical Approach Series

SERIES EDITORS

D. RICKWOOD

Department of Biology, University of Essex
Wivenhoe Park, Colchester, Essex CO4 3SQ, UK

B. D. HAMES

Department of Biochemistry and Molecular Biology
University of Leeds, Leeds LS2 9JT, UK

Affinity Chromatography
Anaerobic Microbiology
Animal Cell Culture (2nd edition)
Animal Virus Pathogenesis
Antibodies I and II
Behavioural Neuroscience
Biochemical Toxicology
Biological Data Analysis
Biological Membranes
Biomechanics—Materials
Biomechanics—Structures and Systems
Biosensors
Carbohydrate Analysis
Cell–Cell Interactions
The Cell Cycle
Cell Growth and Division
Cellular Calcium
Cellular Interactions in Development
Cellular Neurobiology
Centrifugation (2nd edition)
Clinical Immunology
Computers in Microbiology
Crystallization of Nucleic Acids and Proteins
Cytokines
The Cytoskeleton
Diagnostic Molecular Pathology I and II
Directed Mutagenesis
DNA Cloning I, II, and III
Drosophila
Electron Microscopy in Biology
Electron Microscopy in Molecular Biology

Electrophysiology

Enzyme Assays

Essential Developmental Biology

Essential Molecular Biology I and II

Experimental Neuroanatomy

Extracellular Matrix

Fermentation

Flow Cytometry (2nd edition)

Gas Chromatography

Gel Electrophoresis of Nucleic Acids (2nd edition)

Gel Electrophoresis of Proteins (2nd edition)

Gene Targeting

Gene Transcription

Genome Analysis

Glycobiology

Growth Factors

Haemopoiesis

Histocompatibility Testing

HPLC of Macromolecules

HPLC of Small Molecules

Human Cytogenetics I and II (2nd edition)

Human Genetic Disease Analysis

Immobilised Cells and Enzymes

Immunocytochemistry

In Situ Hybridization

Iodinated Density Gradient Media

Light Microscopy in Biology

Lipid Analysis

Lipid Modification of Proteins

Lipoprotein Analysis

Liposomes

Lymphocytes

Mammalian Cell Biotechnology

Mammalian Development

Medical Bacteriology

Medical Mycology

Microcomputers in Biochemistry

Microcomputers in Biology

Microcomputers in Physiology

Mitochondria

Molecular Genetic Analysis of Populations

Molecular Genetics of Yeast

Molecular Imaging in Neuroscience

Molecular Neurobiology

Molecular Plant Pathology I and II

Molecular Virology

Monitoring Neuronal Activity

Mutagenicity Testing

Neural Transplantation

Neurochemistry

Neuronal Cell Lines

NMR of Biological Macromolecules

Nucleic Acid and Protein Sequence Analysis

Nucleic Acid Hybridisation

Nucleic Acids Sequencing

Oligonucleotides and Analogues

Oligonucleotide Synthesis

PCR

Peptide Hormone Action

Peptide Hormone Secretion

Photosynthesis: Energy Transduction

Plant Cell Biology

Plant Cell Culture

Plant Molecular Biology

Plasmids (2nd edition)

Pollination Ecology

Postimplantation Mammalian Embryos

Preparative Centrifugation

Prostaglandins and Related Substances

Protein Architecture

Protein Blotting

Protein Engineering

Protein Function

Protein Phosphorylation

Protein Purification Applications

Protein Purification Methods

Protein Sequencing

Protein Structure

Protein Targeting

Proteolytic Enzymes

Radioisotopes in Biology

Receptor Biochemistry

Receptor–Effector Coupling

Receptor–Ligand Interactions

Ribosomes and Protein Synthesis

RNA Processing I and II

Signal Transduction

Solid Phase Peptide Synthesis

Spectrophotometry and Spectrofluorimetry

Steroid Hormones

Teratocarcinomas and Embryonic Stem Cells

Transcription Factors

Transcription and Translation

Tumour Immunobiology

Virology

Yeast

Plant Cell Biology

A Practical Approach

Edited by

N. HARRIS

Department of Biological Sciences,
University of Durham

and

K. J. OPARKA

Department of Cellular and Environmental Physiology,
Scottish Crop Research Institute,
Dundee

at

OXFORD UNIVERSITY PRESS

Oxford New York Tokyo

Oxford University Press, Walton Street, Oxford OX2 6DP

Oxford New York Toronto
Delhi Bombay Calcutta Madras Karachi
Kuala Lumpur Singapore Hong Kong Tokyo
Nairobi Dar es Salaam Cape Town
Melbourne Auckland Madrid

and associated companies in
Berlin Ibadan

Oxford is a trade mark of Oxford University Press

A Practical Approach ⓢ is a registered trade mark
of the Chancellor, Masters, and Scholars of the University of Oxford
trading as Oxford University Press

Published in the United States
by Oxford University Press Inc., New York

A catalogue record for this book is available from the British Library

Library of Congress Cataloging in Publication Data
Plant cell biology: a practical approach/edited by N. Harris and K. J. Oparka.
(The Practical approach series)
Includes bibliographical references and index.
1. Botanical microscopy—Technique. 2. Plant cytochemistry—Technique. 3. Plant calls and tissues. I. Harris, N. (Nicholas) II. Oparka, K. J. III. Series.
QK673.P58 1994 581.87'072—dc20 93–46555

ISBN 0 19 963398 3 (Hbk)
ISBN 0 19 963399 1 (Pbk)

Typeset by Cambrian Typesetters, Frimley, Surrey
Printed in Great Britain by
Information Press Ltd, Eynsham, Oxford

Preface

Modern plant cell biology encompasses several interrelated research fields, each of which contains a plethora of ever-expanding cellular, sub-cellular, and molecular techniques. When we set out to compile the recipes and protocols most commonly used by plant cell biologists it became apparent to us that we would have to be extremely selective if we were to produce a book that did not run into an inordinate number of chapters. On the one hand we wished to allow newcomers to plant cell biology access to the most useful 'tricks of the trade', while on the other we were keen to include new and exciting techniques which are making, or are about to make, a major impact on the field. In fact, we have attempted to do both within the scope allowed us by the publishers. This book therefore contains a mixture of tried and tested recipes in plant cell biology together with some of the new (at the time of writing) technical advances which are likely to become within the scope of many researchers.

The microscope (whether light, electron, fluorescence, or confocal) continues to be the workhorse of several plant cell biology techniques. Accordingly, the first four chapters of this book are devoted to methods and protocols for viewing, staining, and localizing cell components by a variety of microscopical methods. The subsequent three chapters deal with a range of methods for localizing specific nucleic acid sequences and proteins in plant cells using *in situ* hybridization and immunocytochemistry. Some of the most commonly used methods in protoplast research are given in Chapter 8, while techniques relating to cell wall analysis are given in Chapter 9. Chapters 10 to 12 deal with ways of studying the structure of the plant cytoskeleton, to isolate endo- and plasma-membrane components, and isolate and utilize intact chloroplasts and thylakoids. Mitochondria are not covered, they already have a whole volume to themselves elsewhere in this series, nor are several other components that do not have very specific 'plant' characteristics. Instead, we chose to use the remaining two chapters of the book to take a look at two quite different approaches to the measurement of ions and solutes within plant cells; ion-selective microelectrodes and the emerging technique of single-cell microsampling.

Inevitably we will have omitted someone's 'pet' technique, but we have tried to be as comprehensive as possible in our coverage of the subject within the page allocation. We are grateful to the several researchers who have prepared chapters for the book and allowed us access to the innermost secrets of their lab books. We would also like to thank Denton Prior for his painstaking contribution to the final preparation of *Plant Cell Biology: A Practical Approach.*

K. J. Oparka has also produced a 26 minute narrated educational video entitled *The living plant cell—an introduction to plant cell biology*. The video is particularly suitable for undergraduate teaching, containing several microscope images of plant cells and their organelles. For further information contact Mylnefield Research Services, Scottish Crop Research Institute, Invergowrie, Dundee DD2 5DA.

Durham
Dundee
August 1993

N. H.
K. J. O.

Contents

8. Plant protoplast techniques 177

L. C. Fowke and A. J. Cutler

Contributors

A. J. CUTLER
Plant Biotechnology Institute, National Research Council of Canada, 110 Gymnasium Place, Saskatoon, Saskatchewan, Canada S7N 0W9.

L. C. FOWKE
University of Saskatchewan, Department of Biology, Saskatoon, Saskatchewan, Canada S7N 0W0.

S. C. FRY
University of Edinburgh, Centre for Plant Science, King's Buildings, Edinburgh EH9 3JH, UK.

W. FRICKE
University of Wales, Bangor, School of Biological Sciences, Bangor, Gwynedd LL57 2UW, UK.

K. C. GOODBODY
John Innes Centre for Plant Science Research, Department of Cell Biology, Colney Lane, Norwich NR4 7UH, UK.

N. HARRIS
University of Durham, Department of Biological Sciences, Durham DH1 3LE, UK.

C. HAWES
Oxford Brookes University, School of Biological and Molecular Sciences, Gypsy Lane, Oxford OX3 0BP, UK.

J. S. HESLOP-HARRISON
John Innes Centre for Plant Science Research, Karyobiology Group, Colney Lane, Norwich NR4 7UH, UK.

P. HINDE
University of Wales, Bangor, School of Biological Sciences, Bangor, Gwynedd LL57 2UW, UK.

G. HINZ
Pflanzenphysiologisches Institut der Universität Göttingen, Cytologische Abteilung, Untere Karspule 2, D-37073 Göttingen, Germany.

A. R. LEITCH
University of London, Queen Mary and Westfield College, Mile End Road, London E1 4NS, UK.

C. W. LLOYD
John Innes Centre for Plant Science Research, Department of Cell Biology, Colney Lane, Norwich NR4 7UH, UK.

G. I. McFADDEN
University of Melbourne, Plant Cell Biology Research Centre, Parkville, Victoria 3052, Australia.

A. J. MILLER
Rothamstead Experimental Station, Biochemistry and Physiology Department, Harpenden, Herts AL5 2JQ, UK.

K. OBERBECK
Pflanzenphysiologisches Institut der Universität Göttingen, Cytologische Abteilung, Untere Karspule 2, D-37073 Göttingen, Germany.

K. J. OPARKA
Scottish Crop Research Institute, Department of Cellular and Environmental Physiology, Invergowrie, Dundee DD2 5DA, UK.

J. PRITCHARD
University of Wales, Bangor, School of Biological Sciences, Bangor, Gwynedd LL57 2UW, UK.

D. J. RAWLINS
John Innes Centre for Plant Science Research, Department of Cell Biology, Colney Lane, Norwich NR4 7UH, UK.

N. D. READ
University of Edinburgh, King's Buildings, Edinburgh, UK.

P. RICHARDSON
School of Biological Sciences, University of Wales, Bangor, Gwynedd LL57 2UW, UK.

C. ROBINSON
University of Warwick, Department of Biological Sciences, Coventry CV4 7AL, UK.

D. G. ROBINSON
Pflanzenphysiologisches Institut der Universität Göttingen, Cytologische Abteilung, Untere Karspule 2, D-37073 Göttingen, Germany.

T. SCHWARZACHER
John Innes Centre for Plant Science Research, Karyobiology Group, Colney Lane, Norwich NR4 7UH, UK.

P. J. SHAW
John Innes Centre for Plant Science Research, Department of Cell Biology, Colney Lane, Norwich NR4 7UH, UK.

J. SPENCE
University of Durham, Department of Biological Sciences, Durham DH1 3LE, UK.

D. TOMOS
University of Wales, Bangor, School of Biological Sciences, Bangor, Gwynedd LL57 2UW, UK.

Abbreviations

2-dGal	2-deoxygalactose
2-dGlc	2-deoxyglucose
2-dRib	2-deoxyribose
2-MeXyl	2-*O*-methylxylose
3-MeGlc	3-*O*-methylglucose
α-M	α-mannosidase
A_{240}	absorbance at 240 nm
A/D	analog/digital
AcBr	acetyl bromide
AGP	arabinogalactan protein complex
Ala	alanine
AM	acetoxymethyl
AMCA	7-amino-4-methyl-coumarin-3-acetic acid
ANS	1-anilino-2-naphtol-4-sulphonic acid
AP	acid phosphatase
APAAP	alkaline phosphate-antialkaline phosphatase
Ara	arabinose
as-ni-ATPase	anion-stimulated, nitrate-inhibited ATPase
ATP	adenosine triphosphate
ATPase	adenosine triphosphatase
BAW	butan-1-ol/acetic acid/water mixture
BCECF	2′,7′-bis-(2-carboxyethyl)-5-(and-6)-carboxyfluorescein
BCIP	5-bromo-4-chloro-3-indolyl phosphate
BSA	bovine serum albumin
CAT	catalase
CCR	NAD(P)H-dependent cytochrome c-reductase
CCV	clathrin coated vesicles
cDNA	complementary deoxyribonucleic acid
CF	5(6) carboxyfluorescein
CPD	critical-point drying
cpm	counts per minute
cps	counts per second
cRNA	complementary ribonucleic acid
cs-PPase	cation-stimulated pyrophosphatase
cs-vi-ATPase	cation-stimulated, vanadate-inhibited ATPase
CTP	cytidine triphosphate
DAB	3,3′-diaminobenzidene tetrahydrachloride
DAPI	4′,6-diaminino-2-phenylindole
dATP	deoxyadenosine triphosphate
dCTP	deoxycytidine triphosphate
DEAE	diethylaminoethyl
DEPC	diethylpyrocarbonate
dGTP	deoxyguanosine triphosphate

DIC	differential interference contrast (Nomarski)
DIG	digoxygenin
$DiOC_6$	3,3′-dihexyloxacarboxyanine iodide
DMF	dimethylformamide
DMSO	dimethylsulphoxide
DNA	deoxyribonucleic acid
DNase	deoxyribonuclease
dNTP	deoxynucleotide triphosphate
DTT	dithiothreitol
dTTP	deoxythymidine triphosphate
dUTP	deoxyuridine triphosphate
DW	distilled water
EDTA	ethylenediamine tetra-acetic acid
EGTA	ethyleneglyco-bis(b-aminoethyl ether)*N*,*N*,*N′*,*N′*-tetra-acetic acid
EM	electron microscope
EMF	electro-motive force
EPW	ethyl acetic acid/pyridine/water mixture
ER	endoplasmic reticulum
F	fluorescein
FA-BSA	fatty-acid-free bovine serum albumin
FAD	flavin adenine dinucleotide
Fc	crystallizable fragments
FDA	fluorescein diacetate
Fer	feruloyl ester
FF	freeze-fracture
FITC	fluorescein isothiocyanate
FRAP	fluorescence recovery after photobleaching
Fuc	fucose
Gal	galactose
GalA	galacturonic acid
$GalA_2$	galacturonosyl-α-(1→4)-galacturonic acid
$GalA_3$	galacturonosyl-α-(1→4)-galacturonosyl-α-(1→4)-galacturonic acid
GalNAc	*N*-acetylgalactosamine
GApp	golgi apparatus
GC	gas chromatography
GC–MS	gas chromatography–mass spectrometry
Glc	glucose
Glc_6PDH	glucose-6-phosphate dehydrogenase
GlcA	glucuronic acid
GlcNAc	*N*,-acetyl-glucosamine
Glu	glutamic acid
Gly	glycine
GOT	glutamate oxaloacetate transaminase
GSI/II	glucan synthase I or II
GTP	guanosine triphosphate
GUS	glucuronidase
Hepes	*N*-(2-hydroxyethyl)piperazine-*N′*-(2-ethane sulphonic acid)

His	histidine
HK	hexokinase
HOAc	glacial acetic acid
HPLC	high performance liquid chromatography
HPTS	8-hydroxypyrene-1,3,6,-trisulphonic acid, trisodium salt
Hyp	hydroxyproline
IDPase	inosine diphosphate
Idt	isodityrosine
Ig	immunoglobin
kb	kilobase
kBq	kilobecquerel
kDa	kilodaltons
K_{dis}	dissociation constant
KDO	2-keto-3-deoxy-manno-octulosonic acid
kPa	kilopascals
kV	kilovolts
LM	light microscopy
LR	London Resin
LVDT	linear voltage displacement transducer
LYCH	Lucifer Yellow CH
M	ionic strength (mol $litre^{-1}$)
Man	mannose
ManNAc	*N*-acetylmannosamine
MDH	malate dehydrogenase
Mes	2-(*N*-morpholino)ethanesulphonic acid
Mg-ATP	(Mg)-adenosine 5′-triphosphate
Mn(III)TPPC	5,10,15,20-tetraphenyl-21H,23H-porphin manganase (III) chloride
Mops	3-(*N*-morpholino)propane sulphonic acid
M_r	relative molecular mass (molecular weight)
mRNA	messenger ribonucleic acid
MTDDA.NO3	methyl-tridodecylammonium nitrate
MtSB	microtubule stabilizing buffer
MVB	multivesicular bodies
MW	molecular weight
NAA	naphthalene acetic acid
NAD	nicotinamide adenine dinucleotide (oxidized)
NADH	nicotinamide adenine dinucleotide (reduced)
NADP	nicotinamide adenine dinucleotide phosphate (oxidized)
NADPH	nicotinamide adenine dinucleotide phosphate (reduced)
NCCV	non-clathrin coated vesicles
NHS	N-hydroxysuccinamide
NR	nitrate reductase
NTB	nitroblue tetrazolium
NTP	nucleotide triphosphate
OEC	oxygen-evolving complex
Pa	pascal
PAGE	polyacrylamide gel electrophoresis
PAP	peroxidase-antiperoxidase

PAS	periodic acid–Schiff's
PBS	phosphate buffered saline
PC	personal computer
PCR	polymerase chain reaction
PCR	partially coated reticulum
PEG	polyethylene glycol
PGI	phosphoglucose isomerase
Phe	phenylalanine
pI	isolelectric point
PI	propidium iodide
Pipes	piperazine-*N*,*N*′-bis(2-ethanesulphonic acid)
pK_a	acid dissociation constant
PME	pectin methylesterase
po	phenol oxidase
PPase	pyrophosphatase
PRINS	primed *in situ* hybridization
PTA	potassium phosphotungstic acid
PVC	polyvinyl chloride
PVP-40	polyvinylpyrrolidone-40
R_f	retardation factor
r.p.m.	revs per minute
Rha	rhamnose
Rib	ribose
RNA	ribonucleic acid
RNA polym	DNA-dependent ribonucleic acid polymerase
RNase	ribonuclease
RNasin	ribonuclease inhibitor
rRNA	ribosomal ribonucleic acid
RT	room temperature
rUTP	ribosomal uridine triphosphate
SDS	sodium dodecyl sulphate (sodium lauryl sulphate)
SEM	scanning electron microscope
SLWD	super long working distance
SNARF-1	carboxy-seminaphthorhodafluor
SSC	sodium chloride-sodium citrate mixture
SV	secretory vesicles
TBE	tris/borate/EDTA buffer
TBS	tris buffered saline
TCA	trichloroacetic acid
TE	tris/EDTA buffer
TEA	triethanolamine hydrochloride
TEM	transmission electron microscope
TESPA	3′-aminopropyl triethoxysaline
TFA	trifluoroacetic acid
THF	tetranhydrofuran
TLC	thin layer chromatography
Tm	melting temperature
TP-25	25 kDa tonoplast integral protein

Tris	tris (hydroxymethyl) aminomethane
TRITC	tetramethyl rhodamine isothiocyanate
TTP	thymidine triphosphate
TV	transition vesicles
Tyr	tyrosine
UDPG	uridine 5-diphosphoglucose
UTP	uridine triphosphate
UV	ultraviolet
V	volts
X-gluc	5-bromo-4-chloro-3-indoyl β-D-glucuronic acid
Xyl	xylose
ZIO	zinc iodide/osmium tetroxide mixture

1

An introduction to optical microscopy for plant cell biology

P. J. SHAW and D. J. RAWLINS

1. Introduction

The development of the optical microscope in the seventeenth century opened up many new areas of study in several fields of science. In particular, the observations of plant and animal tissues and micro-organisms can be said to have provided the origins of cell biology, although our modern idea of what a 'cell' actually is arose somewhat later. Some of the data recorded by the early microscopists, using only primitive microscopes and drawing freehand what they saw, are quite remarkable. Hooke, Malphigi, and Leeuwenhoek are probably the best known seventeenth century microscopists, but Nehemiah Grew was the first true specialist in plant microscopy. He produced detailed descriptions of plant microanatomy which have proved remarkably accurate. An example is shown, along with Grew's original legend (1), in *Figure 1*. Although there is only space to show one such picture, the original volumes contain many pages of such detailed, painstaking hand-engravings. With today's access to photography, video cameras, and computer image processing, this should serve as a reminder of how much can be achieved with careful observation and a very simple microscope. Many biologists overlook the useful information that can be rapidly obtained using even a very simple laboratory microscope.

The optical microscope has developed steadily since its invention. Particularly important innovations were the analysis of the imaging process by diffraction theory of Abbe, and the concomitant development of the optimal bright field condenser by Abbe and Zeiss. The development of textile dyes provided many biological stains which were ideal for bright field microscopy. Later Zernike introduced phase contrast optics, which proved particularly useful in visualizing unstained biological material. In fact the early microscopists managed to see unstained biological material to some extent because of the aberrations of their optical components. Paradoxically, as lens design improved, the objects became harder and harder to see. Phase contrast optics reintroduced this contrast in a controlled way. More recently Nomarski

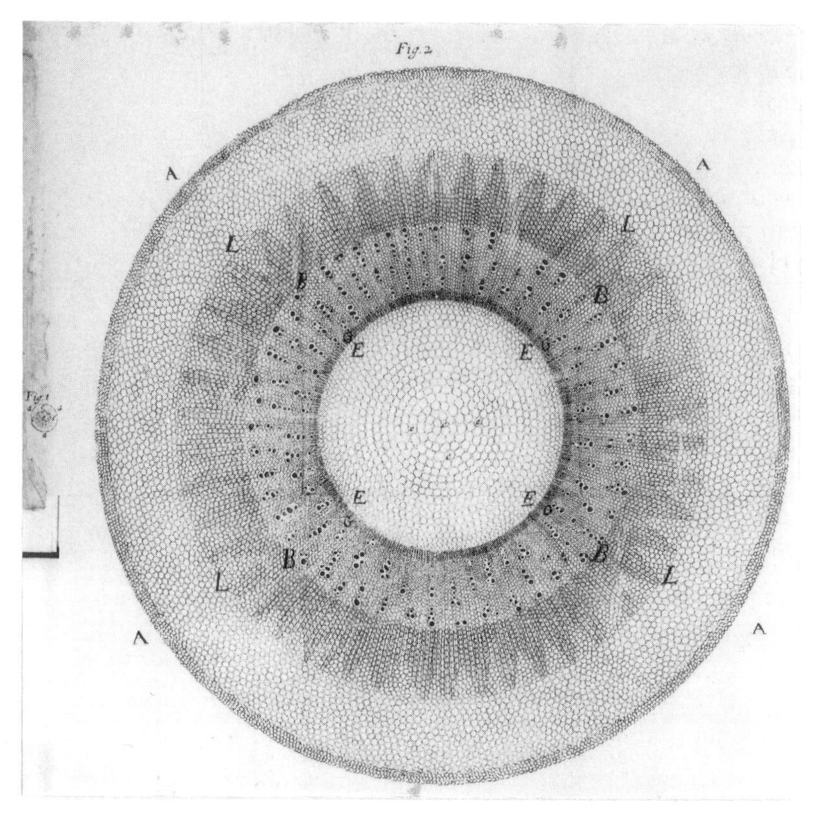

Figure 1. A transverse section of a horse-radish root, hand-drawn and hand-engraved by Nehemiah Grew (1) in 1673. Grew's original legend is as follows:
'Fig 1. A slice of the lower part of the root of Horse-radish cut traversly, as it appeareth to the bare eye. a. The skin. ac. The bark, with the succiferous vessels therein represented by the smaller specks. Within stand the air-vessels represented by the larger and blacker specks. e. The pith.
Fig 2. The same slice, as it appeareth through a microscope. AA The skin. A.B. The bark. B.L. The sufficerous vessels therein postured in the form of a glory. B.G. The air-vessels postured in a thick ring; the several conjugations whereof are radiated. G.E. Other succiferous vessels within the air-vessels postured in a thin ring. E. The pith. ee. The bubles of the pith.' (Reproduced from (1) with permission.)

invented differential interference contrast (DIC) which produces an image whose contrast depends on the changes in refractive index within the specimen. In some types of biological specimen this has proved a very valuable technique.

The progress of development of optical methods has accelerated markedly in the past few years. In the period from 1950 onwards the new technique of electron microscopy provided radically new insights into cell structure and revolutionized our conceptions of subcellular organization, eclipsing optical methods for a while. In the past 10 years optical methods have undergone a renaissance, mainly because of the use of highly sensitive and specific fluorescence probes. These include monospecific antibodies to all types of biological macromolecules (see Chapter 7), specific DNA and RNA segments used as *in situ* probes (see Chapters 5 and 6), and fluorescent dyes for DNA, pH, and particular ions (see Chapter 2). Light is a relatively non-destructive method of probing cells, and this makes optical studies of living cells possible, often coupled with microinjection or other loading of fluorescent markers (see Chapter 2). The use of fluorescently labelled proteins provides a particularly powerful method of analysis—fluorescent analogue cyto-chemisty—allowing the direct visualization of, for example, microtubule dynamics (Chapter 10) or chromatin rearrangements in living cells. A further exciting development in this area is the use of 'caged' fluorochromes—molecules which become fluorescent only after photoactivation by light of a particular wavelength. Thus individual cellular substructures can be marked and followed. This approach may well replace the earlier FRAP (fluorescence recovery after photobleaching) method, since it has the potential of greater sensitivity.

In parallel with these developments in fluorescent probes, there have been radical developments in optical microscopy itself, notably the invention of the confocal microscope. This was suggested by Minsky in 1957 as a method for increasing the resolution of an optical microscope. It was first implemented by Petran for reflection imaging. Although confocal methods have been also devised for transmission imaging, it is with fluorescence imaging that confocal microscopy has had the greatest impact in biological microscopy. It is debatable whether the increase in lateral resolution theoretically obtainable by confocal microscopy is actually normally achieved. However, the most important feature for biological fluorescence imaging is the rejection of 'out-of-focus' flare by the confocal optical system. Fluorescence microscopy is in essence a dark field imaging mode, with very bright structures contrasted against a black background. In conventional epifluorescence imaging the contribution of each part of the specimen to the coarse features in the image extends a long way either side of the focal plane whereas the fine image detail is rapidly attenuated away from the focal plane. This gives rise to a high background out-of-focus contribution which tends to obscure the fine structure and makes the images generally hard to interpret in detail. The confocal arrangement, which is described below, excludes nearly all of the out-of-focus component and thus produces clearer fluorescence images. It is also the best technique for measuring focal section series, and thereby obtaining true three-dimensional reconstructions of cellular and subcellular structures.

In this chapter a brief introduction to the most useful optical imaging modes is given with some examples of the type of image to be expected from plant material. Confocal microscopy is also described with examples of ways in which it is proving useful in plant studies. Rather than give a complete manual on these various techniques, our intention is to show the range of methods available and what they can offer to a plant cell biologist. We refer the reader elsewhere for more detailed instructions on the use of specific techniques (2–4, 7–9).

2. Explanation of terms

Before describing the different types of optical imaging a brief description of some of the technical terms which are encountered in optical microscopy is given, particularly in the description of the optical components. A more complete glossary is given in reference 5. The primary image-forming lens is called the **objective**. It is invariably an assembly of several optical components. Another lens assembly, the **condenser**, focuses illuminating light on the specimen for transmitted light microscopy. The final image is produced by the **eyepiece** or **ocular**. Each objective lens has a type description and some numbers engraved on the side of the barrel; for example 'Plan 25/0.08 160/0.17'. The type description word denotes the level of correction of aberrations of the objective. Optical lenses suffer from various aberrations; these are corrected to different extents in different types of objective. In **achromats**, the chromatic aberration (bringing light of different wavelengths to different focal planes) is minimized for two wavelengths (usually one below 500 nm and one above 600 nm). In **apochromats**, the chromatic aberration is minimized for three wavelengths (generally about 450 nm, 550 nm, and 654 nm). In **plan** objectives the curvature of field (most noticeable at the edge of the field of view) is minimized. In the nomenclature of objectives this is also used as a prefix, e.g. in the terms **plan-achromat** or **plan-apochromat**. Other terms are used by individual manufacturers to describe special features; for example, Zeiss call objectives which are designed for UV transmission **neofluar** and **plan-neofluar**, whereas other manufacturers use different terms such as **fluor** and **UV**. UV transmission is necessary for the excitation of some fluorochromes, such as the DNA dye 4′,6-diamidino-2-phenylindole (DAPI). Plan-apochromats from some manufacturers do transmit light in the near UV and can be used for DAPI, those from others do not. Generally plan-apochromats are the best objectives (and the most expensive). The first number after the type is usually the **magnification**. Objective magnifications range from ×1 to ×100. The most useful magnifications for cell biological work are low (×16 or ×25), intermediate (×40), and high (×63 or ×60) magnification. Next to the magnification is the **numerical aperture**, a number greater than zero and less than 1.5. This is the refractive index of the immersion medium multiplied by the sine of the aperture angle of the lens—in essence it indicates the angle of the cone of scattered light the objective collects. The larger the numerical aperture, the greater the light collection efficiency and the greater the resolution obtainable. **Resolution** is, loosely, the extent to which fine image detail is observable. There are several more rigorous definitions; the one most often used in microscopy is the minimum distance between two points such that they can be recognized as distinct. With the highest numerical aperture objectives, the resolution is approximately 0.25 μm (depending on the wavelength of the light). Some objectives have an adjustment collar to allow the numerical aperture to be changed. The highest available numerical aperture should normally be used, but it is occasionally useful to be able to decrease it, for example for dark field imaging (see below) or to increase the depth of field. Most objectives are designed for a **tube length** of 160 mm, which is usually the next number engraved on the side of the objective. This is the distance between the objective and the eyepiece. Some objectives are termed **infinity-corrected**, which has the advantage of allowing an arbitrarily long distance between the objective and the eyepiece. In general 160 mm tube length objectives are interchangeable between different microscopes, microscopists sensibly having long ago standardized the mounting thread (although some of the correction of chromatic aberration is often in the eyepiece, so eyepiece and objective should strictly be matched for optimal performance). But infinity-corrected objectives **cannot** be interchanged with 160 mm ones, unless a special correction lens is used. A final number may be engraved on the objective, usually 0.17. This is the coverglass thickness. It means the objective has been designed for a coverglass of this thickness, and any other thickness should not be used. A marking of 0.17 corresponds to 1½ thickness, and unless there is a very special reason, it is a good idea never to have any other thicknesses in the laboratory to avoid confusion. Many objective lenses are **immersion** lenses. This means that they are designed to have a liquid of a certain refractive index between the lens and the coverglass. These objectives will have 'Oil', or 'Oel', or 'Imm' written on them. Some objectives are designed for other immersion liquids—usually either glycerol ('Glyc') or water ('W'). Still other objectives have a variable correction collar for different media—make sure the collar is in the right position before use. Objectives which are not designed for immersion are often called 'high dry' lenses. If possible, it is useful not to mix immersion and non-immersion objectives on a microscope. This will avoid confusion and simplify changing from one objective to another. Using the wrong immersion medium or the wrong coverglass thickness will increase the **spherical aberration**. This means that different levels of image detail are brought to a focus at different distances from the objective. In general this will make the images more blurred. In fact in normal biological use spherical aberration is almost always worse than it should be. This is because the objectives are designed to image specimens immediately below the coverglass. In most biological specimens there is a layer of mounting medium between the underside of the coverglass and the specimen, and this extra optical pathlength increases the spherical aberration. This aberration is particularly easy to see with epifluorescence. Focus on a small bright spot and observe how the spot changes either side of focus. In the absence of spherical aberration it should expand and fade equally either side. Almost invariably it is asymmetric, disappearing quickly on one side, but slowly on the other side. An image of a single point such as this is called the **point spread function**. It is an important characteristic of a microscope, especially when there is interest in detailed three-dimensional imaging or image processing. Within certain limits, if the imaging of a single point is known, then the image which should result from any specimen (at least for fluorescence and bright field imaging) can be determined. A final objective characteristic which is often important is the **working distance**. This is the distance between the focal plane and the front of the objective. It is usually very small (perhaps 0.2 mm) for high numerical aperture lenses, since to obtain a large collection angle it is necessary either to have the lens close to the specimen or to have very large diameter lenses, which are expensive. It is possible to obtain objectives with particularly long working distances, of the order of tens of millimetres (for example the Nikon SLWD range, which still have quite large numerical apertures). This type of lens is normally used on inverted microscopes for observation of samples in containers such as Petri dishes, but is also very useful for micromanipulation and microinjection on a non-inverted microscope, since it gives reasonable access to the specimen during observation.

The condenser is crucially important in transmission imaging. Its role is obvious in phase contrast and dark field imaging (see below), but it is often not appreciated that its contribution to the overall resolution is equal to that of the objective in bright field and DIC imaging. For high resolution, a high quality plan-apochromat condenser should be used. If the objective is an oil immersion one, then the condenser should also be an oil immersion lens. A large research microscope will have a number of different condenser positions. A wheel in the condenser is rotated to bring different phase rings, dark field apertures, and DIC compensator prisms into the optical path.

The eyepieces on a microscope are rarely, if ever, changed. Their magnification is calculated so that the image is magnified enough for the human eye to see all the significant detail. This is usually ×10 or ×20. In binocular microscopes there is a focusing collar on one or both eyepieces. If the microscope has a camera attached, one eyepiece will probably have a photoscreen graticule in it, or there may be an eyepiece graticule. If so, focus on the specimen and, looking only through this eyepiece (shutting the other eye), adjust the eyepiece collar until the graticule is sharply in focus. Then the specimen and graticule should both be sharply in focus. Now look through the other eyepiece and adjust it to bring the specimen sharply into focus. This will have adjusted the eyepieces for each individual's eyes, and ensured that when the image looks in focus to the observer, it will be in focus at the camera (provided the camera has been set up correctly).

The overall magnification of the microscope is the product of the eyepiece (or camera projection lens) magnification and the objective magnification. On some microscopes this has to be multiplied by an additional factor; occasionally there is another intermediate magnification-changing lens (sometimes called an **optovar**). To measure the actual magnification of a microscope, a **stage micrometer**—a microscope slide with a scale engraved on it—is required. An **eyepiece graticule** is also useful. This is a glass plate with a dimensionless scale engraved on it, which is inserted in the eyepiece. It is calibrated for each objective magnification using a stage micrometer, and is used to estimate the size of objects through the eyepiece.

3. Photomicrography

For all but very routine microscopy it is a good idea to use photography almost as a matter of course. Somehow it is never possible to find such a good field of view as the one seen when taking a quick look! The moral is—photograph it, and take each photograph as if it was a picture for publication, because it is not known in advance which ones will be required for publication. Most modern exposure meters use centre-weighted average metering. This is usually fine for bright field types of image, since the brightness of the area of interest is about the same as the rest of the field. However, it is not appropriate for other types of image, particularly fluorescence, since the objects of interest are much brighter than the dark background, and the average brightness is near to the background value. Here, the best solution is to use spot metering, where the exposure is set in a small area in the centre of the field. Move the object of interest into the central area to set the exposure. If spot metering is not available trial and error must be used. As a guide, using 800 ASA film, a reasonably bright cytoskeleton immunofluorescence image might need an exposure of 30 sec, while DAPI-stained nuclei might need only 1 sec.

For immunofluorescence, Kodak TMAX film is quite extensively used in this laboratory for black and white photography; it has good contrast and low grain, and is available in a range of speeds (TMAX 400 is popular in many laboratories). Kodak Technical Pan film is much slower and has lower contrast, but is more suitable for bright field or DIC images. Technical Pan film with liquid Technidol low contrast developer can be used for photographing video and computer displays; most other films have too high a contrast for this application. For colour slides, use tungsten colour-rated films if using a tungsten lamp, daylight-rated films for a mercury arc lamp (for example when photographing immunofluorescence specimens in their original colours, or photographing colour video or computer screens). For colour prints, the use of slide film and making prints from these (e.g. Cibachrome prints) is recommended.

4. Microscope imaging modes

Microscope imaging may be divided into transmission modes (such as bright field, dark field, phase contrast, and DIC (Nomarski)) and epi-illumination modes (primarily epifluorescence and reflection contrast). Confocal micro-

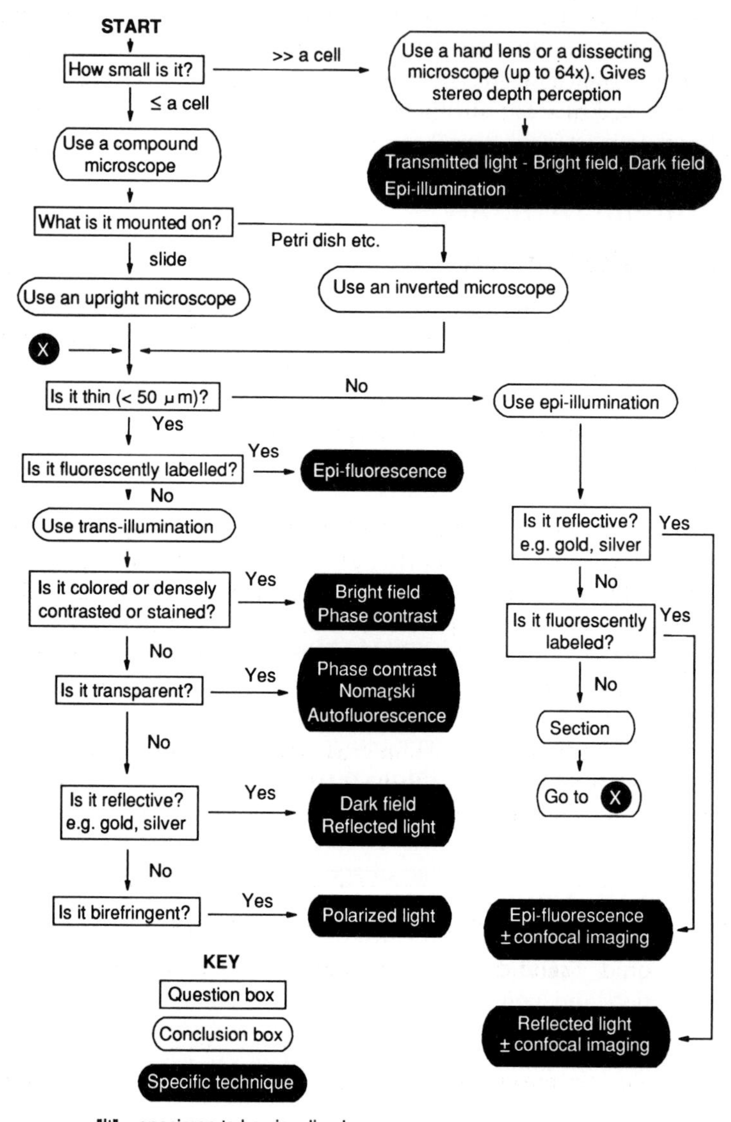

Figure 2. Flowchart for choice of imaging methods. (Reproduced from (2) with permission.)

scopes have been developed which use either type of illumination, but epifluorescence and reflection contrast are much simpler to implement in a confocal arrangement and have been by far the most widely used in biology to date. Deciding what type of imaging to use is sometimes straightforward, sometimes difficult. *Figure 2* gives a flowchart to help in this decision, but this should only be regarded as a guide to the relevant considerations. It is important to be aware of the different imaging modes available and of their capabilities, and to be sufficiently familiar with the microscope to try different imaging methods with a specimen. Some examples are given in the figures of images from typical plant specimens; in some cases the same specimen has been photographed in two or three imaging modes to show the comparison between them.

4.1 Bright field imaging

This is the most basic imaging technique. Adjustment for optimal bright field imaging (Köhler illumination) is fundamental for all the other transmission imaging modes, and familiarization with this process should be the first step in using any microscope. Always check this adjustment before using the microscope for anything else (except epifluorescence, dark field, and reflection) and every time objectives are changed or any other element is inserted into the light path.

Light from each point on the bulb filament is focused by the collector lens to a point at the condenser aperture. At the field aperture this gives a wide area of even illumination. Since the field aperture is at an equivalent focal plane to the specimen, the specimen is evenly illuminated. Light from the specimen is focused by the objective at a plane known as the primary image plane. The eyepiece then forms an image a small distance from the top lens of the eyepiece.

Refer to the handbook for each microscope for a detailed description of its parts and how to adjust them. *Protocol 1* gives a description which should apply in general to any microscope.

Protocol 1. Adjustment for bright field imaging

1. Focus on a specimen with the condenser and field apertures wide open.
2. Close the field aperture until its edges can be seen.
3. Focus the condenser up or down until the edges of the field aperture are sharp.
4. If necessary, centre the image of the field aperture with the condenser centring adjustments.

Protocol 1. *Continued*

5. Close the condenser aperture until the glare around the field aperture just disappears, and then reopen it a little (if in doubt, open the aperture about half-way). Alternatively re-adjust the condenser aperture during observation of the specimen to obtain the best image. Opening the condenser aperture increases the resolution, but decreases image contrast and increases bright flare around objects. Closing it decreases flare and increases image contrast, but decreases resolution.
6. Finally open the field aperture until it just disappears outside the field of view. For lower power objectives (e.g. ×16) the illumination may be not uniform, but brighter in the middle. In this case there is generally an accessory lens in the condenser—swing this in and re-adjust the condenser. In the case of very low power (e.g. ×2.5) even this will not give even illumination. Lower the condenser right down and open up both apertures—this will not be Köhler illumination, but it should be more even, and optimal resolution is not a consideration at such low magnifications.

Examples of bright field images are shown in *Figures 3a*, *4a*, and *5a*.

4.2 Phase contrast

Many specimens, especially biological material, have little inherent direct contrast and are thus nearly invisible in simple bright field. However, different parts of the specimen may have different refractive indices. This causes differing retardations of the light waves but cannot be seen directly by eye. Phase contrast is a method of producing visible differences from these retardations. It works by illuminating the specimen through an annular condenser aperture. The objective has a special plate in it which has a phase ring matching the condenser annulus. Light passing through anywhere in the plate except the phase ring is retarded by a quarter wavelength, that passing through the phase ring is unretarded. Light that is not scattered by the specimen passes through the condenser aperture and through the phase ring. Any light that interacts with the specimen will have had a specimen-dependent phase retardation introduced (generally approximately $\lambda/4$) and will be scattered so that it no longer passes through the phase ring, so it will be additionally retarded by another $\lambda/4$. The total retardation will thus be around $\lambda/2$, and this light will interfere destructively with the unscattered light. The specimen-dependent phase changes will thus be turned into changes in amplitude, and be visible.

Phase contrast therefore requires special objectives with a phase ring and a matching condenser annular aperture—it cannot be done on other types of

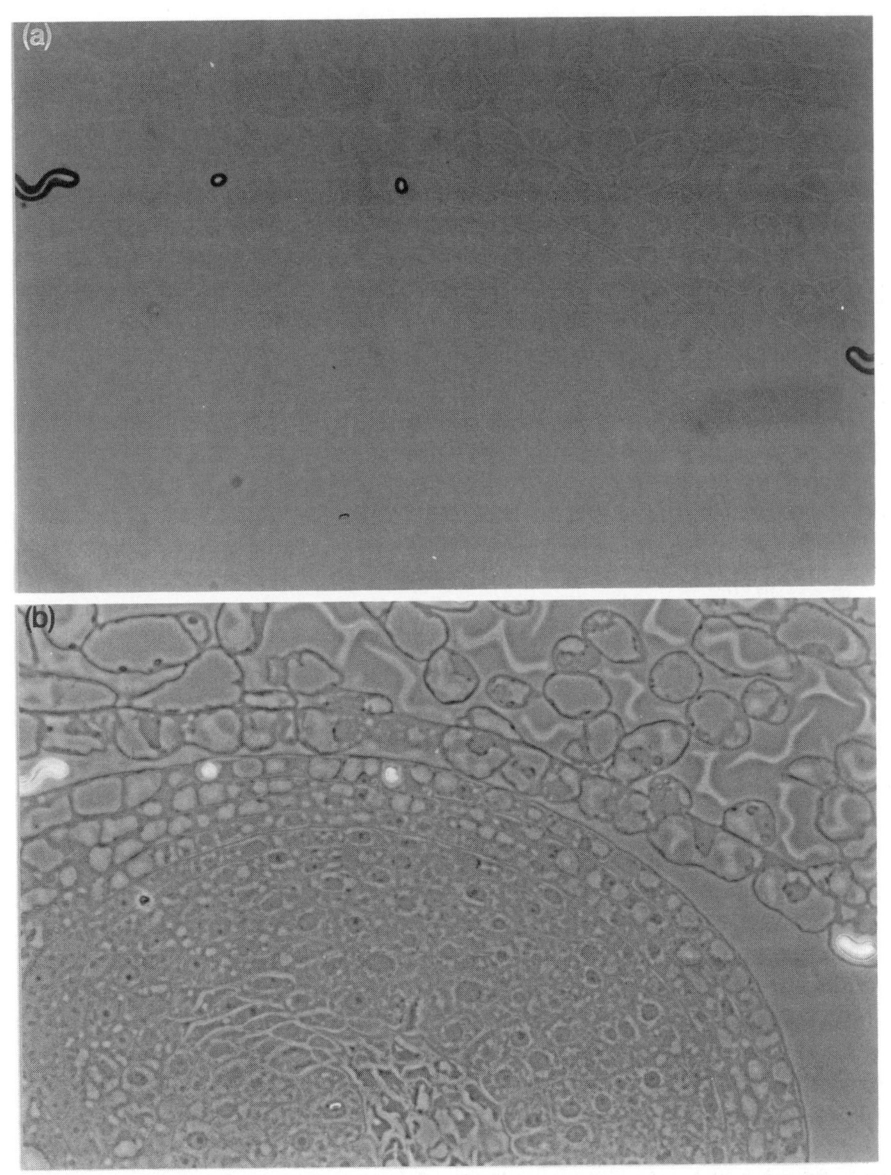

Figure 3. Unstained resin section through an embryo of *Brassica napus*. (a) Bright field image shows very poor contrast in comparison with (b), the phase contrast image.

objective. To adjust a microscope for phase contrast, first set it up for bright field. Next open the condenser aperture fully, and either replace an eyepiece with a phase telescope, or, more usually, move a Bertrand lens into the light path (usually by a rotating wheel somewhere between the lens turret and the eyepiece). This enables the back focal plane of the objective to be seen, which

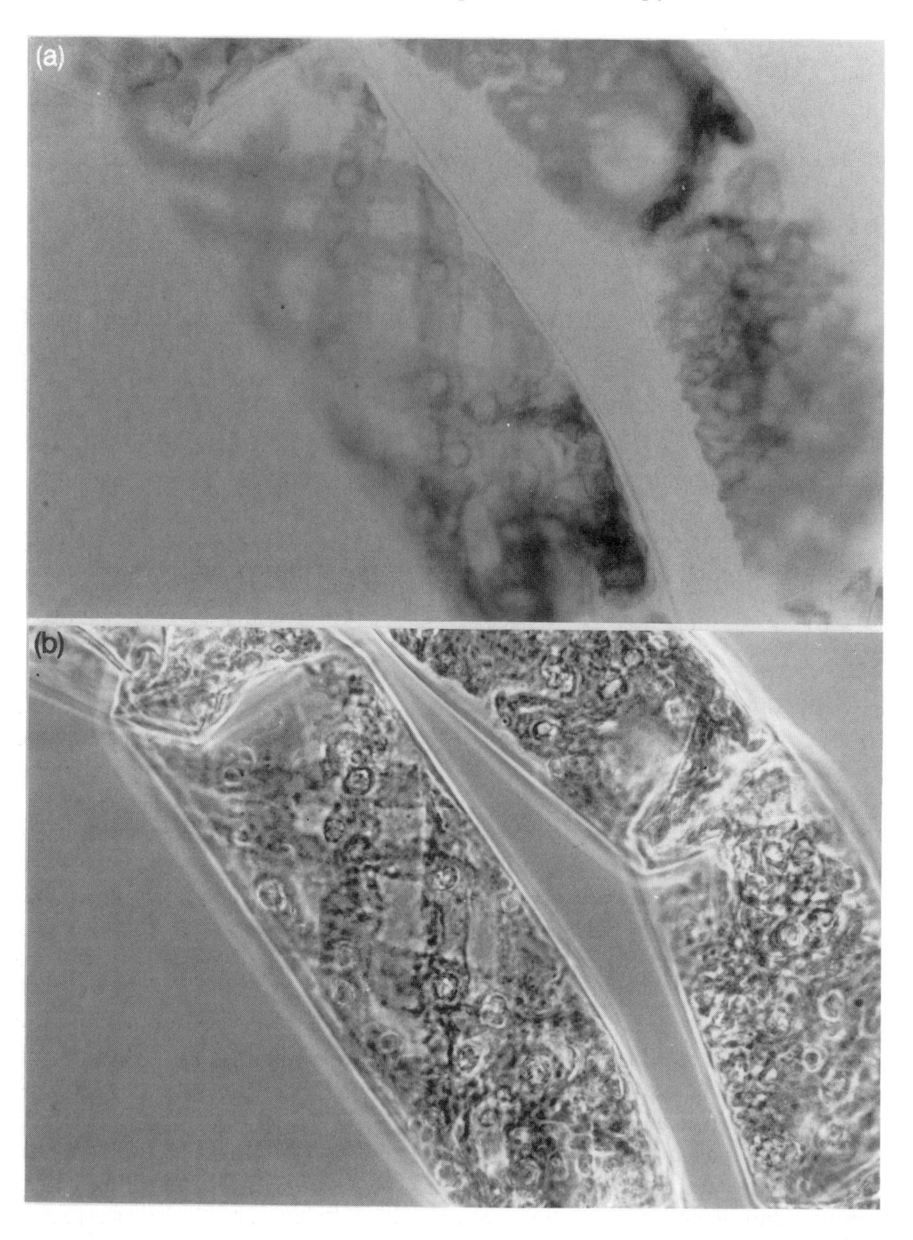

Figure 4. Filament of the alga *Spirogyra grevelliana*. (a) Bright field image. The contrast is fairly low but the characteristic spiral chloroplasts are visible. (b) Phase contrast image of the same cells. The contrast is greatly enhanced, but note the prominent haloes around all the structures, particularly the cell wall. (c) DIC image of the same cells. This is probably the optimal imaging mode for this specimen, showing fine structure inside the cell, and good contrast, without the haloes present in the phase image. (Reproduced from (2) with permission.)

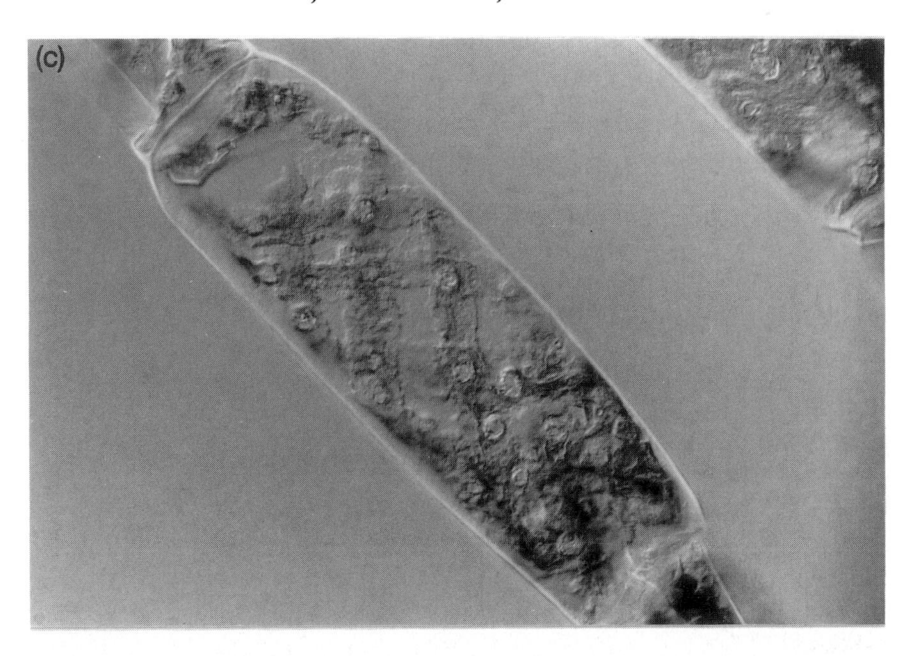

is where the phase ring is located. Two rings should be visible—a bright one corresponding to the condenser annulus and a dark one corresponding to the phase ring. These rings must overlap symmetrically. Generally it is necessary to adjust the position of the condenser annulus, which is accomplished by some sort of centring device (*not* the same as the condenser centring mechanism.) On some modern microscopes the annuli are precentred and should not require adjusting. If the two rings are of very different radii then the condenser annulus is not the right one for the objective. Now move out the phase telescope and return to normal viewing, to observe by phase contrast.

Remember that phase contrast is an interference phenomenon; light and dark areas are related to the refractive index within the specimen. Small structures can appear bright or dark at slightly different focus levels. A dark structure might be a dense object, or it might be something else like a hole or a vacuole. Rings and haloes around structures are also highly visible, and the resolution is less than in bright field. For these reasons, although phase contrast can often be very useful, the resulting images should be interpreted with caution. See *Figures 3b*, *4b*, and *5c* for examples of phase contrast images.

4.3 Polarized light microscopy

In polarized light microscopy a polarizing filter is placed in the light source path—generally on top of the field aperture. Birefringent structures (i.e.

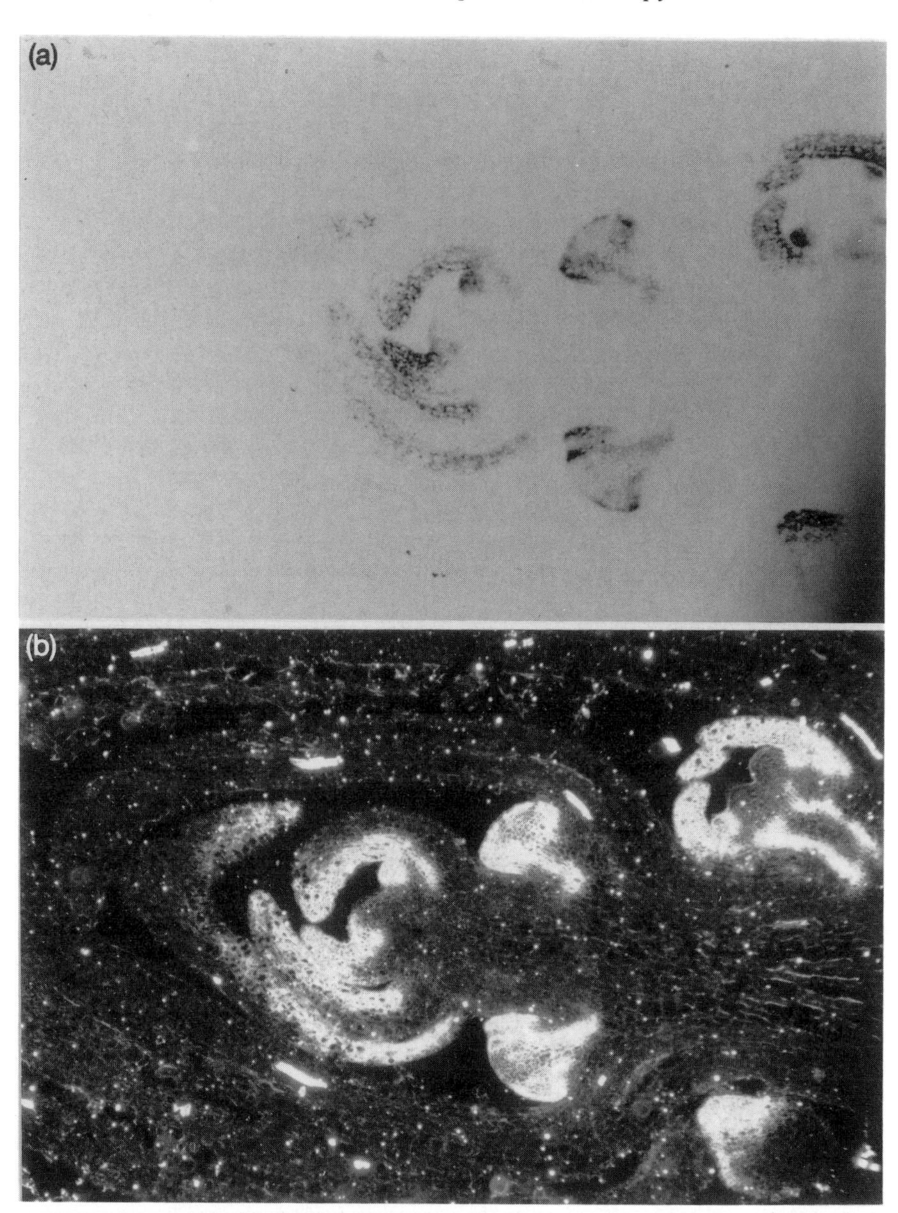

Figure 5. A wax section of wild-type flower of *Antirrhinum majus* labelled with an *in situ* probe to the transcript of the gene *floricaula* detected by alkaline phosphatase-conjugated antibody. (a) Bright field image, showing the distribution of the transcript as revealed by the insoluble dark-coloured product of the alkaline phosphatase reaction, but poor detail of the cell structure. (b) The equivalent dark field image. The crystalline product of the alkaline phosphatase reaction reflects light strongly and is thus seen very clearly. (c) For comparison the phase contrast image of a portion of the same specimen is shown. The contrast is better than that in (a) but poorer than that in (c). (Reproduced from (2) with permission.)

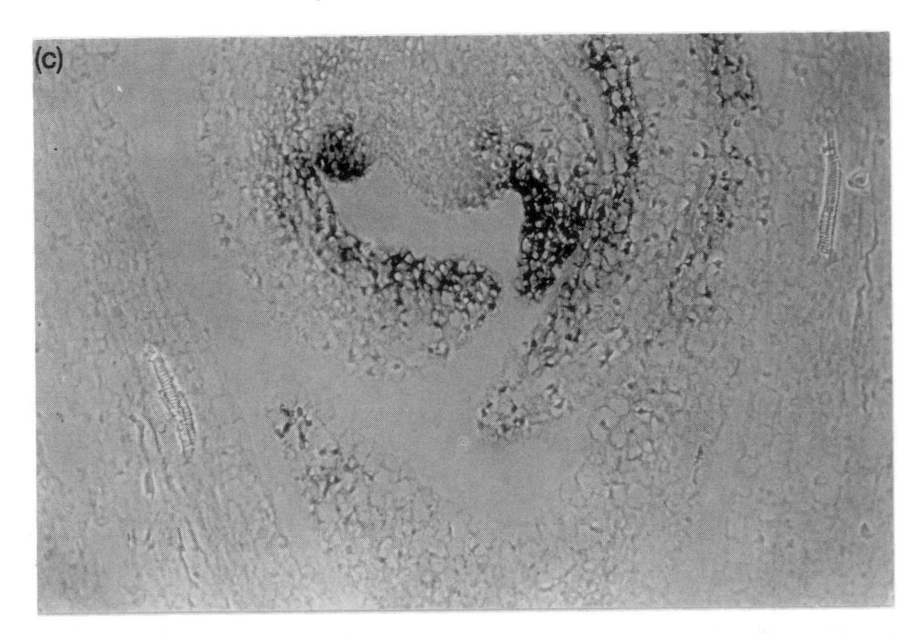

those having different refractive indices in different directions of light polarization) cause alterations in the plane of polarization of the light. A second analyser filter, polarized at right angles to the first and located after the objective in the light path, then converts the changes in plane of polarization to visible amplitude changes. Unrotated light is blocked by the second filter, whereas the more the plane of polarization is rotated by the specimen, the greater the component that is passed by the second filter.

To set up polarized imaging, simply adjust for bright field, then insert the two polarizing filters and rotate one of them until the field is dark; this is called extinction. Polarized light microscopy is most useful for highly birefringent objects, which appear bright against the dark background. This tends to be the case for objects which have a strong inherent directional structure, like cellulose microfibrils in cell walls, or muscle filaments. Polarized light microscopy has not been very extensively used so far in plant studies. An example is shown in *Figure 6*.

4.4 Differential interference contrast (Nomarski)

Differential interference contrast (DIC) was invented by Nomarski and so often goes by his name. Another similar technique is called Hoffman modulation microscopy, which may be less expensive. In DIC two special optical components called Wollaston prisms, each consisting of two crystalline quartz prisms cut parallel to crystallographic axes and cemented together with their respective crystal axes at right angles to each other, are introduced in

Figure 6. A cell wall fragment of *Datura stramonium* imaged in polarized light. The specimen is at 45° to the orientation of the crossed polars and so appears bright against the dark background. Note the characteristic Maltese cross appearance of the starch grains. (Reproduced from (2) with permission.)

addition to the crossed polarizers. The Wollaston prisms have the effect of splitting an incoming beam into two beams polarized at right angles to each other, with a small angular divergence between them. A phase difference is also introduced between the beams, whose value depends on which part of the Wollaston prism the beam hits. One prism is located after the objective in the optical path and splits the image from the objective into two images which are slightly laterally displaced relative to each other—typically by about 0.12 μm. The wavefront reaching the analyser polarizing filter will effectively be the sum of the two images. Where the two points summed have the same phase, the resultant will have the same plane of polarization as the original incident light, and thus be blocked by the analyser filter. However, where there is a phase difference between the two points, the resultant will have a rotated plane of polarization and a component of this light will pass through the analyser. The Wollaston prism will also introduce a specimen-independent relative phase difference between the images; this additional phase difference can be adjusted by displacing the prism, which alters the final image seen. This diplacement is provided via a knurled knob/lead screw device on most microscopes. Because the light source is an extended one, light rays originating at its edge traverse the specimen at an angle to the optical axis and reach different parts of the Wollaston prism from axial rays. They will therefore receive a different relative phase difference from axial rays. This is compensated for by the Wollaston prism in front of the condenser, which introduces an equal and opposite phase change. Large condenser apertures can therefore be used giving excellent resolution and optical sectioning. In principle the beam-splitting prism should be located in the back focal plane of the objective. At least one commercially available system uses this arrangement. However, this is a difficult part of the microscope to reach, and such systems can be difficult to use. It is possible to modify the construction of the prism so as to allow it to be placed further away from the objective in a much more convenient position; this arrangement is much easier to use.

Thus for DIC, a microscope must have provision for two polarizing filters and two prisms in addition to standard bright field optics. In principle DIC should work for any high quality objective lens. It was originally necessary to use specially strain-free objectives, but modern high quality plan-apochromat lenses are quite good enough for DIC. To adjust a microscope for DIC, first obain Köhler bright field illumination with the condenser set to the DIC position (with the compensator prism in place), but with the image-splitting prism out of the optical path. Next open the condenser aperture and insert the polarizing filters. Rotate one of them unil the field is maximally dark, then insert the image-splitting prism and adjust the translation to obtain the 'best' image. It is up to the viewer to judge what the best image is, but it is generally best to use the least contrast required to see the detail of interest (i.e. with condenser aperture as far open as possible), since closing the aperture, while increasing the contrast, will decrease the resolution.

DIC images can be difficult to interpret. The method displays an image of changes in refractive index in the direction defined by the orientation of the prisms. For this reason it is particularly good at revealing edges in biological structures, for example, organelle and nuclear boundaries, cell boundaries, and cell walls. It has also been used very effectively to image fibrous subcellular components such as microtubules, often in combination with video enhancement. One side of an edge will be brighter than the background, the other side will be darker. For this reason the images have a false 'shadowing' appearance. This is visually very attractive and can be very informative. However, it can also be very misleading, since it makes the images seem three-dimensional because of the way the human brain interprets lighting effects. It is important to realize that these images are *not* three-dimensional ones, they merely look that way. To obtain true three-dimensional information focal sectioning must be used. In a simple way this can be achieved by focusing up and down through the specimen. Because DIC can use the highest condenser aperture available it can give smaller depths of field than other conventional imaging modes, and so is better at focal sectioning. For highest resolution with DIC, a condenser with numerical aperture as great as the objective should be used. If the objective is a high numerical aperture oil immersion one, the condenser should be also, and should be used with oil—this is rarely appreciated, still less actually done. A DIC image of *Spirogyra* is shown in *Figure 4c*.

4.5 Dark field

All of the foregoing techniques, with the exception of polarized light microscopy, are essentially bright field methods; that is, darker structure in the specimen is seen imaged against a bright background. This is very familiar and readily interpretable. However, it can be difficult to see small, weakly imaged features, since they represent small changes in a large background, and therefore provide inherently low contrast. In dark field techniqes, the background is low and the specimen structure is bright, and so contrast is much greater. Such techniques are particularly good for visualizing reflective structures such as silver grains in autoradiographs and silver-enhanced, gold-labelled immunocytological specimens. Plant cell walls also show up brightly in dark field.

Dark field transmission images are produced by using oblique illumination. The unscattered light does not enter the objective and thus the background is dark. Any light scattered into the objective by the specimen is seen as bright structure. There are two methods for producing the oblique illumination. Firstly the central portion of the light from the condenser can be blocked with a patch stop (an opaque disc) leaving only the oblique illumination. The second method uses a special condenser with reflective surfaces. The former type is the most common.

To set up dark field with a patch stop condenser, focus on the specimen in bright field. Next bring in the patch stop condenser position. In many modern microscopes the patch stop condenser is precentred, and so the only adjustment necessary is to focus the condenser until a dark background with the specimen shining brightly is seen. When using a patch stop condenser which requires centring, use the phase telescope or Bertrand lens to view the back focal plane of the objective. Focus the condenser up and down until the patch stop is seen as clearly as possible, then centre it. Switch back to normal viewing and adjust the condenser focus as before for the best dark field image. If the condenser is an oil immersion type, use oil and the maximum condenser aperture. It may be necessary to re-adjust the centring of the patch stop; look for the location of the darkest part of the image as the condenser is focused, and adjust if it is not central. In contrast to bright field, the angle of the illuminating cone of light must be *greater* than that of the objective aperture, otherwise the unscattered light will not be excluded from the image. This may mean that dark field will not work effectively with very high numerical aperture objectives, unless the condenser can provide an even greater effective numerical aperture; it may therefore not be possible to get a dark background unless the objective numerical aperture can be stopped down. At low magnification, the largest phase annulus may work as a patch stop. Otherwise it is possible to improvise one with a disc of card underneath the bright field condenser—there is usually a suitable filter holder above the condenser accessory lens.

4.6 Epifluorescence and reflected light microscopy

The previous methods are all transmission imaging modes, i.e. the specimen is illuminated from the side opposite the objective by means of a condenser lens. In epifluorescence and reflection imaging the specimen is illuminated by light introduced from the same side of the specimen as the objective, usually through the objective, which therefore acts as its own condenser. This means that it is not necessary to adjust the condenser for these modes (although it is a good idea to get into the habit of always checking the bright field adjustment before using a microscope.) A dark field image is shown in *Figure 5b*.

Reflection imaging is particularly useful for imaging highly reflective particles, such as silver grains in autoradiographs. This requires an epi-illumination source, a 50% semi-silvered mirror, and two polarizing filters, one in the incident light and a second with its plane of polarization at right angles to the first in the path of the reflected light. The polarized illuminating light is reflected down through the objective by the semi-silvered mirror at 45°. Reflected light from the specimen then passes up and straight through the mirror. Light that is reflected, as opposed to back-scattered light, will have its plane of polarization rotated, and will therefore pass through the second polarizing filter. Back-scattered light will be blocked by the filter. Most epifluorescence microscopes can be converted to reflection imaging simply by substituting a semi-silvered mirror for a dichroic one and adding polarizing filters in appropriate positions. The normal mercury lamp epifluorescence source can be used, but a visible wavelength excitation filter must be included, so as to cut out UV light, as must a heat filter to prevent damage to the polarizer. Commercially available reflection systems use circularly polarized light, which has some advantages but costs more. For very low magnification reflection imaging a dissecting microscope with an annular illumination collar which fits around the objective lens works very well.

Epifluorescence is probably the most important optical technique in use in subcellular imaging studies at the moment. The details are described in later chapters; only the principles will be discussed here. Fluorescent molecules absorb light of particular wavelengths, and then re-emit light at longer wavelengths and in all directions, the energy difference ultimately heating the specimen. Epifluorescence achieves extremely high specificity and low backgrounds (and thus sensitivity) by a double discrimination—in frequency and direction—against the incident light and in favour of the emitted light. A mercury arc lamp is the most usual light source. The light passes through an excitation filter that matches the absorption characteristics of the fluorescent label being used. The light is then reflected down through the objective lens by a dichroic mirror; this is a special type of mirror which has frequency-dependent reflectivity. Light at frequencies higher than the mirror's cut-off is reflected, whereas light at lower frequencies is passed through. This type of mirror is thus both more efficient than a semi-silvered mirror and more frequency-selective. Back-scattered light from the specimen will have the same wavelength as the incident light and will therefore be reflected again by the dichroic mirror and will not reach the eyepiece. Fluorescently emitted light from the specimen passes through the dichroic mirror and then reaches another specific filter—the emission filter—which only passes light of the emission wavelength of the fluorescent marker and further discriminates against non-fluorescent light.

Provided the microscope has the correct parts and provided the mercury lamp has been correctly set up, no alignment is necessary for epifluorescence. Some precautions are advisable with the mercury lamps generally used as the light source. Most mercury bulbs have a rather short life and are also expensive. The life is considerably shortened by frequent switching on and off. It is a good rule not to switch the bulb off for at least 30 min after switching on, and not to switch it on again until at least 30 min after switching off. The recommended life is only 100 h for the older types of bulb, although it is much longer for a more recent mercury bulb (made by Ushio). Resist the temptation to run the bulbs longer than their recommended time; they can explode if this is done, which is highly dangerous since mercury vapour is then released into the air. If this happens, evacuate and close the room until the mercury vapour has dispersed.

It is important to understand the properties of the optical components and fluorescent probes being used. For example, the commonly used fluorescein and rhodamine probes are excited with visible light, and therefore all objective lenses will be suitable. Use plan-apochromats if available. However, other fluorescent dyes, such as the DNA dye DAPI, need near-UV excitation, and not all objectives transmit these wavelengths; plan-apochromats from some manufacturers work for DAPI, while those from others do not, and special UV-transmitting objectives must be used. Use the highest possible numerical aperture objective; because the objective also acts as the condenser, the amount of fluorescence gathered is very highly dependent on the objective's numerical aperture. It is well worth measuring the transmission spectra of all the filters and dichroic mirrors in the microscope, and comparing them with the fluorescent probes to be used. Do not assume that the limited choice provided by the microscope manufacturer necessarily provides the best solution for each particular need, and make enquiries from a specialist filter manufacturer if necessary.

5. Confocal microscopy

Confocal microscopy is currently the subject of much interest among biologists and could easily occupy a volume of its own. A brief description of the principles is given here, with examples of some of the applications to which it is particularly suited, but the interested reader is referred elsewhere for a fuller description (8). As already mentioned, the idea was first put forward by Minsky many years ago, then implemented by Petran. The optical principle is shown in one form in *Figure 7* for epi-illumination. Light from the illumination source passes through a small aperture which is located at a conjugate focal plane to the specimen. If the aperture is small enough, the light will be focused to a diffraction-limited spot in the specimen. The reflected or fluorescently emitted light then passes back up through the objective, as for conventional reflection or fluorescence imaging, but passes through a second aperture, also at a conjugate plane to the focal plane in the specimen. The second set of light rays shows what happens to light coming from other focal planes: originating at a different distance from the objective it is brought to a focus not at the position of the second (detector) aperture, but either in front of or behind it. Thus at the level of the detector aperture it will be spread out, and only a relatively small proportion will pass through to the detector, the rest being blocked. This shows why the confocal microscope has substantially better rejection of out-of-focus light and produces good focal sections of thick specimens. This applies to imaging a single point in the specimen. In order to obtain an extended image of the focal plane the illuminating spot of light has to be scanned across the specimen, or alternatively the specimen must be moved through a stationary illuminating spot.

This is where the designs of confocal microscope differ. In Petran's machine, the scanning was achieved by a Nipkow disc. This is a disc containing many small apertures arranged in spiral arrays. As the disc is rotated, the apertures scan in a raster. In the original design, the apertures were centrosymmetric and so incident light passed through each aperture and back up through an equivalent aperture on the other side of the disc. In more recent modifications, the incident and reflected light or fluorescence pass through the same aperture in opposite directions. The main problem with this design is that an extended light source is needed to illuminate all the apertures on the disc. This means that only a small amount of light passes through any one aperture. This is not a problem for reflection imaging, where the amount of light emitted from the specimen is large, but limits the use of this type of microscope for fluorescence imaging; an image intensifier or low light level camera is necessary. Most confocal microscopes use laser light to illuminate the specimen. In many early prototypes, the specimen was moved through a stationary optical beam. This simplified the optics and its alignment, but

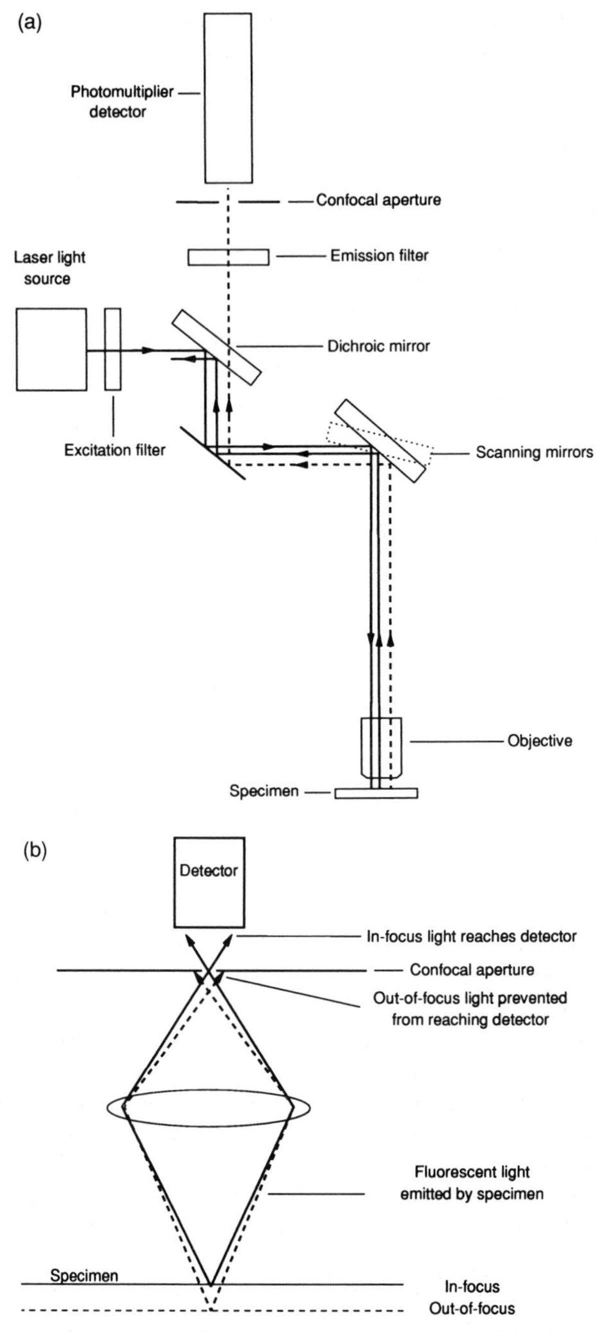

Figure 7. Diagram of a confocal laser scanning microscope. (a) The principle optical components. (b) Ray diagram showing the exclusion of out-of-focus light by the detector pinhole. (Reproduced from (2) with permission.)

made the scanning slow. The first really satisfactory biological confocal microscope, developed by White and colleagues (10) and commercialized by Bio-Rad, scans the light spot across the specimen by a system of vibrating mirrors. This is much faster than scanning the specimen and makes the confocal scanning part a straightforward attachment to most research microscopes. It is very important to have standard fluorescence and other imaging modes available during confocal microscopy of biological specimens, since it is almost invariably necessary to survey large areas of the slide to find a suitable part of the specimen for detailed analysis.

Other commercial confocal microscopes have recently been developed using similar beam scanning systems. One interesting development is the acousto-optical modulator, which is used on at least one instrument. This device causes beam deflection by inducing acoustical waves in a glass plate. This is capable of much faster scanning than the mechanical vibrating mirrors, and can be used at video frame rates. Another development is the use of scanning slit apertures instead of circular apertures; this has the advantage of allowing a much greater proportion of the light from the specimen to contribute to the image, at the expense of a loss of some of the 'confocality'.

Using and adjusting a confocal microscope is beyond the scope of this chapter, and in any case is very specific to the particular model of microscope. However, there are several matters to be considered before beginning any confocal microscopy. The first question is why use a confocal microscope at all? Confocal microscopes are very good indeed for some things, but no better than conventional microscopes for others. They are good at giving clean optical sections and at producing three-dimensional reconstructions at the subcellular and cellular levels. They are not well suited to analysing any structure much larger than a few cells; other methods are probably more efficient for analysing large specimens like an entire root. Although confocal microscopes can be used to analyse structures within a tissue, the depth of penetration possible is usually limited to about 100 μm below the surface by the working distance of the objective lens and by absorption and scattering in the specimen. Most confocal microscopes to be found in biology laboratories are primarily fluorescence microscopes, and so can only visualize fluorescent objects, and only if the fluorescence is excitable by the few wavelengths available from the laser provided. Although some structures are autofluorescent, the vast majority of imaging studies make use of introduced fluorescent probes, as in conventional fluorescence. Thus, for the most part, only consider confocal microscopy once good, or at least strong, conventional fluorescence labelling has been achieved. Then consider confocal microscopy only if the biological questions being asked require it—for example are good three-dimensional images required, or are fine, faint structures in the presence of other more brightly labelled structures being examined, or is ultimate optical resolution required. In spite of the foregoing caveats, the fact remains that confocal microscopy is a very powerful technique and is rapidly

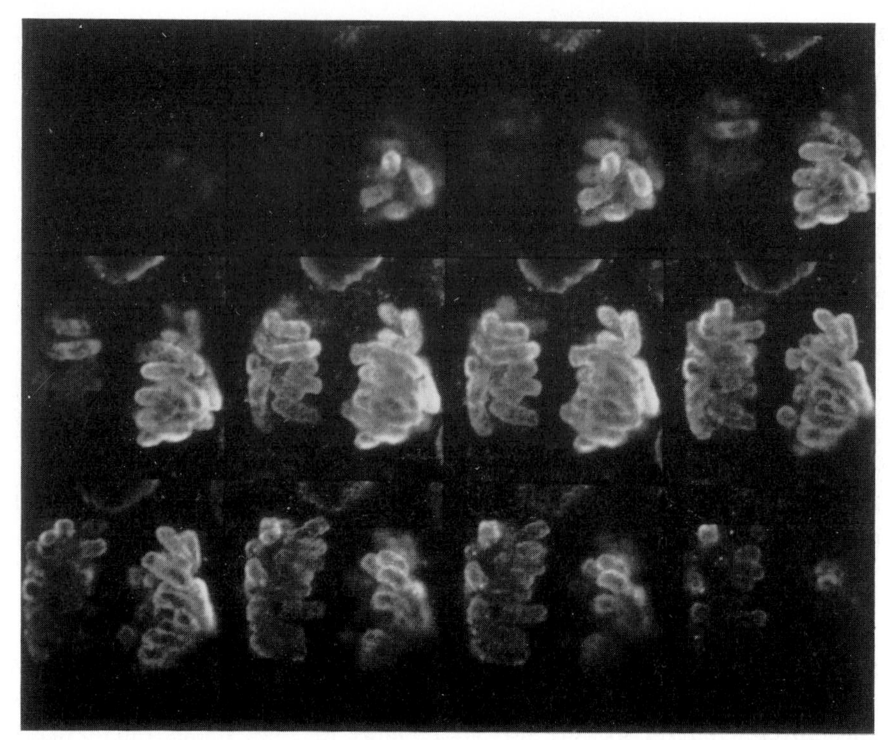

Figure 8. Confocal optical sections through anaphase chromosomes of *Vicia faba*, fluorescently labelled with an antibody that localizes to the outside of the chromosomes.

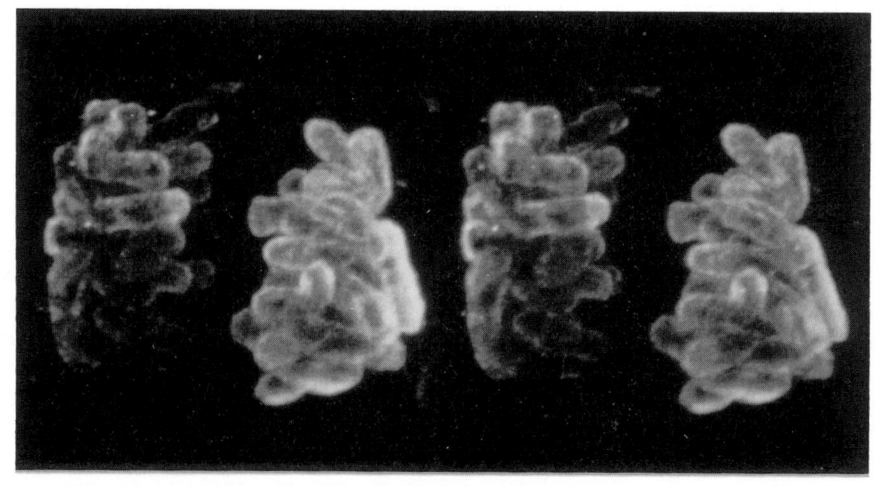

Figure 9. Stereo projection of the entire data stack of optical sections from *Figure 8*, giving an overall impression of the three-dimensional structure.

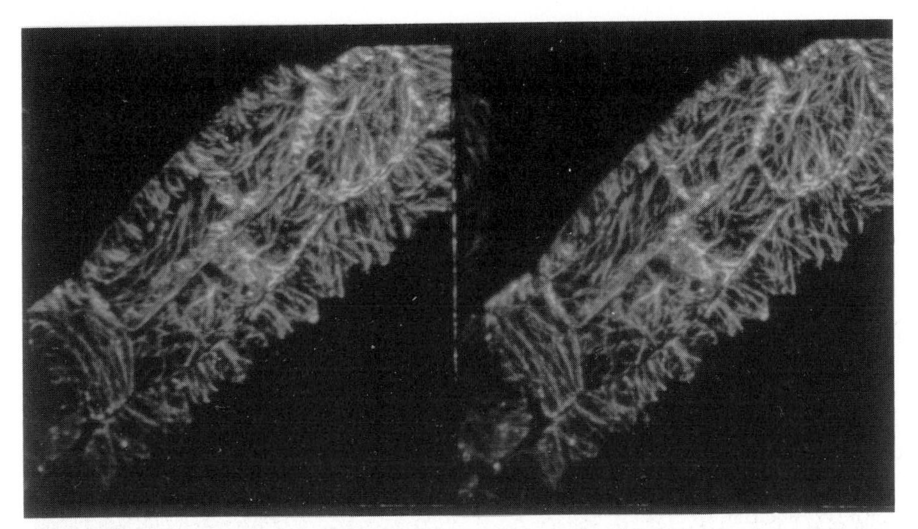

Figure 10. Stereo projection of three-dimensional confocal image of microtubules in epidermal cells of *Tradescantia stramonium*, fluorescently labelled with anti-tubulin. (Reproduced from (11) with permission.)

becoming a standard part of cell biological research. An example of the optical sectioning capability is shown in *Figure 8*, which shows anaphase chromosomes from root tip issue of *Vicia faba*, immunofluorescently labelled with an antibody that reveals the surface of the chromosomes. In *Figure 9* the entire set of sections has been projected to give a calculated stereo pair, giving an overview of the three-dimensional structure. Another, very different, example is shown in *Figure 10*. This shows labelling of the cortical microtubules in some epidermal cells of *Tradescantia*. The cells are very large, which makes it very difficult to analyse the cytoskeletons using conventional fluorescence microscopy. By the use of stereo and rotated reconstructions it is possible to determine the details of the arrangement of the cortical microtubule arrays.

Acknowledgements

We are grateful to Luise Janniche, David Flanders, and Coral Robinson for providing some of the specimens used in this chapter. Much of the work was supported by the Agricultural and Food Council of the UK, via a grant-in-aid to the John Innes Institute.

References

1. Grew, N. (1673). *An idea of a phytological history propounded. Together with a continuation of the anatomy of vegetables, particularly prosecuted upon roots. With*

an account of the vegetation of roots grounded chiefly thereupon. Published by Nehemiah Grew.

2. Rawlins, D. J. (1992). *Light microscopy*. Bios Scientific, Oxford.
3. Lacey, A. J. (1989). *Light microscopy in biology: a practical approach*. IRL Press, Oxford.
4. Bradbury, S. (1989). *An introduction to the light microscope*. Royal Microscopical Society Handbook no. 1. Oxford University Press.
5. *RMS dictionary of light microscopy*. (1989). Royal Microscopical Society Handbook no. 15. Oxford University Press.
6. Spencer, M. (1982). *Fundamentals of light microscopy*. Cambridge University Press.
7. O'Brien, T. P. and McCully, M. E. (1981). *The study of plant structure. Principles and selected methods*. Termacarphi Pty, Melbourne.
8. Taylor, D. L. and Wang, Y.-L. (1989). *Fluorescence microscopy of living cells in culture*, Volumes A and B. Academic Press, London.
9. Pawley, J. B. (1989). *Handbook of biological confocal microscopy*. Plenmum Press, New York.
10. White, J. G., Amos, W. B., and Fordham, M. (1978). *J. Cell Biol.*, **105**, 41.
11. Shaw, P. J. and Rawlins, D. J. (1991). *Prog. Biophys. Mol. Biol.*, **56**, 187.

2

The use of fluorescent probes for studies of living plant cells

K. J. OPARKA and N. D. READ

1. Introduction

The use of fluorescent probes in the field of plant cell biology has undergone phenomenal expansion in the last 10 years. There is now an enormous number of fluorescent probes on the market which are being used for a variety of purposes, including highly specific vital staining of membranes and organelles, Ca^{2+} and pH measurement, and tracing of intercellular transport pathways (for a current update of available probes, see reference 1). Fluorescent probes have also become the 'backbone' of several immunocytochemical techniques (see Chapter 10).

Several of the methodologies utilizing fluorescent probes have become extremely complex, sometimes requiring specialist equipment for the detection of minute quantities of a fluorescent product or ion. It is beyond the scope of this chapter to deal at length with specialized fluorescence techniques, or indeed to catalogue the several hundred probes available to plant cell biologists. Instead, we have selected some of the fluorescent probes which have been most commonly employed in studies of living plant cells (*Table 1*) and have produced a set of protocols for using them which we hope will satisfy the basic requirements of most researchers. Emphasis has been placed on those fluorescence techniques which are within reach of laboratories in possession of a basic fluorescence microscope.

We describe the different ways in which fluorescent probes can be loaded into plant cells and detail some specific protocols for tracing, measuring, and staining with these dyes. Particular attention has been paid to the problems most commonly encountered using fluorescence techniques and how these might successfully be overcome.

2. Methods for loading cells with fluorescent probes

A wide range of methods is available for loading fluorescent probes into plant cells. The ability of a probe to cross the plasma membrane depends on a

Table 1. Properties and staining characteristics of some commonly employed fluorescent probes

Probe selectivity	Fluorescent probe	Mol. wt.	Probe concentration	Filter combination[a]
Symplastic transport	Lucifer Yellow CH	457	10–50 mM[b]	blue
	Cascade Blue hydrazides	548–645	10–50 mM[b]	UV
Phloem transport	5(6) carboxy-fluorescein	376	1 mM	blue
Apoplastic transport	HPTS	524	0.5 mM	UV
	Lucifer Yellow CH	457	20 mM[c]	blue
Lipids and membranes	Nile Red	318	3–300 μM	blue or green
ER and mitochondria	$DiOC_6$	573	1–20 μM	blue
Nuclei	DAPI	350	0.5–30 μM	UV
Lignin and cutin	Acridine Orange	265	4 μM	green
Callose	Aniline Blue	737	7 μM	UV
Cellulose	Calcofluor White M2R	931	1 μM	blue
Free Ca^{2+}	Fura-2	832	100 μM[d]	UV[e]
	Indo-1	840	100 μM[d]	UV[f]
	Fluo-3	855–960	100 μM[d]	blue
	Calcium Green-1	1233	100 μM[d]	blue
pH	BCECF	520	5–50 μM[g]	blue[h]
	carboxy-SNARF-1	453	5–50 μM[g]	blue or green[g]
Live cells	FDA	416	240 μM	blue
Dead cells	propidium iodide	668	150 μM	blue or green

[a] Common filter combinations for visualizing fluorescent dyes are: UV (330–380 nm excitation filter (exf), 400 nm dichroic mirror (dm), and 420 nm barrier filter (bf), blue (450–490 nm exf, 510 nm dm, and 520 nm bf), and green (510–560 nm exf, 580 nm dm, and 580 nm bf). Although these standard filter combinations will generally achieve satisfactory results, slightly different filter combinations may be necessary to increase the dye fluorescence detected and to minimize the problems of autofluorescence.

[b] Relatively high dye concentrations are recommended to minimize volume injected and to maximize fluorescence signal.

[c] Dye concentration recommended for aldehyde fixation. Lower concentrations can be used for freehand sections of fresh material.

[d] Dye concentration recommended for iontophoretic injection.

[e] For dual excitation ratio imaging of $[Ca^{2+}]_c$ by Fura-2, 340 and 380 nm excitation and 510 nm emission wavelengths are recommended (3).

[f] For dual excitation ratio imaging of $[Ca^{2+}]_c$ by Indo-1, 340 nm excitation and 405 and 408 nm emission wavelengths are recommended (3).

[g] Dye concentration recommended for ester loading.

[h] For dual excitation ratio imaging of pH by BCECF, 400 and 490 nm excitation and 535 nm emission wavelengths are recommended (1).

[i] For dual emission ratio imaging of carboxy-SNARF-1, 515 nm excitation and 580 and 640 nm emission wavelengths are recommended.

combination of its dissociation constant (pK_a) and polarity (2). Some stains are intrinsically membrane permeant and will cross the plasma membrane freely by simple diffusion while many others are impermeant molecules which require a specific means of delivery in order to reach their target site within the cell.

2.1 Loading membrane-permeant probes

Simple diffusion may be sufficient to carry permeant probes into plant cells. Some membrane-permeant probes are colourless, except when they react within the cell to form a fluorescent product. The cells can therefore be mounted in the probe without washing. In the case of intrinsically fluorescent compounds (e.g. fluorescein and its derivatives) diffusion may allow sufficient entry of the probe into the cell but the fluorescence of the staining medium will mask its location. In such cases the cells have to be washed before examination to remove unbound or non-internalized probe. Washing steps should be performed with caution as probes which diffuse into cells may also readily diffuse out, causing potential dislocation artefacts. Dye leakage is a common problem with diffusible probes (Section 4.6).

2.2 Loading membrane-impermeant probes

The use of membrane-impermeant probes overcomes the problems associated with diffusive dye leakage but brings the extra problem of by-passing the plasma membrane to introduce the probe into the cell in the first place. A range of delivery strategies is available for impermeant probes, the choice depending on the nature of the experimental system under study. The following methods may be used.

2.2.1 Ester loading

A number of dyes can be loaded as their esterified derivatives which are non-polar and can cross the plasma membrane easily. Once inside the cell the ester groups are cleaved off by the action of intracellular esterases causing the active polar dye to be released. This approach has proved successful for loading some plant cell types, particularly protoplasts, with Ca^{2+} dyes (e.g. Fluo-3, Calcium Green, or Indo-1) and pH dyes (e.g. BCECF (2′,7′-*bis*-(2-carboxyethyl)-5-(and-6)-carboxyfluorescein) or SNARF-1 (semi-naphthorhodafluor).

Ester loading is also the principle behind using fluorescein diacetate (FDA) as a stain for cell viability (see Section 4.3.1).

Protocol 1. Ester loading Ca^{2+} and pH dyes

Reagents

- DMSO (Sigma)
- Pluronic F-127 (Molecular Probes)

Protocol 1. *Continued*

- HPLC grade water (BDH)
- acetoxymethyl (AM) ester of dye (Molecular Probes)

1. Make up a 5 mM stock solution of the acetoxymethyl (AM) ester of the dye in anhydrous DMSO. The stock solution can be stored for up to several months at −70 °C.
2. Dilute the dye in HPLC water to a suitable concentration (typically between 25 and 100 μm for Ca^{2+} dyes and between 5 and 50 μm for pH dyes). To aid solubilization of the dye-ester, 1–5 μl Pluronic F-127 solution (prepared according to the manufacturer's instructions) per ml of dye can be added. Pluronic is a surfactant of low toxicity.
3. Incubate the cells in dye solution either at room temperature at 32 °C in the dark for a suitable period (typically between 30 min and 2 h for Ca^{2+} dyes and between 20 and 45 min for pH dyes[a]). Incubation at 32 °C has been found to increase dye uptake in some cell types.
4. Wash cells of unloaded dye-ester.

[a] The concentration of dye and the time of loading need to be optimized for each cell type used.

2.2.2 Electropermeabilization

Electropermeabilization, sometimes termed electroporation, causes a temporary loss in the semi-permeability of cell membranes by exposing cells to electric pulses of kilovolts per centimetre intensity and of micro- to millisecond duration. This technique has been much used by molecular biologists to transform cells genetically with DNA (for protocols see Chapter 8). Electropermeabilization has also been used to load both living protoplasts and walled cells with various fluorescent probes (e.g. Ca^{2+} dyes and rhodamine-phalloidin). The conditions necessary to permeabilize cells vary considerably between types of cells and the dyes to be loaded. Electropermeabilization variables which have been modified for dye loading include (3):

- concentration of dye in the surrounding medium (e.g. 50 μM to 6 mM for Ca^{2+} dyes; 40 nM for rhodamine-phalloidin)
- field strength (1.0–7.5 kV/cm)
- time constant (15 μsec to 10 msec)
- number of pulses (1–10) and time separating pulses (5–30 sec)

Apparatus for electropermeabilization is available from various manufacturers.

2.2.3 Microinjection

Several of the fluorescent probes currently used in plant cell biology are either too large or too highly charged to cross cell membranes by the dye loading methods described above and, accordingly, these dyes require to be introduced into cells directly. Microinjection provides a means to target fluorescent probes into specific cell types and within these cells to specific compartments (e.g. cytosol or vacuole). It also avoids the washing procedures necessary to remove external dye and is more economical in terms of the amount of dye used. It should be borne in mind that all types of microinjection involve impalement of the cell with a micropipette and inevitably induce some degree of cellular damage, albeit temporary. Thus, cell viability and other indicators of cell health should be assessed routinely (Sections 4.1 and 4.3).

2.2.4 Minimum requirements for microinjection

The degree of sophistication in microinjection systems depends to a large extent on the requirements of the investigator and the difficulty of the cell system being studied. However, a basic set up would comprise the following:

(a) Fluorescence microscope. Inverted microscopes are in common use as the objectives are below the stage, allowing easy access to the specimen from above. An inverted microscope also has the advantage of permitting a sample perfusion chamber to be used easily. If an upright microscope is selected (perhaps essential if plant surfaces are to be injected) long working distance lenses are necessary to allow sufficient clearance between the objective and the specimen during injections.

(b) Microcapillary holder, coupled to a coarse three-directional (x, y, and z movements) variable tilt control unit. This equipment allows coarse positioning of the capillary tip above the tissue and also its angle of attack. Coarse-control systems which are not mounted on the microscope stage are also available but these can lead to increased vibration from the bench.

(c) Micromanipulator that allows fine, three-directional movements (x, y, and z) of the capillary holder. These systems can work by either a mechanical or hydraulic mechanism; hydraulic systems tend to eliminate transmitted vibration from the manipulator to the capillary.

(d) Needle puller to make fine microcapillaries. Several types of needle puller are available, the simplest relying on gravity to form the microcapillary tip. In this system, a glass capillary tube is clamped vertically through a heating coil and a variable weight attached at the lower end of the tube. As current passes through the heating coil the glass becomes hot and is stretched and thinned by gravity acting upon the weights. The glass capillary eventually breaks leaving a fine tapered microcapillary. Variations in temperature and weight produce microcapillaries of differing tip diameter. A common but more expensive system is one that replaces gravity with an electromagnet. The main advantage of the latter is that it allows a much finer control over capillary shank length and tip aperture.

(e) Injection system. Once the target cell has been impaled with the microcapillary, force is required to eject the probe from the capillary tip into the cell. This can be done by the application of a positive- or negative-going current (iontophoresis) or simply by applying physical pressure behind the solution in the needle tip (pressure injection). Several iontophoretic and pressure-injection systems are commercially available. The choice of injection system will be dictated by the system under study and the nature of the compound to be injected into the cell.

2.2.5 Iontophoretic injection

Iontophoretic injection involves using an electrical current to introduce charged dye molecules into cells. It is best suited for injecting dyes with molecular weights up to approximately 10 kDa. Iontophoretic current sources can be obtained commercially but simple devices can be easily made (4, 5). The polarity of the current used is the same as the charge on the dye molecule (e.g. negative-going current is used to inject negatively charged dyes). The time required to iontophorese sufficient dye varies and depends on such factors as pipette tip diameter, dye concentration, the size of the current, sensitivity of the fluorescence detector, and the quantum yield of the dye. For more detailed descriptions of iontophoretic injection methods the reader is referred to references 4–6. The following parameters have been used successfully to introduce Ca^{2+} dyes into cells (3):

- micropipettes with external diameters of 0.1–1.0 μm
- pre-loaded dye concentrations of 50 μm to 1 mM
- currents of < 0.1 to 10 nA
- injection times of 10 sec to 10 min

In general, cells have a better tolerance of low currents for long periods than high currents for short periods of injection. Either continuous or pulsed currents have been used to introduce dyes into plant cells.

2.2.6 Pressure injection

Pressure injection may be the appropriate choice when the compounds to be injected are either uncharged, large ($M_r > 1$–10 kDa), or both. Pressure (typically in the range 200–500 kPa) is generated at the back end of the injection micropipette either through the use of a commercially available pressure-injection system, which delivers a preset pulse of pressure into the injection pipette, or manually through the use of micrometer-based syringes.

Recent advances in pressure injection have allowed the simultaneous measurement of cell turgor pressure during microinjection (7).

2.2.7 Filling micropipettes with fluorescent probes

Several filamented micropipettes are commercially available (e.g. from Clark Electromedical Instruments, Pangbourne, UK). Such micropipettes have a solid glass filament running through the central chamber and can be filled simply by placing the back end into a droplet of the solution to be loaded. The probe moves rapidly to the pipette tip by capillarity. If a completely hydraulic injection system is required (sometimes desirable as air is compressible and leads to loss of sensitivity during injection) then the remainder of the micropipette is filled with 200/100 silicone fluid using a heat-pulled Pasteur pipette with an outer diameter smaller than the internal diameter of the microcapillary being loaded.

In the case of non-filamented micropipettes, capillarity is ineffective as a filling procedure and dye has to be introduced into the pipette tip using the Pasteur pipette method.

3. Fluorescent probes for living cells

3.1 Probes for studying symplastic transport

A number of fluorescent probes are in current use as tracers of symplastic transport in plants. As a general rule, symplastic traces must fulfil a number of criteria. They should:

- be completely membrane impermeant and remain in the compartment to which they were introduced
- be small enough to pass readily through plasmodesmata
- be non-toxic to the cells through which they are moving
- not bind to cell constituents
- possess a high quantum yield, allowing their use at very low concentrations.

Lucifer Yellow CH (LYCH) (8) and Cascade Blue hydrazide (Molecular Probes) are in common use as symplastic tracers and fulfil most of the above requirements. The dyes are microinjected into the cytosol of single cells and their intercellular transport is monitored under a fluorescence microscope. The dyes are usually injected at concentrations of 10–50 mM, although the final choice of concentration is largely dependent on the sensitivity of the fluorescence detector system used.

Considerable advances in the study of plasmodesmata have been made in recent years through the use of fluorescent probes (usually fluorescein isothiocyanate (FITC) conjugated peptides) of differing molecular weights. Such probes can be injected directly into the cytosol of living plant cells to determine the functional size exclusion limit of plasmodesmata connecting differing cell types (9, 10).

Protocol 2. Synthesis and purification of FITC-conjugated peptides

A. *Synthesis* (11)

Reagents:

- acetone
- peptides of differing molecular weights (Sigma)
- $KHCO_3$
- FITC (Sigma)

1. Dissolve 10 mg FITC in 1 ml acetone (this can be difficult to achieve completely).
2. Add this with mixing to a solution containing 12.5 μmol of the peptide (approximately 5 mg for a di- or tri-peptide) and 20 mg $KHCO_3$ in 9 ml water.
3. Mix thoroughly for 10 min then store in darkness for 2 days at room temperature. Freeze dry to a small volume (< 0.5 ml).

B. *Purification* (9)

The following are required:

- Whatman 3MM chromatography paper (BDH)
- UV lamp
- Sephadex G-25 column (Pharmacia)
- $KHCO_3$

1. Apply dye solution as a thick streak across a sheet of Whatman 3MM chromatography paper measuring about 30 cm × 30 cm.
2. Develop by descending chromatography using 50 mM $KHCO_3$ as the eluant. The conjugate is bright yellow and moves away from the orange unreacted FITC.
3. Examine the sheet under UV-A light to reveal fluorescent bands, making sure that the narrow leading orange band is removed.
4. Elute into water and freeze dry to a small volume. At least two paper chromatography purifications may be necessary to achieve good purity. With large FITC conjugates (e.g. FITC-insulin) one paper chromatography separation followed by passage of the conjugate through a Sephadex G-25 column pre-equilibrated with distilled water appears to be successful.

C. *Assessment of purity* (9)

The following are required:

- Kieselgel 60 TLC plates (BDH)
- butan-1-ol
- pyridine
- acetic acid

1. Load spots of purified conjugate onto a Kieselgel 60 TLC plate. Resolve with butan-1-ol:water:pyridine:acetic acid (15:12:10:3 by volume).
2. Observe under UV-A light. FITC has an R_f of about 0.9 in this system[a, b].

[a] Approximate values for other conjugates are as follows, with R_f given first and M_r second. F-Phe[3]: 0.82, 849; F-Gly: 0.77, 464; F-Tyr-Gly-Gly: 0.72, 684; F-Ala[6]: 0.69, 839; F-Gly: 0.68, 536; F-His: 0.59, 544; F-Gly[6]: 0.51, 749; F-Glu-Glu: 0.42, 666.
[b] FITC is usually resolved as two closely spaced spots.

3.2 Probes for monitoring phloem transport

Until recently, phloem transport in plants was difficult to monitor except by the use of radioisotope procedures. Some fluorescent probes have an optimal combination of dissociation constant (pK_a) and oil-to-water partition coefficient which allows them to enter the phloem freely but subsequently to become 'trapped' within the sieve-element lumen. 5(6)-carboxyfluorescein (CF) is a good example of such a phloem-mobile fluorescent probe (12).

Protocol 3. Tracing phloem transport with 5(6) carboxyfluorescein

A. *Preparation*

Reagents:

- 5(6) carboxyfluorescein (Sigma)
- KOH
- HCl

1. Dissolve CF in a minimal volume of 0.3 M KOH.
2. Dilute with water to 1 mM and adjust to pH 6.3 with 1 M HCl.

B. *Application*

Equipment:

- carborundum powder 300 grit (BDH)
- paint brush

Protocol 3. *Continued*

- cotton cloth
- Parafilm (BDH)
- glycerol

1. Abrade 1 cm^2 of adaxial leaf surface (avoiding large veins) using carborundum powder gently rubbed into the leaf surface with a small paint brush.
2. Apply CF as an approximately 20 μl drop onto the abraded region. For continuous applications of CF, a plastic well (1 cm diameter) can be affixed to the abraded region with silicone grease and the reservoir continuously replenished with CF. CF can be applied to the petiole or stem by abrasion of the epidermis with carborundum followed by application of the dye in a cotton cloth soaked in CF solution and sealed with Parafilm ribbon.
3. At varying periods after dye application (one to several hours) mount freehand sections of petiole, stem, or root in water: glycerol (1:9 (v/v)) and observe under a fluorescence microscope with blue excitation (*Table 1*). Unlike fluorescein, CF remains confined to the phloem and does not leak readily from tissue sections. Examine untreated control sections for tissue autofluorescence (Section 4.4).

3.3 Apoplastic tracers

Several fluorescent compounds have been used to trace the plant apoplast. Apoplastic tracers should be non-toxic and able to move as freely as water through cell walls. In practice, both criteria may prove difficult to meet. Negatively charged dyes of low pK_a are to be preferred as these are not bound by the prevailing positive charge of the cell walls and are sufficiently dissociated at apoplastic pH not to cross the plasma membrane and stain cell contents (for a discussion on the use of apoplastic tracers in plants see reference 13). The following probes have been used successfully to trace apoplastic transport pathways in plants.

Protocol 4. Tracing the apoplast with HPTS (14)

1. Prepare a 0.02% (w/v) aqueous solution of HPTS (8-hydroxypyrene-1,3,6,-trisulphonic acid, trisodium salt, Molecular Probes).
2. Dip intact plant roots or cut stems into the dye solution and allow the plant to transpire for one to several hours.

3. At the conclusion of dye treatment, rinse the root for several seconds under flowing tap water to remove excess dye from the outer surfaces, blot dry with absorbent tissue, and hand section with a clean, dry razor blade.
4. Examine sections quickly using UV excitation (*Table 1*). Use non-fluorescent immersion oil to avoid lateral spread of dye from the apoplast into the cytoplasm of cut cells.

We have also found the fluorochrome LYCH to be an excellent apoplastic tracer (15). It is non-toxic and highly fluorescent and has an exceptionally low pK_a. Although more expensive than some other anionic probes, LYCH has the advantage that it can be retained in the apoplast by aldehyde fixation and its distribution between cells examined at high resolution using semi-thin plastic sections.

Protocol 5. Tracing the apoplast with Lucifer Yellow CH

Reagents:

- LYCH (Molecular Probes, Sigma)
- paraformaldehyde
- glutaraldehyde
- sodium cacodylate
- graded ethanol series (25–100%)
- London Resin White, medium grade (Agar Scientific)
- plastic embedding capsules (Agar Scientific)
- Fluoromount (Gurr, BDH)

1. Prepare a 1% (w/v) solution of LYCH and apply to roots or stems as described in *Protocol 4* for one to several hours.
2. Cut freehand sections and examine immediately under a fluorescence microscope with blue excitation (*Table 1*), or, alternatively:
3. Rinse tissue in three rapid changes (20 sec) of fixative (3% (w/v) paraformaldehyde, 2% (v/v) glutaraldehyde in 50 mM sodium cacodylate buffer, pH 6.8). Fix tissue *en bloc* with fresh fixative overnight at 5 °C.
4. Rinse in several changes of 50 mM cacodylate buffer, dehydrate in a graded ethanol series (25–100%), and leave overnight on a rotator in 50:50 dry ethanol:London resin. Transfer to fresh resin at room temperature and rotate for a further 8 h.

Protocol 5. *Continued*

5. Allow to polymerize in plastic capsules at 60 °C overnight, eliminating as much air as possible from the capsule. Section tissue on glass knives at 1–2 μm thickness.
6. Collect sections on a glass slide, dry down, and mount without further staining in Fluoromount or another suitable non-fluorescent mountant. Examine sections with blue excitation (*Table 1*).
7. Prepare control tissue as above but in the absence of LYCH. Cells walls containing LYCH will appear a highly fluorescent yellow. The walls and cytoplasm of control sections appear a dull green-yellow, easily distinguishable from those containing the fluorescent probe. If required, parallel LYCH-containing sections can be counterstained with a variety of water-soluble stains.

3.4 Quantifying uptake of fluorescent probes by plant cells

In several circumstances it may be necessary not only to observe the subcellular distribution of a fluorescent probe but also to quantify the amount taken up. The latter allows an examination of the uptake kinetics of probe in response to a range of experimental treatments, as well as an evaluation of the transport processes involved in its uptake (16). Several probes can be extracted from plant cells following thorough washing of the cells and lysis in an appropriate detergent (e.g. Triton X-100 or SDS). This approach is applicable to a wide range of probes but is obviously most suited to those which do not bind appreciably to cell contents. A general protocol for the extraction of fluorescent probes from plant protoplasts is given below.

Protocol 6. Measurement of fluorescent probes in leaf protoplasts

Reagents required:

- 5(6)-carboxyfluorescein (Molecular Probes, Sigma) or appropriate non-binding probe
- mannitol
- Mes–KOH buffer
- Triton X-100

1. Isolate leaf protoplasts by standard methods (see Chapter 8) and resuspend in buffered osmoticum (e.g. 400 mM mannitol, 20 mM Mes–KOH, pH 5.5, but may vary depending on protoplast type).
2. Dissolve the dye in a suitable solvent (e.g. buffered osmoticum or ethanol if required). Mix the protoplasts (10^5/ml final dilution) with the dye stock solution.
3. Incubate the protoplasts for varying time intervals at 25 °C[a].
4. During the incubation period wash 1 ml aliquots of the protoplast suspension three times in 10 ml of buffered osmoticum by centrifugation at 100 *g* for 2 min. Resuspend in buffered osmoticum and monitor the accumulation of dye within the protoplasts using a fluorescence microscope.
5. To quantify dye uptake, lyse the protoplast pellet that results from centrifugation by the addition of buffered osmoticum plus 0.1% Triton X-100 (release of dye is immediate).
6. Measure the fluorescence of the lysate using a spectrofluorimeter, at appropriate excitation and emission wavelengths for the dye being examined (for CF the appropriate values are 490 nm and 515 nm for excitation and emission respectively), with reference to a standard curve of dye prepared in lysis buffer.

[a] Controls are essential to measure staining of damaged protoplasts. This can be done by including a brief (30 sec) incubation in dye which allows dead cells to become stained without significant entry of probe into living cells. This value is subsequently subtracted to calculate net uptake rates.

3.5 Organelle stains

A number of fluorescent probes bind to plant cell organelles specifically and are non-toxic at those concentrations which produce optimal staining reactions. In recent years the number of organelle-specific probes which has come on the market is overwhelming (1). Here, we will restrict ourselves to a number of organelle-specific stains in common use in plant cell biology (*Table 1*).

3.5.1 Lipids and membranes

Nile Red (9-diethylamino-5H-benzo[a]phenoxazine-5-one; Molecular Probes, Sigma) is an extremely sensitive fluorescent probe for lipids, especially neutral lipids, which frequently allows the visualization of membranes as well. By being hydrophobic, Nile Red is readily membrane permeant and preferentially dissolves in lipids within cells (17).

Protocol 7. Staining lipids and membranes with Nile Red

1. Make up a 10 mg/ml stock solution by dissolving Nile Red (Molecular Probes) in acetone. The stock solution can be kept for several months at 4 °C in the dark.
2. Apply the fluorochrome to isolated cells or tissues after diluting it to a concentration of 1–100 μg/ml depending on the intensity of staining required.
3. Wash the cells or tissues briefly in distilled water or examine, while still immersed in stain, under blue excitation (*Table 1*). The lipids fluoresce yellow-gold.

3.5.2 Endoplasmic reticulum and mitochondria

Mitochondria and endoplasmic reticulum (ER) can be stained up with potential-sensitive dyes of which the most common is $DiOC_6$ (3,3-dihexyloloxacarbocyanine), a cationic, cyanine dye which accumulates inside mitochondria and ER in a potential-sensitive manner. It is readily permeant and non-toxic and provides a non-invasive means with which to examine the relative membrane potentials of cells and organelles too small to be impaled with microelectrodes.

Protocol 8. Staining mitochondria and ER with $DiOC_6$

1. Dissolve $DiOC_6$ (Molecular Probes, Sigma) in a minimal quantity of DMSO or absolute ethanol and make up a stock solution of 5 mg/ml in distilled water.
2. Apply the fluorochrome to cells or tissues at a concentration of 0.5–10 μg/ml.
3. Wash the tissue briefly in distilled water or examine, while still immersed in the stain, under blue excitation (*Table 1*). Mitochondria and ER will fluoresce green[a].

[a] Mitochondria stain more rapidly than the ER and may swell up with prolonged exposure to blue irradiation. As a general rule, low dye concentrations (< 0.5 μg/ml) favour mitochondrial staining while higher dye concentrations (2–5 μg/ml) stain the ER as well as mitochondria. The nuclear membrane may also become stained at high dye concentrations.

3.5.3 Nuclei

DAPI (4′,6-diamidino-2-phenylindole) is a very sensitive probe which stains AT-rich regions of DNA. Cells are permeant to the dye and it does not fluoresce until bound to DNA. Besides nuclear DNA, mitochondrial and chloroplast DNA can sometimes be visualized with DAPI staining. DAPI can also stain other cell components (e.g. polyphosphates).

Protocol 9. Staining nuclei with DAPI

1. Make up a 1 mg/ml stock solution of DAPI (Molecular Probes, Sigma) dissolved in distilled water. The stock solution can be kept for several months at 4 °C in the dark.
2. Apply the fluorochrome to isolated cells or tissues at a concentration of 0.1–10 μg/ml and stain for 5–10 min.
3. Wash the tissue in distilled water and examine with UV excitation (*Table 1*). DNA fluoresces blue.

3.6 Ca^{2+}-selective probes

A large number of fluorescent dyes which are highly selective for free Ca^{2+} have recently become available (Molecular Probes). In studies of plant cells, the most commonly used of these fluorochromes have been Indo-1, Fura-2, Fluo-3, and Calcium Green-1. These dyes have different properties from each other, particularly in relation to their spectral characteristics, affinities for free Ca^{2+}, and quantum yields, and are best suited for measuring cytosolic free Ca^{2+} ($[Ca^{2+}]_c$) in the 50 nM–1 μM range. The Ca^{2+} dye of choice is dependent on the application. On binding to Ca^{2+}, Indo-1, and Fura-2 exhibit shifts in their emission and excitation spectra, respectively, allowing them to be used for ratio analysis and therefore achieve more precise quantification of $[Ca^{2+}]_c$ (3).

Cells are loaded with sufficient dye to obtain a reasonable signal-to-noise ratio with the detector being used. It is important to keep the dye concentration as low as possible to avoid significant buffering of the $[Ca^{2+}]_c$. The Ca^{2+} dyes can be loaded in different ways and unfortunately no single method works for all cell types. The dyes can be loaded in their charged free acid forms either by iontophoretic microinjection (Section 2.2.5). Alternatively, pressure injection (Section 2.2.6) can be used to load high molecular weight dye–dextran conjugates which, in many cases, do not readily compartmentalize within organelles or leak out of cells (Section 4.6). The other main method of introducing Ca^{2+} dyes into cells is as their uncharged AM esters (Section 2.2.1). In our hands, however, ester loading has proved routinely successful only with protoplasts (3).

A number of critical controls must be employed when using Ca^{2+} dyes. It is particularly important to determine whether the dyes are actually reporting $[Ca^{2+}]_c$ because dye compartmentation within organelles is common (Section

4.6). Organelles such as mitochondria, ER, and vacuoles possess free Ca^{2+} concentrations above 5 μm, which is significantly higher than in the cytosol. Because the dyes have high affinity for free Ca^{2+} (most have a K_{dis} between 100 and 500 nM), any dye sequestered within these organelles will be Ca^{2+}-saturated and thus maximally fluorescent. It is important to demonstrate that artificially raising $[Ca^{2+}]_c$ levels causes a corresponding change in dye fluorescence. This can be done in a variety of ways. The most common approach is to increase $[Ca^{2+}]_c$ levels by adding a Ca^{2+} ionophore (e.g. 10 μm Br-A23187 or 100 μm ionomycin) in the presence of external Ca^{2+}. The latter is often present within the medium but if not 10 mM Ca^{2+} can be added with the ionophore.

An explanation of how $[Ca^{2+}]_c$ is calibrated by using fluorescent dyes is beyond the scope of this chapter. Instead, the reader is referred to Thomas and Delaville (18) for a detailed practical account of what is involved. Protocols for the calibration of $[Ca^{2+}]_c$ in plant cells have been reviewed in (3).

3.7 pH-dependent probes

A range of pH probes is available from Molecular Probes. The pH dyes which are currently recommended for use with plant cells are BCECF and carboxy-SNARF-1 which are dual excitation and emission ratio dyes, respectively. These dyes are best suited for measurements in the 6.0–8.0 pH range.

Similar loading protocols to those used for the Ca^{2+} dyes (Section 3.6) are used for the pH dyes. In general, however, it seems to be much easier to ester-load (Section 2.2.1) both protoplasts and walled plant cells with pH dyes than it is with Ca^{2+} dyes.

3.8 Stains for cell wall components

The following stains are in general use for localizing cell wall components. While this chapter is devoted to living cells, some of the protocols below are also particularly suited to fixed and embedded material and are included here for completeness.

3.8.1 Cellulose

Calcofluor is used to stain cellulosic cell walls (19). It is best used with fresh sections cut either by hand or by cryotome, although it may also be used effectively with dewaxed sections cut from wax-embedded blocks or with semi-thin sections of material embedded in London resin. Calcofluor White M2R (American Cyanamid) is used at a concentration of 0.1 mg/ml (aqeuous) for 30 sec to 5 min. After water wash, sections are mounted in water or aqueous mountant and viewed with UV excitation (*Table 1*); the pale blue signal is generally very strong and anti-fade reagents (Section 4.5) may not be required.

3.8.2 Callose

Aniline Blue (BDH) can be used with fresh (20) or aldehyde-fixed tissue sections (beware, glutaraldehyde can induce the formation of callose (21)) and may also be used to stain cleared whole mounts. Use freshly made stain (0.5 mg/ml in water or phosphate buffer (pH 8.5)) with tissue observed in the staining solution (UV excitation, *Table 1*) to give a blue/green localization of callose. The stain works particularly well with material embedded in London resin.

3.8.3 Lignin and cutin

Acridine Orange (BDH; Sigma) can be used with fresh or dewaxed sections from fixed and embedded blocks. Use freshly made stain (0.1 mg/ml in water) for 5–10 min. After washing, mount in aqueous anti-fade mountant (Section 4.5) and observe under green excitation (*Table 1*). Lignified tissue stains very bright yellow and cutinized tissue a bright red; autofluorescence results in a pale orange background showing general tissue structure.

4. Problems and overcoming them

4.1 Cytotoxicity

Many fluorescent probes are potentially cytotoxic. Cytotoxicity will depend on the concentration of the probe employed, the duration of its application, and its interaction with cellular constituents. There may be little point in proceeding with an experiment if the cell one is trying to monitor is dead! Although this may seem obvious, the plant cell biology literature is replete with examples of studies in which the cytotoxicity of the fluorescent probe has not been considered. Ideally, all probes should be tested routinely for potentially adverse effects. The following check list may aid in the assessment of the tissue being examined:

(a) Is the cell viable? Use viability tests (Section 4.3).
(b) Do the cells divide and grow normally following application of the probe? Continue to examine individual cells or cell populations for prolonged periods following treatment.
(c) Are cytoplasmic streaming and organelle motion unaffected by the probe? Observe cells, preferably with differential interference contrast optics, before, during, and after application of the probe.
(d) Do the cells continue to function normally? This can often be easily assessed if one is examining a physiological process which can be continuously monitored or even quantified (e.g. the kinetics of stomatal closure).

4.2 Test systems

Ideally, cytotoxicity tests should be performed on the cell type under investigation. However, some plant cells are either too small or too inaccessible to observe some of the above parameters readily. In such instances it may be desirable first to examine the effects of a given probe (e.g. staining specificity or cytotoxic effects) on a suitable test system before embarking on a further detailed study. The inner epidermis of onion storage leaves has become a classical test system in plant cell biology. The cells are large, contain prominent nuclei, and undergo vigorous cytoplasmic streaming within a complex system of trans-vacuolar strands. Unfortunately, when held under coverslips for long periods of time, anaerobiosis may lead to cessation of cytoplasmic streaming and eventual loss of viability. Handling procedures also cause considerable cell damage. The following protocol describes an easy and reliable method for obtaining large areas of viable onion epidermal peels which can be subsequently stained, without damage due to handling, by a wide range of fluorescent probes (22).

Protocol 10. Preparation of viable onion epidermal peels for fluorescence staining and cytotoxicity testing

Equipment:

- 0.5 mm plastic sheeting (easily obtained from empty storage bottles for non-toxic chemicals)
- 1 cm diameter cork borer
- double-sided adhesive tape (Agar Scientific)
- scalpel and paint brush

1. Cut the plastic sheeting into 15 mm × 15 mm pieces. Remove a central, inner core with the cork borer, leaving a plastic holder with a hollow centre.
2. Firmly secure one side of the holder to double-sided adhesive tape, trim the tape flush with the edges of the holder, and remove the tape covering the central hole by cutting round the edge of the hole with a scalpel.
3. Separate the storage leaves of an onion bulb by hand, taking care not to cause damage that will lead to the wetting of the inner epidermis with sap.
4. Press the plastic holder, with the remaining protective layer of the double-sided tape now removed, firmly against the inner epidermis and use a scalpel to cut the tissue around the outer edge of the holder.

5. Lift the holder off. This removes a section of epidermis which adheres firmly to the plastic holder[a].
6. Place the floats directly on top of fluorescent probes in Petri dishes or retain floating on distilled water until ready for use. Viable epidermal floats can be maintained in this way, without loss of cytoplasmic streaming, for several hours, or days in some instances. Occasionally, a gelatinous cell wall material accumulates on that side of the epidermis in contact with the water. This may cause uneven uptake of fluorescent probe and may reduce the optical quality of the specimen. It can be removed by gentle brushing with a soft paint brush.
7. After staining, wash the epidermal floats in several changes of buffer or distilled water, cut the central portion of epidermis from the holder with a scalpel, and examine under a coverslip. Alternatively, fill the central region with water or buffer, place the coverslip directly over the epidermal float, and examine the central region using long working distance lenses[b].

[a] Such epidermal 'floats' can then be handled with ease, without damage to the exposed central region, by gripping the plastic holder with a pair of forceps. The above procedure takes seconds.

[b] The above method is ideal for preparing large numbers of viable epidermal peels (e.g. for class practicals) and has also proved satisfactory for removing the lower epidermis from leaves of a range of plant species.

4.3 Stains for testing cell viability

One could debate at length on the precise definition of a 'viable' plant cell. In general, different investigators have used different criteria for determining viability. A wide range of fluorescent probes is in common use as viability stains. Some of these (e.g. propidium iodide) test for the integrity of the plasma membrane (the ability either to exclude or to allow penetration of fluorescent molecules) while others (e.g. fluorescein diacetate (FDA)) are based on the ability to detect specific enzyme reactions within the cytosol. Further probes detect the maintenance of a pH gradient across the tonoplast. It is undoubtedly unwise to depend exclusively on a single viability test. Ideally, a range of tests should be performed and where possible these should be combined with the diagnostic observations described in Section 4.1. The following viability tests are in common use.

4.3.1 Fluorescein diacetate (FDA)

This is a popular viability stain which works on the same principle as the process used to ester load fluorescent dyes (Section 2.2.1). FDA is essentially a non-fluorescent, non-polar molecule which is passively taken up into the cell. If the cell is living it possesses esterase activity and the FDA is de-esterified to polar fluorescein which is fluorescent and accumulates within the cell. If the cell is dead, esterase activity is lacking and there is no cell fluorescence after adding FDA.

Protocol 11. 'Live' staining with FDA

1. Prepare a stock solution of FDA (Molecular Probes, Sigma; 1–5 mg/ml in acetone) and store at 0 °C in the dark.
2. Add one drop of the stock solution to 10 ml distilled water or appropriate buffer (incubation medium in the case of protoplasts) and apply to cells or tissue.
3. After 5 min examine under a fluorescence microscope with blue excitation (*Table 1*). The % viability = (number of fluorescing cells/total number of cells) × 100.

4.3.2 Propidium iodide

A number of viability stains are based on the exclusion of fluorescent dyes, usually due to their large size or charge, by the plasma membrane. In damaged or dead cells, however, the plasma membrane is permeable to such dyes which then enter the cells and stain them. Propidium iodide is a dye of this type which only stains the DNA of damaged or dead cells. Care needs to be taken in handling propidium iodide because it is a known mutagen.

Protocol 12. 'Dead' staining with propidium iodide

1. Make up a 1–5 mg/ml stock solution of propidium iodide (Molecular Probes, Sigma) dissolved in distilled water. The stock solution can be kept for several months at 4 °C in the dark.
2. Apply the fluorochrome to cells or tissues after diluting to a concentration of 100 μg/ml with distilled water.
3. After 5 min examine the stained material, while it is still immersed in the stain, under blue or green excitation (*Table 1*). Dead or damaged cells fluoresce red.

4.3.3 Simultaneous 'live and dead' staining

In some instances it may be desirable to stain dead cells as well as living cells by the use of two probes applied sequentially. When FDA and propidium iodide are used in this way (Sections 4.3.1. and 4.3.2) live cells fluoresce green and dead cells fluoresce red, after excitation with blue light.

A number of 'live and dead' stains are now commercially available as kits. For example, Molecular Probes Inc. supply a live/dead viability kit which consists of Calcein AM (acetoxymethyl), a fluorogenic esterase substrate that is cleaved only in viable cells to form a green fluorescent membrane-impermeant product, and ethidium homodimer, a red fluorescent DNA stain that is only able to pass through the compromised membranes of dead cells.

4.4 Autofluorescence, background fluorescence, and stray light

Autofluorescence, background fluorescence, or stray light can seriously mask dye fluorescence within plant cells. Autofluorescence may result from the plant material itself, the mounting medium, or various other sources such as the glass slide, coverslip, or immersion oil used. Another source of background fluorescence is dye which has leaked out of cells (Section 4.6) or dye which has not been adequately washed out of the medium after loading. It is essential to examine unstained plant material routinely with the same filter combination as used for the stained material in order to determine whether any of these problems are significant.

Most plant tissues autofluoresce to varying degrees. Common sites of autofluorescence are xylem vessels, chloroplasts, starch grains, and vacuoles. Phenolic-containing cells are a common source of autofluorescence. Complex growth media, in which the plant cell is mounted, can also autofluoresce. Autofluorescence can be reduced or even eliminated by changing the filter combination (Section 5) or by using a probe with different spectral characteristics. Some types of autofluorescence can be quenched by the application of bright-field stains which bind specifically to the autofluorescent component. For example, autofluorescent starch can be rendered opaque by first staining the tissue in iodine/potassium iodide before application of the fluorescent probe. In such cases it is essential to check that the bright-field reaction does not interfere with either the fluorochrome or its specific tissue reaction.

It is important always to select slides, coverslips, and immersion oil which exhibit low autofluorescence.

It is strongly recommended that all fluorescence microscopy is performed in the dark or under subdued illumination. The problems of stray light are often only fully appreciated after using long exposures to photograph a fluorescent preparation.

4.5 Dye photobleaching and irradiation damage

A common problem in fluorescence microscopy is that fluorochromes fade, often rapidly, during illumination. This photobleaching is mainly due to photochemical reactions induced by the excitation light. It can be minimized in a number of ways:

(a) Reduce the irradiation exposure of the tissue to a minimum. Use a shutter to cut off the excitation light when not imaging fluorescence.
(b) Reduce the intensity of excitation light with neutral density filters, by using low numerical apertures and/or by initially examining the specimen with low magnification objectives.
(c) Reduce the bandwidth of the excitation light.
(d) Use an anti-fade mountant. Several mountants, both temporary or permanent, are now available which retard the photochemical fading of several fluorescent probes. These include sodium dithionite (0.02 M sodium dithionite, $Na_2S_2O_4$, in 0.01 M phosphate buffer, pH 7.2 (23)), Citifluor (City University, London), and Mowiol 4–88 (Calbiochem). Suitable controls must be performed to ensure that the mountants do not have cytotoxic effects (Section 4.1).

4.6 Dye sequestration and leakage

Some fluorescent dyes can be sequestered by cell organelles other than those of interest or, alternatively, can leak out of cells. In both cases, the dye will become diluted, or even lost altogether, from the cell compartment under study. Sequestration of dye into the wrong compartment may also mask the dye fluorescence of interest.

Dye sequestration has been a particular problem with the Ca^{2+} dyes, pH dyes, and a range of anionic fluorochromes (2, 3, 22). Significant sequestration can occur within seconds or over several hours and it is sometimes only a problem with certain cell types or methods of dye loading. The organelles which take up the dye can also vary among cell types and include vacuoles, nuclei, plastids, and ER.

If dye sequestration is a problem, critical controls need to be done to assess the location of the dye and in particular to determine whether sufficient dye remains in the cell compartment of interest to be useful for localization or measurement purposes. Often the compartmentalized dye can be easily visualized, especially by confocal microscopy. When attempting to image or measure $[Ca^{2+}]_c$ it is always essential to demonstrate clearly that sufficient dye is present in the cytosol to respond to artificially induced changes in $[Ca^{2+}]_c$ (Section 3.6). It is sometimes possible to find a window of time within which one can image a dye before too much is sequestered.

Another, and often related problem to dye sequestration, is dye leakage from cells. A number of methods are available to prevent or reduce dye sequestration or leakage. These include:

(a) Use of a different dye which reports the same cell component but which does not compartmentalize to the same extent. Sometimes closely related derivatives of the dye do not sequester significantly or leak out of cells.

(b) Use of a dye conjugated to a high molecular weight dextran. A range of reporter dyes coupled to dextrans of different molecular weights (3–7 kDa) is commercially available (Molecular Probes). Their high molecular weights render such dyes membrane impermeant but usually necessitate their introduction into cells by microinjection.

(c) To prevent leakage of aqueous, diffusible dyes, cells can be mounted in a non-aqueous medium such as glycerol, immersion oil, or silicone oil for short periods of time.

5. Choice of filters

Although the standard filter sets (*Table 1*) recommended in the various protocols in this chapter will mostly give satisfactory results, further optimization is usually possible. There is often considerable scope to match a filter set with the spectral characteristics of a fluorescent dye and, at the same time, reduce sample autofluorescence to a minimum. The aim is to maximize both the brightness and the signal-to-noise ratio of the image. In doing this, however, one often needs to make compromises because there may be considerable overlap between the excitation and emission spectra of the dye and those of autofluorescence. In addition, the excitation and emission spectra of the dye alone may also significantly overlap. If autofluorescence is not a problem, the maximum dye signal will be achieved by using a wide bandwidth excitation filter centred on the absorption peak of the dye and a long pass emission filter which transmits all light above a wavelength sufficient to allow all of the emitted dye fluorescence to be detected. If autofluorescence or spectral overlap are problems, one or more of the following may provide a solution:

- use of a wide bandwidth emission filter centred on the emission spectrum of the dye instead of a long pass emission filter
- use of a narrow bandwidth excitation and/or emission filters centred on the absorption and emission spectra of the dye
- use of either narrow or wide bandwidth excitation and/or emission filters which are not quite centred on the absorption and emission spectra of the dye

These options may provide improved spectral discrimination and thus a greater signal-to-noise ratio. However, this needs to be traded off against a reduction in signal brightness which will be unacceptable below a certain level.

The choice of filters may be further complicated by the need to excite dyes or to detect fluorescence at more than one wavelength. This is significant when staining a sample with more than one dye or when performing ratio analysis. Nevertheless, one still follows the same rule of thumb described above when choosing filters.

Molecular probes provide filter sets optimized for a wide range of fluorescent dyes used in studies of living cells. These filter sets, manufactured by Omega Optical Inc., are of very high quality and can be configured for most types of fluorescence microscope.

References

1. Haugland, R. P. (1992). *Molecular probes—handbook of fluorescent probes and research chemicals*. Molecular Probes Inc., Eugene, OR.
2. Oparka, K. J. (1991). *J. Exp. Bot.*, **2**, 565.
3. Read, N. D., Allan, W. T. G., Knight, H., Knight, M. R., Mahlo, R., Russell, A. *et al.* (1992). *J. Micros.*, **166**, 57.
4. Purves, R. D. (1981). *Microelectrode methods for intracellular recording and iontophoresis*. Academic Press, London.
5. Callaham, D. A. and Hepler, P. K. (1991). In *Cellular calcium* (ed. J. G. McCormack and P. H. Cobbold), pp. 383–403. IRL Press, Oxford.
6. Warner, A. E. and Bate, C. M. (1987). In *Microelectrode techniques: the Plymouth workshop handbook* (ed. N. B. Standen, P. T. A. Gray, and M. J. Whitaker), pp. 169–86. Company of Biologists, Cambridge, UK.
7. Oparka, K. J., Murphy, R., Derrick, P. M., Prior, D. A. M., and Smith, J. A. C. (1990). *J. Cell Sci.*, **98**, 539.
8. Stewart, W. W. (1981). *Nature*, **292**, 17.
9. Goodwin, P. B. (1983). *Planta,* **157**, 124–30.
10. Tucker, E. B. (1982). *Protoplasma*, **113**, 193.
11. Simpson, I. (1978). *Anal. Biochem.*, **89**, 304.
12. Grignon, N., Touraine, B., and Durand, M. (1989). *Am. J. Bot.*, **76**, 871.
13. Canny, M. J. (1990). *New Phytol.*, **114**, 341.
14. Peterson, C. A., Emanuel, M. E., and Humphreys, G. B. (1981). *Can. J. Bot.*, **59**, 618.
15. Oparka, K. J. and Prior, D. A. M. (1988). *Planta*, **176**, 533.
16. Wright, K. M. and Oparka, K. J. (1989). *Planta*, **179**, 257.
17. Greenspan, P., Meyer, E. P., and Fowler, S. D. (1985). *J. Cell Biol.*, **100**, 965.
18. Thomas, A. P. and Delaville, F. (1991). In *Cellular calcium* (ed. J. G. McCormack and P. H. Cobbold), pp. 1–35. IRL Press, Oxford.
19. Hughes, J. and McCully, M. E. (1975). *Stain Technol.*, **50**, 319.
20. Smith, M. M. and McCully, M. E. (1978). *Protoplasma*, **95**, 229.
21. Hughes, J. E. and Gunning, B. E. S. (1980). *Can. J. Bot.*, **58**, 250.
22. Oparka, K. J., Murant, E., Wright, K. M., Prior, D. A. M., and Harris, N. (1991). *J. Cell Sci.*, **99**, 557.
23. Gill, D. (1978). *Experientia*, **35**, 400.

3

General and enzyme histochemistry

N. HARRIS, J. SPENCE, and K. J. OPARKA

1. Introduction

Fluorescent probes and stains have proved very valuable in identifying cell components and activities (Chapter 2). However, not all laboratories have access to flourescence microscopes; this chapter therefore contains various protocols for both general and specific staining of plant tissues, cell types, and organelles which may be used with bright field microscopy. Generally such protocols have the added benefit of providing permanent mounts which may be viewed on numerous subsequent occasions. Only a brief selection is provided here including some of the more useful for general staining, for general light counterstaining, and for localizing some of the major cell components. Many such stains are available; used carefully the methods, though simple, can be very informative. More comprehensive lists and protocols can be found in references 1–5. Cross reference to Chapter 2 is recommended: fluorescent probes, which in many cases are not cytotoxic and may be used with living tissues, are available for many cell components.

Stains can be used with whole mounts, perhaps after 'clearing' to remove pigments and expose inner tissue components, or with sections of tissue cut freshly, after cryofixation (see Chapter 4, *Protocol 10*; Chapter 5, *Protocol 3*), or after embedding in wax, polyethylene glycol, or resin.

Pieces of tissue in which secondary thickening has not occurred may be rendered semi-translucent by treating with ethanol:glacial acetic acid (3:1(v/v)) at 60 °C for 15–60 min. Samples initially 'clear' but remain intact; longer treatment is used to macerate tissue. Cleared samples can subsequently be stained, for example with phloroglucinol/HCl (see Section 2.2.3 below), to demonstrate the distribution of lignified vascular elements and their complex inter-relationships at and between nodes. The three-dimensional inter-relationships are visible within the whole mount more readily than by reconstruction from stained sections. In many investigations, however, whole mounts are not suitable and the required details are only available following sectioning.

1.1 Tissue embedding and sectioning

Tissue blocks may be sectioned 'freehand' (<25 μm thick) or using various commercial tissue choppers which slice fresh tissue (to *circa* 15 μm) or microtomes which require that the tissue is first embedded in a matrix (for 0.5–10 μm thick sections). The matrix may be the frozen cell sap (for cryosectioning, see Chapter 5), wax, or a water-soluble medium such as polyethylene glycol (this chapter, *Protocol 1*), or one of various resins (this chapter, *Protocol 2*; Chapter 4, *Protocol 3*; Chapter 5, *Protocol 5*; Chapter 6, *Protocol 2*). Cryosectioning of vacuolate plant tissues containing numerous intercellular air spaces may present problems; meristematic tissues are more amenable.

Protocol 1. Tissue processing and embedding in wax or polyethylene glycol (PEG)

- crystalline paraformaldehyde
- 25% glutaraldehyde stock solution
- 0.5 M phosphate or cacodylate buffer pH 7.0 (stock)
- ethanol
- Histo-Clear or similar[a]
- wax (Paraplast or similar) or PEG 1000 or 1500
- specimen vials and embedding moulds
- rotator or shaker for specimen vials[b]
- oven

1. Prepare primary fixative (typical: 1.5% (w/v) formaldehyde, 2.5% (v/v) glutaraldehyde in 0.05 M phosphate buffer pH 7.0[c]).
2. Cut tissue blocks, with at least one dimension a maximum of 5 mm, in the primary fixative as soon as possible after removing from plant.
3. Fix overnight at 4 °C (volume of fixative >> 10 × volume of sample).
4. Wash in buffer (two changes with 30 min between changes).
5. Dehydrate through a graded ethanol series (10%, 25%, 40%, 60%, 75%, and 95%) with two 15–30 min changes in each step and three 15–30 min changes in dry ethanol (ethanol kept over molecular sieve).

For polyethylene glycol embedding:

6. Infiltrate with 1:1 ethanol:PEG[d] overnight at 40 °C.
7. Infiltrate with PEG for 48–72 h at 56 °C, with changes to fresh PEG each evening and morning.

8. Place in prewarmed embedding moulds with fresh PEG, orient, and cool on ice or at 4 °C.

For wax embedding:

6. Wash with 2:1 ethanol:Histo-Clear[a] for 2 h at room temperature; repeat with 1:1 and 2:1 ethanol:Histo-Clear, and leave in Histo-Clear overnight.
7. Infiltrate with Histo-Clear:wax at 1:1 for 8 h at 56 °C.
8. Infiltrate with wax for 96 h at 56 °C, with changes of wax every 24 h.
9. Place in prewarmed embedding moulds with fresh wax, orient, and cool on ice or at 4 °C.

[a] Xylene, toluene, or benzene are commonly replaced with Histo-Clear or a similar, less hazardous clearing reagent.
[b] During fixation, washing, dehydration, and infiltration with resin, the samples should be constantly but gently moved: several manufacturers produce suitable rotators which hold 10 ml glass specimen vials.
[c] Make formaldehyde freshly from crystalline paraformaldehyde; see Chapter 4, *Protocol 1*
[d] PEG 1000 gives softer blocks which require cooling to 4 °C before sectioning

Protocol 2. Embedding tissue in LR White resin

- as in *Protocol 1* except LR White resin (London Resin Company) in place of wax
- sealable polypropylene embedding moulds

A. *Ambient temperature embedding*

1. Cut tissue blocks of (2–3 mm)^3, in the primary fixative as soon as possible after removing from plant.
2. Fix for 2–6 h at 4 °C (volume of fixative $>>$ 10 × volume of sample).
3. Wash in buffer (two changes with 15 min between changes).
4. Dehydrate through a graded alcohol series (10%, 25%, 40%, 60%, 75%, and 95%) with two 15 min changes in each step and three 15–30 min changes in dry alcohol (alcohol kept over molecular sieve).
5. Infiltrate with 2:1 ethanol:resin overnight at 4 °C, with 1:1 ethanol: resin for 4 h, and with 1:2 ethanol:resin for 4 h at room temperature.
6. Infiltrate with resin for 12–48 h with changes of fresh resin every 12 h[a].
7. Place in polypropylene or gelatin capsules, fill and seal capsules with only minimum trapped air, and allow to polymerize at 50 °C for 24 h[b].

Protocol 2. *Continued*

B. *Low temperature embedding*

Preliminary comments and steps 1–3 are as above.

4. Dehydrate through a graded alcohol series with 30 min changes in each step (10% and 25% ethanol at 0 °C; 40%, 60%, 75%, and 100% ethanol at −20 °C.

5. Infiltrate with 2:1 ethanol:resin overnight, with 1:1 ethanol:resin for 4 h, and h, and with 1:2 ethanol:resin for 4 h, all at −20 °C.

6. Infiltrate with resin with catalyst (0.5% (v/v) benzoin methyl ether) for 12–48 h at −20 °C with changes of fresh resin every 12 h[a].

7. Place in polypropylene or gelatin capsules, fill and seal capsules with only minimum trapped air, and polymerize by irradiating with UV light for 24 h at −20 °C and 24 h at room temperature.

[a] The time taken for resin infiltration is dependent upon the type of tissue; small pieces of non-woody tissues will be infiltrated within 12 h whereas woody tissues or seed tissues packed with starch and other storage reserves require up to 48, or exceptionally, 72 h.
[b] LR White may be 'cold cured' using an accelerator—the reaction is highly exothermic and not recommended for immunological work.

Sections are collected on pretreated glass slides. Pretreatments to ensure adhesion of samples include gelatin (Chapter 5, *Protocol 1*), poly-L-lysine (Chapter 10, section 3.3), or TESPA. To silanize with TESPA (3′amino-propyl triethoxylsilane, Sigma A3648), wash slides with dilute detergent, rinse with distilled water, immerse in 2% TESPA in acetone for 30 sec, rinse twice in acetone and twice in distilled water, and leave to dry in a dust-free environment. For RNase-free slides water treated with diethylpyrocarbonate (DEPC) is used where appropriate (see Chapter 5).

Preparation of cryosections and sections from resin and from wax are described in Chapter 5. *Protocols 3*, *7*, and *8* respectively. Sections from PEG-embedded material are cut at 5–15 μm using standard microtomes with steel or disposable knives. Sections can be transferred in a ribbon to the slide which is subsequently warmed very gently so that the PEG just begins to melt. Slides with sections are then cooled and stored at 4 °C.

2. General histo- and cytochemistry

Toluidene Blue is suitable for staining structures in sections of fresh or fixed and embedded tissues; it is particularly good as a general stain of sections from resin-embedded tissues. Use at 0.01% in 0.1% aqueous sodium tetraborate: if the tissue is not overstained Toluidene Blue is metachromatic:

cellulose cell walls, cytoplasm, and plastics stain blue, nuclei stain blue with the nucleolus purple, and lignified walls are green/blue.

Alternatively, two dyes may be used to contrast various components within sections, for example haematoxylin/Orange G (*Protocol 3*).

Protocol 3. Haematoxylin/Orange G for histological sections

- haematoxylin solution (Sigma)[a]
- ethanol
- Orange G (Sigma 07252)
- clove oil
- 0.5% acid alcohol
- reagents for permanent mounting[b]

1. Prepare Orange G solution: dissolve 0.5 g Orange G in 100 ml ethanol and add 100 ml clove oil. Mix and allow ethanol to evaporate until 100 ml solution remains.
2. Dewax and rehydrate sections. (see Chapter 5, *Protocol 8*, steps 1–4)
3. Stain with haematoxylin solution for 10–15 min.
4. Differentiate (remove excess stain) with 0.5% acid alcohol for 1 min.
5. Wash with tap water until the haematoxylin is blue.
6. Dehydrate through 50, 75, and 100% ethanol.
7. Stain with Orange G for 2–5 min.
8. Differentiate in a solution of equal parts clove oil, ethanol, and xylene (or Histo-Clear).
9. Rinse in xylene (or Histo-Clear) and mount permanently[b]. Organelles stain dark blue; cell walls and cytoplasm are orange.

[a] Haematoxylin solution is available commercially or can be prepared as follows. Dissolve 1 g haematoxylin in 50 ml ethanol and then add, in this order, 50 ml glycerol, 50 ml water, 50 ml acetic acid, and 1 g potassium aluminium sulphate. The solution should be left to 'ripen' in sunlight for several weeks before use.

[b] For permanent mounting, samples are air dried, or dehydrated in ethanol and xylene or Histo-Clear; mounting medium (Canada balsam, DPX, Histomount, Euparal, or similar), is then added and a clean coverslip placed on top, taking care to avoid trapping any air bubbles. The mountant may be left to set at ambient temperature or at 30 °C.

2.1 Counterstains

To contrast with the distribution of intensely stained specific tissues a variety of counterstaining methods are available to stain the general distribution of tissues lightly. Light Green only, Safranin and Light Green, or Safranin and Fast Green are suitable for counterstaining.

Protocol 4. Counterstaining with Safranin and Fast green

- 1% (w/v) Safranin (Sigma S8884) in 95% aqueous ethanol
- 0.5% (w/v) Fast Green (Sigma F7258) in 95% aqueous ethanol
- ethanol
- xylene, Histo-Clear (or similar) (see *Protocol 1,* footnote *a*)
- permanent mounting reagents (see *Protocol 3,* footnote *b*)

After primary staining[a] to give high contrast/high colour to specific components within the tissue section:

1. Stain with Safranin for 10–15 min.
2. Destain with water until Safranin is no longer lost from the specimen.
3. Stain with Fast Green for 5–30 sec[b].
4. Dehydrate with 50% ethanol for 1 min.
5. Rinse with xylene or equivalent.
6. Mount permanently (see *Protocol 3,* footnote *b*).

Lignified walls stain red; nuclei and plastids stain red/pale pink.

[a] It is critical that the primary stain product is not alcohol-soluble.
[b] If green staining masks the Safranin, dilute the Fast Green stain 5- to 10-fold with 95% ethanol.

2.2 Cell wall components

Cell well components may be localized by 'classical' staining methods such as that in *Protocol 5*, or by techniques for individual components resulting in either bright field or fluoresence microscopy examination: fluorescent stains are listed in Chapter 2, Section 3.

Protocol 5. Chlor-zinc-iodide (Schutze's reagent) for localization of cell wall components

- zinc chloride
- potassium iodide
- resublimed iodine

1. Dissolve 2.5 g potassium iodide in 3 ml water and add 0.5 g resublimed iodine.

2. Add 7 ml water and 15 g $ZnCl_2$ and allow to dissolve.
3. Use small volumes to stain tissue sections.
4. Wash with water, cover with coverslip, and examine.

Cellulose walls stain blue/violet, lignified and suberized walls yellow/brown, starch also stains blue.

2.2.1 Cellulose

For chloride–iodide staining, dissolve 30 g zinc chloride, 5 g potassium iodide, and 1 g iodine in 15 ml distilled water, and stain fresh tissue or sections for 5–15 min; examine the sections while in staining solution; cellulose is stained blue. The stain solution should be stored in the dark. Where fluorescence microscopy is available use Calcofluor (see Chapter 2, Section 3).

2.2.2 Pectins

Pectins can be stained with 0.02% aqueous Ruthenium Red: incubate fresh tissue sections for 2–10 min, wash with distilled water, and examine by bright field microsopy. Pectins stain red (RNA may also stain). Alternatively the hydroxylamine hydrochloride–ferric chloride method may be used (*Protocol 6*).

Protocol 6. Hydroxylamine hydrochloride–ferric chloride staining for pectins

- sodium hydroxide
- hydroxylamine hydrochloride
- ethanol
- ferric chloride
- concentrated HCl

1. Prepare hydroxylamine hydrochloride stain (A) using equal volumes of 14% (w/v) sodium hydroxide in 60% aqueous ethanol and 14% (w/v) hydroxylamine hydrochloride in 60% aqueous ethanol.
2. Prepare ferric chloride stain (B) by mixing 10% (w/v) ferric chloride in 60% aqueous ethanol with 1% 0.1 M HCl.
3. Incubate sections of fresh tissue sections in 1 ml solution A for 5–10 min.
4. Add 1 ml ethanolic HCl (95% ethanol:conc. HCl, 2:1 (v/v)) and leave for 30 sec.

Protocol 6. *Continued*

5. Remove excess liquid and stain with solution B.
6. Wash with 60% ethanol and examine by bright field microscopy. (Pectin esters stain red.)

2.2.3 Lignin

Acidic phloroglucinol is widely used as a stain for lignin. Staining involves carefully adding a solution of 10% w/v phloroglucinol in 95% ethanol to an equal volume of concentrated HCl and staining sections of fresh or fixed tissues, or cleared whole mounts, for 3–30 minutes. Wash the samples thoroughly in water and examine by bright field microscopy; lignin is stained bright red.

An alternative is aniline sulphate: stain fresh or rehydrated sections in 1% aniline sulphate in 60% aqueous ethanol with 0.005 M sulphuric acid (final concentration) for 5–10 min; mount in water and examine by bright field microscopy. Lignin is stained bright yellow.

0.01% aqueous Acridine Orange can be used as a fluorescent stain for lignin and also cutin (see Chapter 2, Section 3.8.3).

2.2.4 Callose

Callose is most conveniently stained with the fluorochrome Analine Blue (see Chapter, Section 3.8.2). When only bright field microscopy is available, stain tissue sections or stigma squashes with Resorcinol Blue (Sigma). Make stock solution by dissolving 3 g white resorcinol in 200 ml distilled water, add 3 ml '0.88' ammonia, and heat in a water bath at 90 °C for 10 min. The red-brown solution will turn blue after storage; heat again for 30 min, filter, and store as stock. Use freshly diluted stock (1:50 in distilled water) and stain for 10–20 min. Wash with 0.1M citrate phosphate buffer at pH 3.2; callose remains blue but other wall components change from blue to red (6).

2.3 Nuclei and nucleic acids

Fluorescent staining of nuclei, chromosomal components, and RNA are covered in Chapter 2, Section 3.5.3, in Chapter 6 (*Protocol 8*), and in Chapter 5 (*Protocol 9*) respectively. For permanent mounts viewable without fluorescent optics use Methyl green/pyronin Y for DNA and RNA (*Protocol 7*), or the general staining methods described above.

Protocol 7. Methyl Green–Pyronin for localization of DNA and RNA

- Walpole's buffer pH 4.8 (60 ml 0.2 M sodium acetate and 40 ml 0.2 M acetic acid (11.54 ml glacial acetic acid in 1 litre)

- 0.2 M phosphate buffer pH 6.0
- RNase
- Methyl Green (Sigma M5015)
- Pyronin Y (Sigma 6653)
- glycerol
- permanent mountant (see *Protocol 3*, footnote *b*)
- coverslips

1. Dewax and rehydrate sections.
2. Incubate control sections in 0.05 mg/ml RNase in 0.2 M phosphate buffer pH 6.0 for 1 h at 37 °C; incubate test sections in buffer only.
3. Wash well in distilled water.
4. Wash in Walpole's buffer pH 4.8.
5. Stain with Methyl Green–Pyronin for 25 min (9 ml 2% (w/v) aqueous Methyl Green, 4 ml 2% (w/v) aqueous Pyronin Y or G, 14 ml glycerol and 23 ml Walpole's buffer pH 4.8).
6. Wash in Walpole's buffer and gently blot dry.
7. Dehydrate and mount permanently (see *Protocol 3*, footnote *b*).
8. Examine by bright field microscopy. RNA is stained pink, DNA green/blue.

2.4 Endomembrane system

Components of the endomembrane system in living cells can be stained with the fluorochrome $DiOC_6$ (see Chapter 2, Protocol 8). The zinc iodide–osmium tetroxide reagent is used for contrasting the endoplasmic reticulum and Golgi apparatus in sections of resin-embedded tissue examined by optical or electron microscopy (see Chapter 4, *Protocol* 5).

2.5 Proteins

Toluidene Blue (Sigma T7029) is a valuable metachromatic stain (see page 54) which gives good general staining of protein in fresh tissues and sections from wax- and resin-embedded tissues. For protein staining make a stock of 0.1% (w/v) Toluidene Blue in 1% (w/v) aqueous sodium tetraborate; fresh and rehydrated sections from wax/embedded tissues require approximately 1 min of staining prior to distilled water washing (if staining is too intense dilute stock 1:10 or 1:100 with water). Staining of sections of resin-embedded material may be accelerated by gentle warming on a hot plate, although it is critical that the stain does not start to precipitate.

2.6 Carbohydrates and starch

Periodic acid–Schiff's reaction (*Protocol 8*) is used to localize total carbohydrates in histological sections.

Protocol 8. Periodic acid–Schiff's (PAS) stain for carbohydrates

- 0.5% (w/v) aqueous periodic acid
- Schiff's reagent (Sigma 395-2-016)
- 2% (w/v) aqueous sodium bisulphite

1. Rehydrate sections from wax-embedded or freeze-dried tissue.
2. Incubate with 0.5% (w/v) aqueous periodic acid for 5 to 30 min.
3. Wash in running water for 10 min.
4. Stain with Schiff's reagent for 15 min.
5. Rinse in water and incubate in 2% (w/v) aqueous sodium bisulphite for 2 min.
6. Wash in running water.
7. Counterstain lightly (see *Protocol 4*) if required.
8. Dehydrate and mount permanently (see *Protocol 3*, footnote *b*) if required.

Polysaccharides, including starch, stain deep purple/red.

Iodine in potassium iodide is the classic stain for starch: made as 0.5% (w/v) iodine in 5% (w/v) aqueous potassium iodide, the stain should be stored in the dark. When used with fresh or fixed material it gives a very strong reaction within a few minutes. The deep blue/black colour indicates long chain starch, with shorter chain molecules staining red/brown. Starch may also be identified, without staining, by the 'Maltese cross' pattern given by starch grains when viewed under polarized light (see p. 13).

2.7 Lipids

Lipids may be stained with Sudan Black B (Gurr) or Sudan III (Gurr): the stains are made as saturated solutions in 70% (v/v) ethanol and sections are stained for extended periods, usually in excess of 30 min. The dye preferentially partitions from the staining solution into the specimen lipids which is stained black or blue with the former, or red with the latter. Nile Blue (Gurr) may be used to distinguish between neutral and acidic lipids: unfixed tissue is stained with 1% (w/v) aqueous Nile Blue for 30–60 sec at

37 °C, sections are rinsed briefly with 1% (v/v) acetic acid and washed with distilled water. Nile Blue is, however, a general basic stain and may give high levels of staining of non-lipid components in some tissues: it is useful where high levels of free fatty acids or phosopholipids are present.

2.8 Tannins

Tannins are precipitated by 1% (w/v) aqueous osmium tetroxide to give a dense deposit which is visible by light and electron microscopy. Older, and still valuable, staining methods for tannins use ferric salts or the nitroso reaction. After incubating fresh sections in 1.0% (w/v) ferric chloride in 0.1 M HCl tannins give a blue precipitate. If a permanent mount is required, fix tissue in 2% (w/v) ferric sulphate in 10% (v/v) formalin, dehydrate, embed in wax (see *Protocol 1*), and section.

The nitroso reaction gives red deposition of tannins: use fresh sections and add equal volumes of (i) 10% (w/v) sodium nitrate, (ii) 20% (w/v) urea, (iii) 10% (v/v) acetic acid, stain for 5 min, add 2 volumes of 2 M sodium hydroxide, and observe by bright field microscopy.

3. Enzyme histochemistry

Cryosections from fresh frozen tissues are recommended for all histochemical techniques. Delicate enzymes may be damaged or lost entirely during the harsh processing procedures involved in embedding the tissue. Increasing the incubation times is usually necessary when using embedded material.

Enzyme cytochemistry for localization at the electron microscope level has recently been comprehensively and excellently reviewed by Sexton and Hall (7), and is not repeated here.

3.1 Acid and alkaline phosphatases

Phosphatases are involved in the hydrolytic cleavage of organic phosphate esters and are active under alkaline or acidic conditions. As well as indicating important physiological activities within plant tissues alkaline phosphatase is also used in immunolocalization studies where it is conjugated to a secondary antibody. Alkaline phosphatases are not as widely distributed in plant tissues as in animal ones; the conjugated calf intestinal alkaline phosphatase may be distinguished from any endogenous activity by the use of the inhibitor levamisole which, at millimolar concentrations, does not inhibit the calf enzyme. Several methods are available, based on the Gomori (8) reactions. The colour of the final precipitated reaction product may be blue (with the substrates NBT and BCIP) or red (using naphthol AS-MX phosphate and Fast Red TR).

Staining artefacts are readily produced when testing for phosphatases and a comprehensive range of controls must be used, particularly with electron microscopic studies (see Chapter 4 and reference 7).

Protocol 9. Localization of acid and alkaline phosphatases

- acetate buffer pH 5 (30 ml of 0.2 M acetic acid and 70 ml of 0.2 M sodium acetate)
- naphthol AS-MX phosphate (Sigma N4875)
- Fast Red TR (Sigma F2768)
- Tris–HCl buffer pH 8
- Dimethylformamide (DMF)

A. *Acid phosphatase*

1. Prepare a 100× stock solution of naphthol AS-MX phosphate (A) by dissolving 20.8 mg naphthol AS/MX phosphate in 1 ml DMF (store frozen).
2. Prepare a 100× stock solution of Fast Red TR (B) by dissolving 51.4 mg Fast Red TR in 1 ml 70% (v/v) DMF (store frozen).
3. Prepare substrate solution by dissolving 10 μl solution A and 10 μl solution B in 5 ml acetate buffer.
4. Incubate test sections in substrate solution for 10–60 min at 37 °C in darkness.
5. Incubate control sections in the above solution from which the substrate has been omitted.
6. Wash in distilled water.
7. Air dry and mount permanently.

Red deposits that are absent from control sections indicate the sites of enzyme activity.

B. *Alkaline phosphatase*

As in *Protocol 9* A but substitute Tris–HCl buffer pH 8 for the acetate buffer.

3.2 ATPase

Adenosine triphosphatases catalyse the hydrolysis of ATP to give inorganic phosphate: in animal and plant cells the reactions are often linked to important membrane-associated physiological processes which are energy dependent.

Protocol 10. Localization of ATPase

- 0.04 M Tris–maleate buffer pH 7
- ATP
- calcium nitrate
- lead nitrate
- H_2S water

1. Prepare substrate solution containing 2 mM ATP, 2 mM calcium nitrate, 3.6 mM lead nitrate in 0.04 M Tris–maleate buffer pH 7.
2. Incubate test sections in substrate solution for 10–30 min at room temperature in darkness.
3. Incubate control sections in the above solution from which the substrate has been omitted.
4. Wash in buffer.
5. Place sections in freshly prepared H_2S water for 2 min.
6. Wash in distilled water.
7. Air dry and mount permanently.

Brown/black deposits that are absent from control sections indicate the sites of enzyme activity.

3.3 Esterase

A range of esterase activities is present within plant tissues. Within this broad range, specific categories may be defined (5) by application of inhibitors: since these inhibitors include organophosphates many investigators simply refer to the localization of non-specific esterases. Esterase has been used as a marker of early developmental changes that occur during vascularization within plant tissues (9), and also as a test of cell viability (see p. 46).

Protocol 11. Localization of general esterase

- 0.2 M Tris–HCl buffer pH 6.5
- naphthol AS-D acetate (Sigma N2875)
- Fast Blue BB (Sigma F3378)
- DMF

1. Prepare substrate solution as follows. Dissolve 5 mg naphthol AS-D acetate in 0.5 ml DMF. Slowly add this, while mixing, to 25 ml 0.2 M

Protocol 11. *Continued*

Tris–HCl pH 6.5 and 25 ml distilled water. Add 20 mg Fast Blue BB, shake well, and filter into a dark bottle.

2. Incubate test section in substrate solution for 10–20 min at room temperature or 37 °C in darkness.
3. Incubate control sections in the above solution from which the substrate has been omitted.
4. Wash in distilled water.
5. Air dry and mount permanently.

Blue deposits that are absent from control sections indicate the sites of enzyme activity.

3.4 Glucuronidase (GUS)

GUS is widely used as a reporter gene to examine the temporal and spatial regulatory roles of nucleic acid sequences expressed transgenically (10). GUS activity is localized using, as substrate, naphthol AS-B1-β-D-glucuronide (*Protocol 12*), or X-gluc (5-bromo-4-chloro-3-indolyl-β-D-glucuronic acid) which gives a bright blue, insoluble reaction product. This may be viewed after staining of fresh tissue (*Protocol 13*) by bright field microscopy, or may be fixed and viewed, with higher resolution, after sectioning of stained tissue which has been embedded in methacrylate (see *Protocol 13*). Where ultrastructural resolution is required GUS is localized by its antibody and electron microscopic immunocytochemistry (see Chapter 7).

Protocol 12. Localization of glucuronidase (GUS) using naphthol AS-B1-β-D-glucuronide

- 0.2 M acetate buffer pH 4.5 (51 ml 0.2 M acetate acid and 49 ml 0.2 M sodium acetate)
- naphthol AS-B1-β-D-glucuronide (Sigma N1875)
- DMF (Sigma D4252)
- Fast Blue BB (Sigma F3378)
- 0.1 M phosphate buffer pH 7.5

1. Prepare stock naphthol solution (A) as follows. Dissolve 11.4 mg naphthol AS-B1-β-D-glucuronide in 1 ml DMF. Add this to 49 ml 0.2 M acetate buffer pH 4.5.
2. Prepare substrate solution by adding 0.8 ml solution A to 3.2 ml 0.2 M acetate buffer pH 4.5 and 4 ml distilled water.

3. Incubate test sections in substrate solution for 15–60 min at 37 °C in darkness.
4. Incubate control sections in the above solution from which the substrate has been omitted.
5. Prepare post-coupling solution. Dissolve 1 mg Fast Blue BB in 5 ml phosphate buffer pH 7.5.
6. Incubate sections in post-coupling solution for 5 min at 4 °C.
7. Wash in distilled water.
8. Air dry and mount permanently.

Blue deposits that are absent from control sections indicate the sites of enzyme activity.

Protocol 13. Localization of GUS using X-gluc

- X-gluc (5-bromo-4-chloro-3-indolyl-β-D-glucuronic acid)
- 50 mM phosphate buffer pH 7.2
- humid chamber (see Chapter 7, *Protocol 3*)

A. *For semi-permanent mounting*

- fixative (see *Protocol 1*)
- ethanol
- methacrylate JB4 (Polysciences)
- catalyst for JB4
- specimen moulds

1. Cut sections of fresh tissue into 1 mM X-gluc in 50 mM phosphate buffer pH 7.2.
2. Incubate for 2–12 h at 37 °C.
3. Rinse with water.
4. Examine by bright field microscopy (GUS activity indicated by blue staining absent in controls without substrate).

B. *If a permanent mount is required*

Follow steps 1–4 above then proceed as follows.

5. Fix with buffered aldehyde for 4–12 h at 4 °C.

Protocol 13. *Continued*

6. Dehydrate through ethanol series (see *Protocol 1*, step 5).
7. Infiltrate with increasing concentrations of methacrylate JB4 in ethanol over 24 h.
8. Infiltrate with methacrylate JB4 for 24 h.
9. Embed in methacrylate JB4 with catalyst and allow to polymerize in specimen moulds at room temperature.
10. Section at 2–10 μm, mount on glass slides, and examine using bright or dark field illumination without counterstaining.

3.5 Catalase, peroxidase, and phenoloxidase

The oxidases have major roles in both primary and secondary metabolic pathways. Protocols are given here for catalase, peroxidase, and phenol-oxidase; other methods available include these for localization of cytochrome oxidase (11), uricase (12), and glucolate oxidase (13). NADI reagents have been used for the localization of cytochrome oxidase; there are, however, several disadvantages to these including the migration of the stain product to adjacent lipid components within the tissues, a false positive from xylem elements, and occurrence of the reaction in the absence of enzymic activity. Cataslase activity can be discriminated from peroxidaxe activity using the catalase inhibitor aminotriazole.

Protocol 14. Localization of catalase (after Frederick (14))

- 2-amino-2-methyl-1, 3-propanediol buffer pH 10
- diaminobenzidine (DAB) Sigma D5637 or, with special packaging for carcinogen, D9015)
- 30% hydrogen peroxide

1. Prepare substrate solution, as follows. Dissolve 10 mg DAB in 5 ml of 2-amino-2-methyl-1,3-propanediol buffer pH 10. Add 0.1 ml freshly prepared 3% hydrogen peroxide, adjust to pH 9, and filter into a dark bottle.
2. Incubate a control section (i) in the catalase inhibitor 3-amino-1,2,4-triazole (20 mM) for 30 min.
3. Incubate test and control sections in substrate solution for up to 1 h at 37 °C in darkness.
4. Incubate control sections (ii) in the above solution from which the hydrogen peroxide has been omitted.

5. Wash in distilled water.
6. Air dry and mount permanently.

Brown/black deposits that are absent from control sections indicate the sites of enzyme activity.

Protocol 15. Localization of peroxidase (after Graham and Karnovsky (15))

- 0.05 M Tris–HCl pH 7.6
- diaminobenzidine (DAB) (as *Protocol 14*)
- 30% hydrogen peroxide

1. Prepare substrate solution as follows. Dissolve 5 mg DAB in 10 ml 0.05 M Tris–HCl pH 7.6. Add 0.2 ml freshly prepared 1% hydrogen peroxide.
2. Incubate test sections in substrate solution for 5–10 min at room temperature in darkness (after optional preincubation with 1% (w/v) phenylhydrazine hydrochloride to block endogenous peroxidase).
3. Incubate control sections (i) in the above solution from which the hydrogen peroxide has been omitted, and (ii) as in *Protocol 13*, step 2.
4. Wash in distilled water.
5. Air dry and mount permanently.

Brown/black deposits that are absent from control sections indicate the sites of enzyme activity

Protocol 16. Localization of phenoloxidase (after Vaughn and Duke (16))

- phosphate citrate buffer pH 4.5 (18.7 ml 0.5 M sodium phosphate, 10.65 ml 0.5 M citric acid, diluted to 100 ml)
- naphthol (Sigma N1000)
- 4-aminodiphenylamine (*N*-phenyl-*p*-phenylenediamine) (Sigma 5379)
- DMF (as *Protocol 11*)

1. Prepare substrate solution as follows. Dissolve 5 mg naphthol and 5 mg 4-aminodiphenylamine in 0.5 ml DMF. Add slowly while mixing to 10 ml of phosphate citrate buffer pH 4.5.

Protocol 16. *Continued*

2. Incubate test sections in substrate solution for 10–20 min at room temperature or 37 °C in darkness.
3. Incubate control sections in the above solution from which the substrate has been omitted.
4. Wash in distilled water.
5. Air dry and mount permanently (see *Protocol 3,* footnote *b*).

Blue deposits that are absent from control sections indicate the sites of enzyme activity.

References

1. Jensen, W. A. (1962). *Botanical histochemistry*. W. H. Freeman, San Francisco.
2. Pearse, A. G. E. (1974). *Histochemistry: theoretical and applied.* Churchill Livingstone, Edinburgh.
3. O'Brien, T. P. and McCully, B. (1981). *The study of plant structure*. Termarcarphi Pty, Melbourne.
4. Clarke, G. (ed.) (1981). *Staining procedures*. Williams and Wilkins. Baltimore, MD.
5. Gahan, P. B. (1984). *Plant histochemistry and cytochemistry: an introduction*. Academic Press. London.
6. Eschrich, W. and Currier, H. B. (1964). *Stain Technol.*, **39**, 303.
7. Sexton, R. and Hall, J. L. (1991). In *Electron microscopy of plant cells* (ed. J. L. Hall and C. Hawes), pp. 105–80. Academic Press, London.
8. Gomori, G. (1941). *Arch. Pathol.*, **32**, 188.
9. Gahan, P. B. (1981). *Ann. Bot.*, **48**, 769.
10. Jefferson, R. A., Kavanagh, T. A., and Beven, M. W. (1987). *EMBO J.*, **6**, 3901.
11. Seligman, A. M., Karnovsky, M. J., Wasserkrug, H. L., and Hanker, J. S. (1968). *J. Cell Biol.*, **38**, 1.
12. Kaneko, Y. and Newcombe, E. H. (1987). *Protoplasma*, **140**, 1.
13. Kaughn, A. P. (1987). In *Handbook of plant cytochemistry*, vol. 1, (ed. K. C. Vaughn), pp. 25–36. CRC Press, Boca Raton, FL.
14. Frederick, S. E. (1987). In *Handbook of plant cytochemistry*, vol. 1 (ed. K. C. Vaughn), pp. 2–23. CRC Press, Boca Raton, FL.
15. Graham, R. C. and Karnovsky, M. J. (1966). *J. Histochem. Cytochem.*, **14**, 291.
16. Vaughn, K. C. and Duke, S. O. (1981). *Protoplasma*, **108**, 319.

4

Electron microscopy

C. HAWES

1. Introduction

Despite the recent resurgence in the use of light microscopy due to the introduction of confocal and video-enhanced microscopy coupled with the development of new image processing techniques (see Chapter 1), the unparalleled resolution offered by the electron microscope (EM) will maintain its importance as a basic tool in cell biology. Recent developments in preparative protocols, such as ultra-rapid freeze-fixation and freeze-substitution have permitted the preservation of cellular ultrastructure with great fidelity. Thus, many of the old arguments regarding the artefactual nature of EM preparations are no longer valid.

This chapter covers the basic protocols for the preparation of plant material for both transmission and scanning electron microscopy (TEM and SEM) including freeze-fixation. For specific staining and labelling techniques such as enzyme cytochemistry, immunocytochemistry, and *in situ* hybridization histochemistry the reader is referred to chapters 3, 5, 6, and 7 in this volume. It is of course an impossible task to cover the techniques of TEM and SEM comprehensively in one short chapter, so a range of protocols considered to be of most use to the average plant cell biologist have been selected; some omissions are inevitable. For detailed discussion into the merits and theory of the various techniques the reader is referred to some of the numerous books and articles in the literature on electron microscopy (1–5). Suppliers of standard reagents used in the protocols are not listed individually since all are available from a wide range of companies including, for example, Agar Scientific, Balzers, Bio-Rad, Pelco International and Polysciences Inc.

2. Transmission electron microscopy

2.1 Conventional methods

In this section the conventional methods of fixation, dehydration, and embedding will be described.

2.1.1 Fixation and dehydration for thin sectioning

As with all microscopy the ultimate aim of the electron microscopist is to observe cells and tissues preserved in a state that reflects as closely as possible the structure of living cytoplasm. However, due to the hostile environment within the microscope, prior fixation of the material to be observed is required. The specimen may be preserved by conventional chemical fixation, by rapid freezing techniques (see Section 2.2.1), or by a combination of freezing and chemicals (freeze-substitution, Section 2.2.2).

For chemical fixation there is no simple protocol suitable for the majority of plant tissues and experiments have to be carried out with regards to the most appropriate buffer to be used, the correct pH of the fixative, and the strength and combination of fixatives and other additives to the fixative such as Ca^{2+} and sucrose. The two major properties required of a fixative are to create stable chemical cross-linkages in the tissue and to penetrate rapidly into the specimen. In most morphological studies a double fixation protocol will be used with a primary aldehyde fixation followed by post-fixation in osmium tetroxide. For primary fixation a combination of paraformaldehyde (a monoaldehyde) for rapid penetration and glutaraldehyde (for good protein cross-linking) is normally preferred. Post-fixation is carried out in osmium tetroxide which is a strong oxidant of ethylenic bonds and therefore cross-links both proteins and lipids. It is thus a good fixative for cell membranes and any components with a high fat content. For full discussions on the choice of fixatives and buffers for plant tissues the reader is referred to the excellent articles by Coetzee (6) and Roland and Vian (7). There are, however, several basic rules that should be followed for all fixations:

- if possible the osmoticum of the fixative should reflect that of the tissue to be fixed
- initial fixation should be as rapid as possible
- if tissue has to be excised from larger pieces this should be carried out under fixative
- fix as small a piece of tissue as possible i.e. 1 mm^3

The commonest buffers used are phosphate, sodium cacodylate (arsenic-based and very toxic), and Pipes (6).

Protocol 1. Preparation of a typical double aldehyde fixative (1% paraformaldehyde/2% glutaraldehyde)

- calcium chloride
- glutaraldehyde (EM grade)
- paraformaldehyde powder

- sodium cacodylate or Pipes
- 1.0 M and 0.1 M sodium hydroxide
- sucrose
- ultrapure water

1. Prepare buffer stock of either sodium cacodylate or Pipes as follows.
   (a) 0.2 M sodium cacodylate, pH 6.9: add 8.56 g sodium cacodylate to 100 ml ultrapure water, dissolve the powder, adjust pH with 0.2 M HCl, and make up final volume to 200 ml.
   (b) 0.2 M Pipes, pH 6.9: to 6.04 g Pipes add 50 ml ultrapure water; dissolve the powder with 8 M NaOH and adjust pH with 0.1 M NaOH, then dilute to 100 ml.
2. Prepare a 10% stock solution of paraformaldehyde.
   a) Add 10 g paraformaldehyde powder to 60 ml distilled water.
   (b) Heat to 70 °C.
   (c) Slowly add drops of 1 M NaOH until all the powder has dissolved.
   (d) Make up final volume to 100 ml.
3. Add together 25 ml 0.2 M buffer, 4 ml EM grade 25% glutaraldehyde, 5 ml 10% paraformaldehyde stock solution, and 16 ml ultrapure water.
4. Optionally add 1–3 mM $CaCl_2$ and 1% sucrose to optimize quality of fixation.

The speed of fixation of many plant specimens is often restricted by the hydrophobic nature of the surface of the material such as waxy or hairy cuticles. In these cases wetting of the specimen can be facilitated either by fixing under a low vacuum or by adding some surfactant to the fixative such as Brij 35 or Tween 20. Particulate specimens such as cell suspensions or protoplasts can also cause problems of handling during the fixation procedure. Various methods can be used to overcome this, such as centrifugation through all the steps of the fixation, dehydration, and embedding procedure or encapsulation of the suspension in agar or gelatin after osmium post-fixation. Further steps can then be carried out on small blocks of gel.

Dehydration of specimens is required before infiltration with resin, unless a water-miscible embedding medium is to be used. It is during the dehydration steps that major changes in tissue volume can take place. Therefore, it is preferable to use a water/ethanol series rather than a water/acetone series as recommended in many texts and if possible as many different alcohol concentrations as possible should be used. The other major problem with dehydration is the extraction of lipid from the sample. This is also reduced by using alcohol instead of acetone and by having post-fixed the specimen with osmium.

Protocol 2. Fixation and dehydration

- snap-top specimen vials (glass, 10 ml)
- rotator to hold specimen vials
- water/ethanol series
- dried absolute ethanol
- preprepared aldehyde fixative (see *Protocol 1*)
- 2% (w/v) buffered osmium tetroxide
- 0.5% aqueous uranyl acetate

1. If possible excise a small (1 mm^3) portion of tissue under fixative, transfer to fresh fixative in a vial, and agitate or rotate gently. Fix for approximately 1 h at room temperature. Optimum fixation times will vary with different tissues. If tissue does not wet or floats on the surface of the fixative, penetration can be aided by the addition of a surfactant (e.g. Brij 35) or by fixing under vacuum[a,b].
   - cell suspensions and protoplasts can be fixed in small centrifuge tubes and spun down in between treatments
   - for the fixation of protoplasts add suitable osmoticum to the primary fixative
2. Wash three times for 10 min each in buffer.
3. Post-fix in 2% buffered osmium tetroxide for 2 h at room temperature (shorter times can be used for smaller specimens)[b].
4. Wash in buffer followed by distilled water (at least two 10 min washes). After osmication cell suspensions and protoplasts can be embedded in either 2% (w/v) aqueous agar (40 °C) or 10% (w/v) gelatin prior to further processing.
5. Block stain in 0.5% (w/v) aqueous uranyl acetate for 12 h at 4 °C. This is an optional staining step.
6. Dehydrate in a water/ethanol series for 10–15 min each step as follows: 10%, 20% 30%, 50%, 70%, 90%, absolute ethanol, and absolute ethanol stored over a drying agent.

[a] Treatment with 1% tannic acid either in the primary fixative or as a separate 1 h step after primary fixation greatly enhances the contrast of structural proteins such as cytoskeletal elements and clathrin coats.
[b] Osmium tetroxide vapour is toxic, so carry out all fixation procedures in a fume cupboard.

2.1.2 Resin embedding

Ideally an embedding medium for electron microscopy should have low viscosity, will polymerize uniformly without shrinkage, will permit the use of various stains, be easy to section, and be stable in the electron beam (8). The ideal medium would be a water-soluble one such as the melamine resin Nanoplast, but this type of medium has yet to be routinely used with plant material. There is, however, a wide range of embedding media available and the choice will depend on the nature of the specimen and any post-sectioning treatments that may be undertaken such as immunocytochemical localization. This Section will be restricted to the use of epoxy resins. Acrylic resins such as LR White and Lowicryl for immunocytochemistry are described in Chapters 5 (*Protocol 5*), 6 (*Protocol 2*), and 7.

Epoxy resins are multi-component mixtures comprising the resin, hardening agents, flexibilizer, and accelerator. All epoxy resins are immiscible with water and only Spurr resin can be used without the use of an intermediate solvent such as propylene oxide. Great care should be taken when handling epoxy resins as many of the components are toxic, carcinogenic, and allergenic. This is particularly true of the low viscosity Spurr resin. Carry out all handling including polymerization in a fume cupboard, wear gloves when handling resins, dispose of excess resin by polymerization, and if possible use disposable beakers and instruments when mixing resins. Some companies supply resins in pre-weighed quantities which are simply mixed by pouring the components into one bottle and shaking. This bypasses any messy weighing procedures. Remove any contamination from the skin immediately using resin removal cream or with soap and water. The use of solvents will encourage absorption of the resin into the skin.

The choice of resin will often depend upon the nature of the specimen to be embedded. Plant material, due to the nature of the cell walls, is often difficult to infiltrate and low viscosity resins such as Spurr (*Table 1*) are often used for

Table 1. Spurr's resin mix

Component[a]	**Quantity (g)**
ERL/VCD 4206	10.0
DER 736	6.0
NSA	26.0
S-1	0.4

The Spurr mixture has a pot life of 3–4 days (or can be stored at −20 °C) and should be polymerized at 70 °C for 8–10 h. The hardness of the block can be varied by increasing the amount of DER to 8 g for soft blocks and reducing it to 4 g for hard blocks.

[a] Resins in component forms or kits are available from suppliers of EM consumables.

this reason alone. However, it should be noted that better contrast is often obtained with Epon (note that commercially available supplies are limited) or Epon substitutes (*Tables 2* and *3*).

Table 2. Typical two-part Epon mix

Mix A: 15g Epon 812, 15.45 g DDSA, and 0.5 ml BDMA
Mix B: 17.25 g Epon 812, 14.25 g MNA, and 0.5 ml BDMA

Combine the mixes in a 1:1 ratio or vary the ratio to alter the hardness of the final blocks—increase the proportion of A for soft blocks, increase the proportion of B for hard blocks.
Polymerization is best carried out at 45 °C for 10–20 h and 60 °C for 24–40 h.

Table 3. Epon/araldite mix

Component	**Amount (g)**
Araldite CY212	22.3
Epon 812	31.0
DDSA	60.0
BDMA	2.0

Polymerize for 24–48 h at 60 °C.

Protocol 3. Resin embedding

- snap-top specimen vials
- specimen moulds
- dried absolute ethanol
- propylene oxide (if using Araldite, Epon, or Epon substitutes)
- resin of choice

1. Dehydrate specimens through to dried absolute alcohol.
2. Infiltrate with pure ethanol/Spurr[a,b] resin mixture as follows for 1 h each in 3:1, 1:1, and 1:3 ethanol:Spurr. If using other Epoxy resins such as Araldite, Epon, or Epon substitutes take specimens through to propylene oxide and then infiltrate with propylene oxide/resin mixtures of the same proportions as the ethanol/Spurr mixtures.
3. Infiltrate with pure resin for 1 h and then in a fresh batch of pure resin for a further 12 h[c].

4. Embed in resin in suitable moulds and allow to polymerize at 70 °C for 9–11 h depending on the resin mix.

[a] Spurr resin is highly toxic and all procedures including polymerization must take place in a fume cupboard.
[b] This protocol is also suitable for acrylic resins such as LR White.
[c] Times for resin infiltration and the number of resin changes will vary with the viscosity of the resin used and the size and hardness of the specimens.

2.1.3 Sectioning

It is outside the scope of this chapter to consider ultrathin sectioning in detail and the reader is referred back to some of the numerous texts on this topic (3, 9). However, it is prudent to mention that the whole sectioning process is facilitated by the correct choice of embedding moulds for each particular specimen. It is often the case with plant material such as roots and leaves that correct orientation of the specimen in the resin block is vital and that flat dish embedding or flat embedding in coffin-shaped silicone rubber moulds is to be recommended instead of BEEM or gelatin capsules.

2.1.4 Staining techniques

The prime reason for the use of stains on resin-embedded specimens is to increase the scattering of the electron beam by the specimen so that sufficient contrast is generated to permit microscopy and photography (10, 1, 7). To increase the contrast the stain has to have a higher atomic weight than the atoms of the specimen that it is staining and it will have to attach to the specimen. In general heavy metal salts are used in TEM and stains can be divided into two categories, general cytoplasmic stains and selective stains.

Over the years a wide variety of metal salts has been used as stains; the commonest are given in *Table 4*.

Table 4. Common heavy metal stains for TEM

Stain	Effect on image contrast
Osmium tetroxide	Stains lipids and proteins
Uranyl acetate	Stains nucleic acids and proteins
Lead citrate	Stains reduced osmium, membranes, proteins, glycogen, and RNA
Phosphotungstic acid	Stains polysaccharides, glycoproteins, and proteins

i. En bloc *staining*

When to stain is an important choice to be made in a TEM investigation. Stains can be introduced during the fixation and dehydration protocol, so-called *en bloc* staining, or staining can be carried out on the sections. Various factors have to be taken into account when choosing when and what to stain as follows:

- stain specificity
- duration of staining
- rate of penetration of stain
- future use of the *en bloc* specimen
- possible effects of multiple staining

There are various stages during a conventional fixation and dehydration procedure when a staining step can be introduced into the protocol. The most important staining step is post-fixation in osmium tetroxide. However, a mordant such as tannic acid can be introduced with or after primary fixation (see *Protocol 2*). Aqueous uranyl acetate at 0.5–2.0% can be used after osmication to stain the specimen slowly overnight at 4 °C in the dark or, made up as a 0.5% solution in 70% ethanol, as part of the dehydration procedure where staining is more rapid and may only take 30 min.

ii. Section staining

The majority of thin sections require some staining after they have been collected on grids. Each laboratory has developed its own particular protocol for staining and many small gadgets have been devised to aid in the handling of grids during the staining procedure and can be purchased from EM accessory suppliers. If a large number of grids is to be routinely stained then automatic stainers are available.

For conventionally fixed material it is usual to stain with a 0.5–2.0% solution of aqueous uranyl acetate (if *en bloc* staining has not been performed), in the dark for 5–10 min followed (after thorough washing in ultra-pure water) with a 2–10 min stain in lead citrate followed by a final wash. A simple method is to float the grids (with the sections facing downwards) on drops of stain solution on Parafilm or dental wax in Petri dishes and then to wash them on drops of filtered ultra-pure water. Pellets of potassium hydroxide should be included in the dish with the lead citrate to adsorb free CO_2 which will rapidly precipitate out lead carbonate in the stain and contaminate the sections. In between transfer from one drop to the next grids can be dried by carefully dabbing the edge on to filter paper.

Protocol 4. Preparation of Reynold's (11) lead citrate

- acid-washed 100 ml flask
- filter funnel
- glass-stoppered 50 ml volumetric flask

- lead nitrate
- sodium citrate
- 1 M sodium hydroxide

1. Add 1.33 g of lead nitrate and 1.76 g sodium citrate to 30 ml ultra-pure water in an acid-washed flask.
2. Shake vigorously for 1 min.
3. Allow to stand for 30 min with intermittent shaking.
4. Add 8.0 ml 1 M sodium hydroxide, dilute to 50 ml, and mix by inversion.
5. Filter into a glass-stoppered flask, leaving as little air space as possible. The stain is now ready to use[a].

[a] Any further turbidity can be removed by centrifugation

iii. Selective staining

Many different techniques have been devised to highlight specific components of cells. It is often useful to be able to stain selectively various membrane-bounded compartments or polysaccharide-rich components of the cell (7). These techniques can be particularly valuable in combination with thick sectioning and stereo microscopy in order to study the three-dimensional aspects of cellular organization (12, 13).

Osmium impregnation techniques

Osmium impregnation techniques are perhaps the most popular way of increasing the contrast of the endomembrane system and some other membranes such as chloroplast thylakoids. They are based on the reduction of osmium tetroxide to leave a deposit of osmate or osmium on the surface of membranes or within the cisternal space of membrane-bounded organelles such as the endoplasmic reticulum. Prolonged soaking of tissue in osmium tetroxide can give the desired results, but it is preferable to aid the reduction chemically with the addition of either zinc iodide (13, 14) or potassium ferrocyanide (15) to the stain.

The zinc iodide/osmium tetroxide (ZIO) technique (see *Protocol 5*) is suitable for most plant, algal, and fungal tissues and will impregnate the nuclear envelope, endoplasmic reticulum, Golgi apparatus (sometimes showing gradation of staining across the stack), the tonoplast (in some cell types), plastid thylakoids, and occasionally the stroma of mitochondria. It will not selectively enhance the plasma membrane or plastid envelope. After aldehyde fixation the specimens are soaked in a fresh unbuffered ZIO mixture. Any precipitation in the staining mixture can be ignored and the time of impregnation will vary from tissue to tissue. For most higher plant tissues impregnation lasting between 4 and 18 h normally gives satisfactory results although there is often a gradation in the degree of staining across a tissue.

Protocol 5. Zinc iodide/osmium tetroxide (ZIO) impregnation

- aldehyde fixative of choice (see *Protocol 1*)
- 2% aqueous osmium tetroxide
- powdered zinc
- resublimed iodine

1. Fix specimens in aldehydes[a].
2. Wash in buffer (2 × 10 min).
3. Wash in distilled water (2 × 10 min)[b].
4. Prepare a fresh mixture of zinc iodide/osmium tetroxide as follows: add 3 g of zinc powder and 1 g of resublimed iodine to 20 ml distilled water, stir for 5 min, and filter. Mix a small aliquot with an equal volume of 2% OsO_4 and use immediately.
5. Impregnate specimen with ZIO. Time and temperature of impregnation will vary with the specimens, initially try 4–12 h.
6. Wash in distilled water, dehydrate, and embed (*Protocol 3*).

[a] See *Protocols 1* and *2*.
[b] It is particularly important to wash thoroughly if phosphate buffers have been used in the primary fixative to prevent precipitation of zinc phosphate.

Plasma membrane staining

Both phosphotungstic acid and silicotungstic acid when used at low pH are excellent at staining selectively the plasma membrane of plant cells (7) and can also be used to enhance the contrast of secretory vesicle membranes. Stain solutions are acidified with either HCl or chromic acid and staining is carried out directly on thin sections. If the specimens have been post-fixed with osmium then bleaching with periodate will be required.

Protocol 6. Phosphotungstic acid staining of plasma membranes

- chromic or hydrochloric acid
- *dodeca*-tungstophosphoric acid
- 10% H_2O_2

1. Collect ultrathin sections on gold grids.
2. If osmium post-fixed, bleach for 30 min in 10% H_2O_2 or 1% periodic acid.
3. Float grids for 2–15 min on a 1% solution of phosphotungstic acid made in either 10% chromic acid or 10% hydrochloric acid.
4. Wash twice in distilled water.

Polysaccharide staining

The periodic acid–thiocarbohydrazide-silver proteinate (PATAg) developed by Thiery (16) is an excellent method for staining polysaccharides in resin-embedded sections. Developed from the classical periodic acid–Schiff's test it is based on the oxidative cleavage of free glycols on sugars, the condensation of these by thiocarbohydrazide or thiosemicarbazide, and the addition of silver proteinate which is precipitated leaving an electron-opaque precipitate of silver over the reaction site (7).

Protocol 7. PATAg staining of carbohydrates

- acetic acid
- 1% periodic acid
- 1% aqueous silver proteinate
- 0.2% thiocarbohydrazide (or thiosemicarbazide) in 20% acetic acid

1. Collect ultrathin sections of epoxy resin-embedded material on gold or nickel grids, in plastic rings, or in platinum wire loops.
2. Float sections or submerge grids in 1% periodic acid for 30 min[a,b].
3. Treat with 0.2% thiocarbohydrazide (or thiosemicarbazide) in 20% acetic acid at room temperature for 3–24 h.
4. Wash in decreasing concentrations of acetic acid (20%, 10%, 5%, and 1%) and twice for 10 min each in distilled water.
5. Stain with 1% aqueous silver proteinate (made at least 12 h previously) in the dark for 30 min.
6. Wash twice with distilled water.
7. If sections are in loops or rings, mount on grids. If needed, post-stain the sections with uranyl acetate and lead citrate.

[a] As a control treat the sections with 5–10% H_2O_2 for 20–30 min. This will not oxidize free glycols to aldehyde groups thus the thiocarbohydrazide reaction will not link to the carbohydrates.
[b] This step can be omitted if non-osmicated material is used.

The staining of polysaccharides in a dimethylsulphoxide extracted cell wall of an unosmicated preparation of a maize coleoptile epidermis is demonstrated in *Figure 1*.

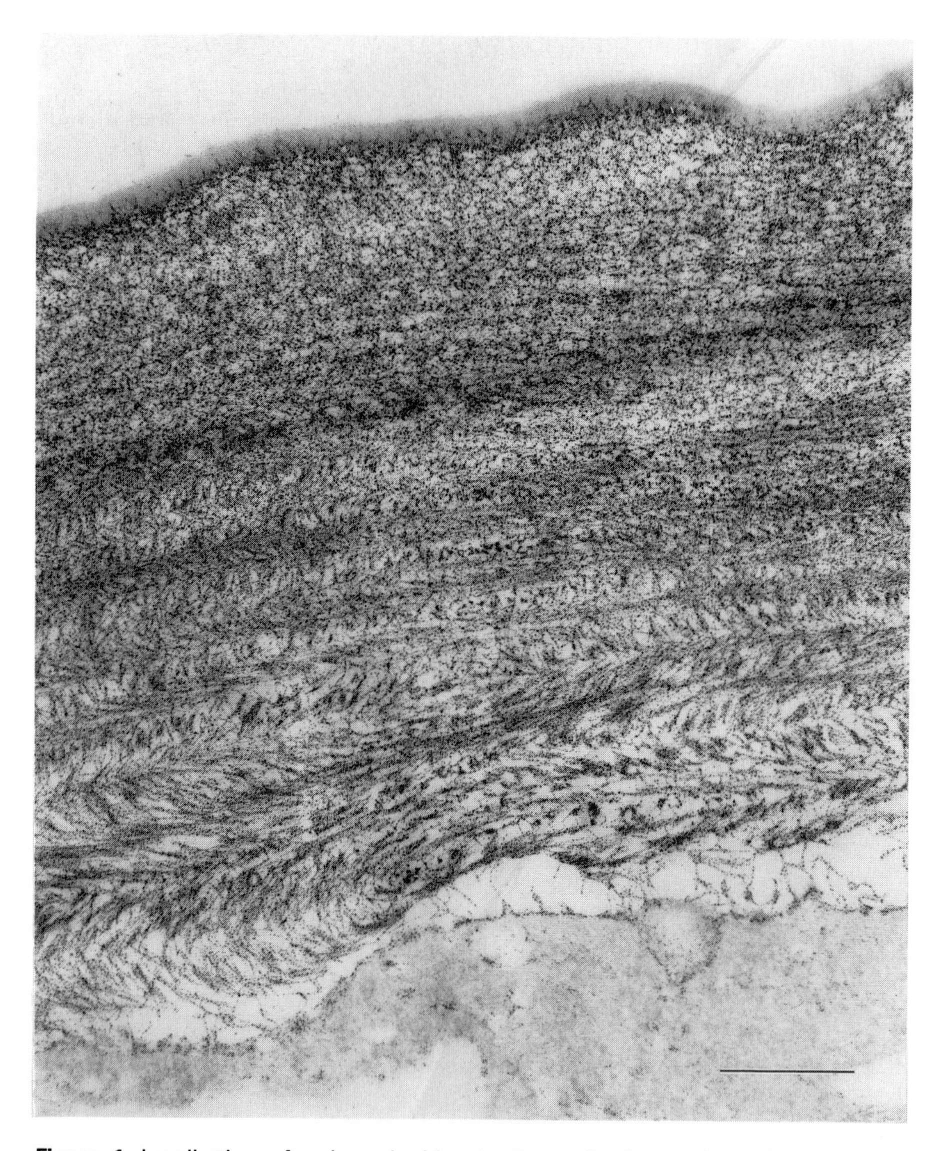

Figure 1. Localization of polysaccharides in the wall of a maize coleoptile outer epidermal cell. The wall matrix was partially extracted in DMSO for 12 h after aldehyde fixation and staining was by the periodic acid, thiocarbohydrazide silver proteinate technique (see Section 2.1.4). Bar = 1.0 μm. (Unpublished micrograph of Dr B. Satiat-Jeunemaitre.)

iv. Negative staining

Negative staining is a very quick technique for preparing samples of particulate specimens (e.g. viruses), small single cells (e.g. bacteria), organelle fractions (e.g. mitochondria, membrane vesicles, Golgi and cytoskeletons), and macromolecules (e.g. nucleic acids and proteins). It is based on the selective scattering of electrons by a heavy metal stain that is dried around the specimen thus giving a negative image of the specimen. Although the basic procedures for negative staining can be quick and simple, combined with the correct use of the electron microscope such as low dose imaging, to reduce radiation damage and movement of the stain, extremely high resolution data can be achieved. Full details of negative staining techniques are presented in two excellent chapters in previous volumes of the *Practical Approach* series (17, 18), so just one simple protocol will be given here.

Protocol 8. On-grid negative staining

- phosphotungstic/silicotungstic acid[a] or
- uranyl acetate[a] or
- ammonium molybdate[a]

1. Prepare Formvar- (available from EM consumables suppliers) or carbon-coated grids on which to carry out the staining. Grids may be 'glow discharged' in a vacuum coating unit prior to staining to reduce their hydrophobicity.
2. Apply a drop (5–10 μl) of suspension of the specimen to the grid. Grids may be held in forceps or placed on dental wax or Parafilm. If the latter technique is used then make sure there is a 'free drop' of the specimen from the pipette so that the grid is not pulled up to the pipette tip by surface tension.
3. Leave to settle for a few seconds or longer (depending on the nature and concentration of the specimen) and then pull off excess liquid by touching the edge of the grid to a piece of filter paper.
4. Apply a drop of negative stain[a] as above and after a few seconds remove it with filter paper.
5. Leave the grid to dry at room temperature prior to observation with the TEM.

[a] Common negative stains are as follows: phosphotungstic or silicotungstic acid at 0.25–2% (w/v) solutions adjusted to pH 6–8 with 1 M KOH or NaOH; 0.5–2% (w/v) aqueous uranyl acetate used at its natural pH of 4.2; ammonium molybdate at 1–3% (w/v) solution adjusted to pH 7.0.

2.2 Low temperature methods

Although chemical fixations is still the most popular method for the preservation of tissue for electron microscopy there is the ever-present problem of artefacts that may be introduced at various stages of the preparative protocols. Chemical fixation by its very nature is slow and there will always be a gradient of fixation across tissues, so rapid events such as cytoplasmic streaming and vesicle movement will rarely be captured in a state reflecting the *in vivo* situation. Induced artefacts can include movement of cellular components from their *in vivo* positions, extraction of various components especially soluble molecules, and chemical alteration of molecules which may for instance destroy antigenic sites. Thus, ultra-rapid freeze-fixation techniques have been seen as way of overcoming these problems, although it should not be assumed that freezing technology produces preparations that are always artefact-free.

The ultimate goal of the cryobiologist is to view the cell in a frozen state where freeze-fixation has been so rapid that all the cellular water is vitrified (i.e. no ice crystals have formed) and no chemicals have been added to the specimen during subsequent processing or used to induce contrast in the electron microscope. Such technology is just becoming available with the combination of high-pressure freezing, cryosectioning, and cryomicroscopy. Images have been obtained of unstained cryosectioned leaf material which show the normal complement of cellular organelles (19, 20). However, although the necessary equipment is expensive and not available in the majority of laboratories, many simple freezing devices whether home-made or commercial are available and techniques such as freeze-substitution should become routine in the average EM laboratory.

2.2.1 Rapid-freeze fixation

The most critical step in any cryopreparation technique is the initial freeze-fixation of the specimen. In an ideal situation heat would be removed from the cells of the specimen at a rate which will result in the formation of vitreous ice, i.e. ice with no crystalline structure. In reality this is a difficult state to achieve in cells, and especially in tissues. However, an acceptable goal is to produce ice crystals that are below the resolution produced by the subsequent preparative techniques so that any ice crystal damage is not recognizable. It is also important to remember that as frozen specimens are heated up cubic ice will turn into hexagonal ice and cause serious damage to specimens; frozen specimens must therefore be retained below −100 °C if ice crystal regrowth is to be avoided. This phenomenon is a potential cause of damage to specimens during freeze-substitution protocols. The other main limitation to all the freezing techniques with the exception of high-pressure freezing is that the depth of good freezing into a tissue is limited to a few tens of micrometres before ice crystal damage becomes unacceptable.

There are many excellent texts and reviews on the fine detail of the ultra-rapid freezing of biological material for electron microscopy and the reader is referred to these for further detailed discussion on this topic (21–24).

i. Choice of cryogen

The choice of cryogen for freezing will to some extent depend on the freezing apparatus to be used. In increasing order of efficiency the commonest cryogens are liquid nitrogen, liquid Freon, liquid propane, liquid ethane, and liquid helium, the latter being used to cool a metal freezing surface. All these cryogens present problems in their use. It is important to realize that liquid nitrogen is by far the worst cryogen in terms of rapid heat exchange when the specimen is plunged into the cryogen. This is due to the well known 'Leidenfrost' phenomenon which arises from film boiling where an insulating layer of gas builds up between the specimen and cryogen, drastically reducing the cooling rate. Thus, liquid nitrogen is only useful for freezing well cryoprotected specimens such as sucrose-infused tissue for cryosectioning and immunocytochemical labelling or for cold block freezing. Freon can be considered to be environmentally unfriendly, while propane and ethane due to the risk of explosion need to be handled with great care and suitable methods of disposal or burning off must be available.

Plunge freezing

The simplest method for cryofixing specimens is to hold them in fine forceps and plunge them into a pot of cryogen. The cryogen such as Freon or propane is kept liquefied by being immersed in a bath of liquid nitrogen. Cell suspensions can be stuck on to coverslips coated with poly-L-lysine (mol. wt. 35 000, 5 mg/ml in distilled water) and pieces of tissue held directly in forceps. Specimens are then stored in liquid nitrogen prior to further processing. However, this technique is obviously uncontrolled and for good freezing a rapid and controlled insertion of specimen into the cryogen is desirable (> 1.0 M/s). To facilitate this there are various plunge freezing devices on the market (24); alternatively, simple devices as described by Lancelle *et al.* (25) allowing controlled plunging can be made in the laboratory. In this latter case the specimen holder was a thin Formvar-coated wire loop on to which cells were stuck.

Jet freezing

Instead of rapidly introducing the specimen into the cryogen, jet freezers fire two synchronized jets of propane either side of the specimen. The technique is best applied to small specimens which are normally sandwiched between two small copper plates or hats. However, both root and leaf segments have been successfully preserved by this technique with satisfactory freezing reported up to a depth of 100 μm (26). For safety reasons this technique should be restricted to commercially available instruments.

Cold block slamming

As heat transfer across a solid metal interface is faster than that from a specimen into a liquid, various devices have been manufactured to permit the rapid slamming of a specimen against a copper block cooled with a liquid helium or liquid nitrogen. This technique has been very successful when combined with freeze fracturing and deep-etching of animal cells, tissues, and isolated macromolecules (27–29), but as yet has not been widely applied to plant cells, although excellent results can be achieved (30, 31).

High pressure freezing

Becoming increasingly popular is the technique of high-pressure freezing as this permits the satisfactory freezing of much thicker specimens to a depth of about 0.5 mm when frozen from both sides. In the (expensive) apparatus specimens are subjected to a pressure of 2100 bar which depresses the melting point of water, reduces the formation of ice nuclei, and retards the growth rate of ice crystals. The net result is that the high pressure has an equivalent effect of adding a chemical cryoprotectant to the tissue. The result is that much improved freezing is achieved deeper into the tissue. Freezing is mediated by jets of liquid nitrogen, under the same high pressure, being fired at the specimen which is sandwiched between metal plates. Tissue is only exposed to the high pressure for approximately 20 msec before freezing to minimize possible pressure-induced artefacts (32, 33) and specimens are usually mounted in non-penetrative cryprotectants such as 1-hexadecene or 15% aqueous dextran to optimize the transfer of pressure to and heat from the specimen (20, 35). To date the results that have been published of high-pressure frozen plant material, either freeze-fractured (34) or sectioned (19, 20, 36), indicate that this technique is likely to prove invaluable for the cryopreservation of whole tissues.

2.2.2 Freeze-substitution

Freeze-substitution is the replacement of ice within the specimen with another solvent whilst keeping the specimen at low temperatures (24). Acetone and methanol are the most common solvents and some other fixative such as an aldehyde or osmium is normally included in the substitution medium. The key to successful freeze-substitution is to hold the specimen at −80 to −90 °C for several days to allow for the slow penetration of the substituting medium into the tissue. If complete substitution does not take place before warming up is initiated then ice crystals can regrow, leading to damage in the specimen that is often mistakenly attributed to poor freezing. The low temperature substitution period is then followed by slowly bringing the specimen to a temperature above 0 °C and infiltrating it with resin (25). For immunocytochemical work osmium can be omitted from the substitution medium and low temperature embedding resins can be used (see Chapter 7).

Although apparatus for freeze-substitution is manufactured by several companies it is a relatively inexpensive technique to carry out in the laboratory. Initial substitution can be carried out in a low temperature deep-freeze and subsequent steps in a combination of freezers and fridges. However, in our laboratory we carry out the final stages of the substitution procedure in the chamber of a small cryostat (Bright Starlet) which gives better temperature control from −50 °C to 0 °C.

Protocol 9. Typical freeze-substitution schedule

- anhydrous acetone
- 2–4% osmium tetroxide in anhydrous acetone[a]
- 0.5% uranyl acetate in acetone
- epoxy or acrylic resin

1. Rapid-freeze specimens by method of choice. Store until required in liquid nitrogen.
2. Make substitution medium of choice. e.g. 2–4% osmium tetroxide in anhydrous acetone[a].
3. Cool to −80 °C in freezer in small vials and rapidly add frozen specimens.
4. Substitute at −80 °C for 3–5 days.
5. Slowly bring to room temperature over 12–48 h[b].
6. Wash three times in acetone.
7. Stain with 0.5% uranyl acetate in acetone for 2 or 12 h at 4 °C.
8. Wash in acetone.
9. Embed in resin of choice[c].

[a] Alternative substitution media after Ding *et al.* (26). Freeze-substitute with 0.1% tannic acid in acetone at −90 °C for 2 days and then with 2% osmium tetroxide/2% uranyl acetate in acetone at −20 °C for 1 day, raise temperature to 4 °C overnight and wash in pure acetone at 4 °C for 2–3 h prior to embedding.

[b] This procedure can easily be carried out in a cryostat by reducing the temperature to −50 °C, closing the lid and turning the machine off.

[c] As in *Protocol 2*, from step 2; ethanol is replaced by acetone.

2.2.3 Freeze-fracture and replication

The so-called freeze-etching technique has been around for many years and has become a standard EM procedure. However, before the advent of the various ultra-rapid freezing techniques, discussed above, specimens were usually fixed and cryoprotected prior to freezing which could result in the

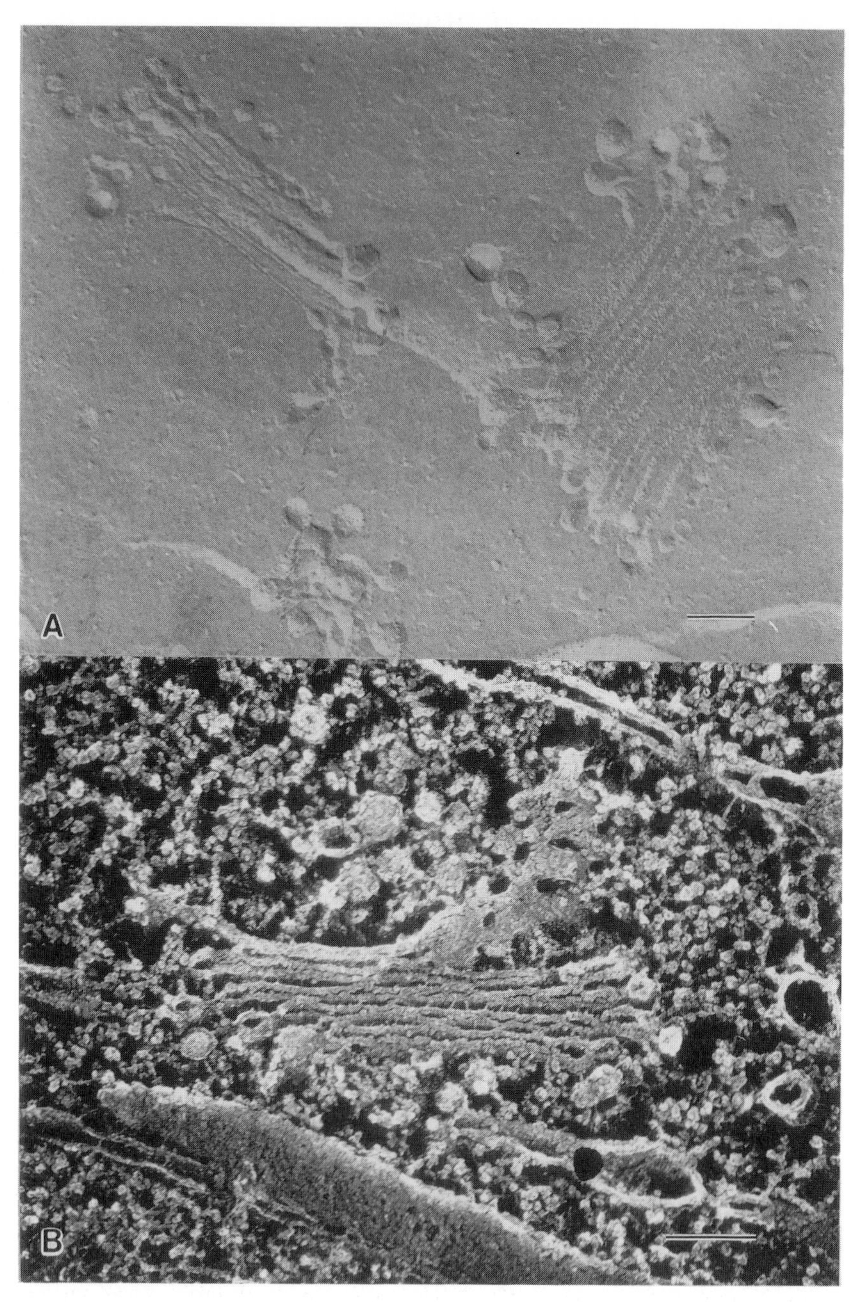

Figure 2. (A) Conventional freeze-fracture replical of a suspension cultured carrot cell, cryofixed on a helium cooled copper mirror at 12 K, showing the stacks of cisternae of the Golgi apparatus (see Section 2.2.3). Bar = 200 nm. (B) Deep-etch rotary shadowed replica of a maize suspension culture cell cryofixed on a helium cooled copper mirror at 12 K showing much greater detail of Golgi cisternae than in a conventional freeze-fracture repica (see Section 2.2.3). Bar = 200 nm.

induction of various artefacts (23, 37, 38). The basic freeze-fracture technique permits study of the macromolecular organization of the hydrophobic inner faces of membrane bilayers whilst partial freeze-drying of the specimen after fracturing (etching) permits the production of rotary shadowed replicas to facilitate high resolution imaging of cytoplasmic components and macromolecules. For full details of the various freeze-fracture techniques refer to Robards and Sleytr (21) and to an excellent series of protocols on the technique presented by Fujikawa in a previous volume of this series (39).

i. Specimen preparation, fracturing, and replication

The fine details of the process of freeze-fracture and replication will vary with the particular machine to be used. However, the major steps are given in *Protocol 10* and an example of a freeze-fractured Golgi apparatus is shown in *Figure 2A*.

Protocol 10. Basic protocol for freeze-fracture and replication

- 70% aqueous sodium hypochlorite
- 70% concentrated sulphuric or chromic acid

1. Prepare specimen for freezing. If possible avoid the use of cryoprotectants and chemical fixatives.
2. Fracture with cold knife or double replica device in freeze-fracture unit.
3. Briefly remove any surface water by sublimation at −100 °C.
4. Shadow with platinum/carbon (95%/5%) at about 45 ° to the fracture plane.
5. Replicate with carbon perpendicular to the fracture plane.
6. Float replicas off the specimen in distilled water and clean in hypochlorite and acid (Section 2.2.3, subsection (iii)).
7. Wash in filtered distilled water and pick up on grids.

For a full detailed protocol of the freeze-fracture technique refer to Fujikawa (39).

ii. Deep-etching and rotary shadowing

The introduction of the rapid-freeze deep-etch technique by Heuser (27) has in the past decade revolutionized the study of animal cell cytoplasm and the constituent macromolecules (28). The improved resolution of cellular ultrastructure over conventionally thin-sectioned material combined with the use of stereo imaging has been invaluable in the study of the cytoskeleton and vesicle trafficking (29, 40). However, the technique has yet to be widely used by plant cell biologists although very striking images of the cytoskeleton, clathrin-coated vesicles, and cell walls have been published (30, 41, 42). Although best for structural elements, deep-etching can also give excellent detail of membrane-bounded organelles such as the Golgi apparatus (*Figure 2B*); however, large sheets of membrane tend to get damaged during prolonged etching.

Protocol 11. Rapid-freeze, deep-etching of plant cells

- hypochlorite (70%)
- sulphuric or chromic acid

1. Ultra-rapidly freeze cells or tissue on suitable stub or pins for the freeze-fracture apparatus to be used. Remove as much water from the specimen as possible just prior to freezing.
2. Fracture specimen in freeze-fracture unit with cold knife at −110 °C.
3. Manoeuvre cold knife over specimen and etch at −100 °C for 10–30 min.
4. Low angle rotary shadow at 26 ° with platinum/carbon at −100 °C[a].
5. Rotary replicate at 70 °C with carbon at −100 °C[b].
6. Remove specimens from freeze-fracture apparatus and float replica on to distilled water.
7. Clean replica in 70% H_2SO_4, followed by 70% hypochlorite over 2–3 days[c]. Replicas can be transferred between solutions by picking up with a platinum loop or by picking up on the end of a glass pipette that has been heat fused to form a solid 'blob' of glass.
8. Wash on filtered distilled water (3 × 5 min).
9. Pick up on Formvar- or carbon-coated grids[d].

[a] Determine film thickness with a quartz crystal monitor. 1.5–2.0 nm should be adequate.
[b] Carbon film should be 10–35 nm thick.
[c] Replicas of deep-etched material can take much longer to clean than conventional freeze-fracture replicas.
[d] Deep-etch replicas are more fragile than conventional freeze-fracture replicas so great care must be taken in handling.

iii. Replica cleaning

Replicas of plant material can often prove very difficult to clean. Cell wall material can often prove to be resistant to hypochlorite or weak acids. However, perseverance and strong solutions of sulphuric acid and chromic acid will eventually dissolve most material. Final washing after transfer through weak acid to water can be in 70% hypochlorite followed by several washes with filtered distilled water.

2.3 Rotary shadowing of proteins

Negative staining techniques can be used for the visualization of isolated proteins and can provide very high resolution images. However, for many slender proteins the contrast provided by negative staining is not adequate and rotary metal shadowing offers an excellent alternative. In this technique proteins are dissolved in buffers containing high concentrations of glycerol, sprayed or dried on to mica, dried, and finally shadowed and replicated with metals and carbon (41, 43). Laboratories tend to devise their own particular technique for spraying the protein-containing buffer on to mica. In our laboratory the buffer is atomized into a cloud in mid-air using a chromatography spray gun with a 100 μl micropipette tip attached. The cleaved mica sheet is then passed through the cloud and droplets adhere to the surface. This is then dried and replicated in a vacuum coating unit.

Protocol 12. Rotary metal shadowing of isolated proteins

- ammonium acetate
- dithiothreitol
- EDTA
- glycerol
- mica sheets

1. Prepare protein to be shadowed and bring to a final concentration of 10–100 μg/ml in the replication buffer made by mixing 0.1 M EDTA, 10 μM dithiothreitol, and 155 mM ammonium acetate with glycerol to give a final glycerol concentration of 70%.
2. Spray 20 μl of protein and buffer on to a freshly cleaved mica sheet.
3. Dry mica in vacuum coating apparatus.
4. Rotary shadow with platinum/carbon (95%/5%) at 20 °C[a].
5. Replicate with carbon.
6. Remove from coating unit and float replica on to filtered distilled water.
7. Pick up on to plastic-coated grids.

[a] Tantalum/tungsten can be used to give a higher resolution shadow (43); a coating system with electron beam guns is the most appropriate for consistent shadow quality.

3. Scanning electron microscopy

Preparation of material for SEM presents a different set of problems than does TEM. For the majority of studies faithful preservation of surface detail is the goal. Processing of specimens for conventional SEM has been shown in many cases to result in a wide variety of artefactual changes which may be induced by fixation, dehydration, and drying protocols. Cryo-techniques have to a certain extent alleviated some of these problems and also permit the easy examination of internal tissue structure. This section will be restricted to some of the commonest protocols for the preparation of plant material for SEM.

3.1 Ambient temperature SEM

It is possible to observe fresh untreated material in the SEM, especially at low accelerating voltages (45). However, the conventional procedure is to fix, dehydrate, dry, and metal-coat the specimen before observation.

3.1.1 Fixation and dehydration

Fixation and dehydration protocols for the SEM are very similar to those used for the TEM. Post-fixation in osmium tetroxide is not always necessary but can be useful in increasing the specimen conductivity and thus reducing the possibility of charging in the microscope.

Protocol 12. Basic preparation of material for ambient temperature SEM

- 2–4% buffered glutaraldehyde
- 2% osmium tetroxide (aqueous or buffered)
- water/ethanol series (see *Protocol 2*)
- anhydrous ethanol

1. Fix specimens in 2–4% glutaraldehyde in buffer of choice.
2. Wash in buffer (2 × 10 min).
3. Post-fix in 2% osmium tetroxide (buffered or aqueous)[a].
4. Wash in distilled water and dehydrate in a water/ethanol series to dried absolute ethanol[b].
5. Critical-point dry from absolute ethanol[c].
6. Mount specimens on to SEM stubs using an adhesive such as double-

sided tape, silver dag, or carbon paste which will not degrade the microscope vacuum.

7. Sputter coat with gold.

[a] Post-fixation is optional but increases specimen rigidity and conductivity.
[b] See *Protocol 2*.
[c] See *Protocol 13*.

3.1.2 Drying techniques

i. Air drying

Air drying can lead to major shrinkage, collapse, and distortion of specimens and is not recommended for most SEM specimens. However, extremely robust material (e.g. some seed coats and wood) does not need to be fixed before coating and can be observed without processing.

ii. Freeze-drying

Freeze-drying techniques allow the preparation of cryofixed material for the SEM and although commercial freeze-driers are available the technique can easily be performed in any EM laboratory with a vacuum coating unit. Frozen specimens are held in wells drilled into a metal block which is cooled in liquid nitrogen in a polystyrene container. The block still in the container is then put in the chamber of the coating unit and held under rotary pump vacuum. Residual liquid nitrogen boils and solidifies and is left to sublime slowly overnight until the metal block has warmed to room temperature. Specimens are then taken out of vacuum and mounted on stubs for coating. The key to good freeze-drying is to remove the water slowly from the specimen over a few days; the ultimate in this is molecular distillation drying which allows very slow removal of water at high vacuums and low temperatures (46).

iii. Critical-point drying

Critical-point drying is the most popular technique for the drying of biological specimens for the SEM (47). It is based on the principle that for every liquid there is a unique combination of pressure and temperature (the critical point) at which the transition from liquid to the vapour phase takes place without an interface or change in volume as the densities of the liquid and the gas are equal. Therefore, the specimen is instantaneously dried without being subjected to the surface tension forces associated with liquid/vapouring boundaries. Liquid CO_2 is a suitable medium for critical-point drying with a critical point of 31 °C and 7.3×10^6 Pa (74 bar). As liquid CO_2 is not miscible with water an intermediate fluid such as ethanol, acetone, or amyl acetate is required, so specimens have to be dehydrated before drying.

Protocol 13. Critical-point drying

1. Fix and dehydrate specimens through to absolute alcohol[a] (see *Protocol 2*).
2. Transfer to pre-cooled chamber of critical-point drying apparatus in a small quantity of ethanol making sure that the specimen remains wet at all times.
3. Quickly seal the pressure chamber and fill with liquid CO_2.
4. Flush out the chamber with liquid CO_2 to replace all the ethanol in the specimen with CO_2 whilst maintaining a chamber temperature of 0–5 °C. The time of flushing will depend on the make of drier and the size of the specimen. Typically three 1 min flushes with 5–10 min between flushes will suffice.
5. Heat up the critical-point drying apparatus to take the liquid CO_2 through the critical point.
6. Let the temperature rise above the critical point (31 °C for CO_2) and then very slowly bleed out the pressure[b].
7. If specimens are not mounted on stubs and coated immediately, store them in an anhydrous conditions over desiccant.

[a] Specimens may be transferred to the carrier of the critical-point drying apparatus early on in the dehydration procedure.
[b] If the pressure is reduced too quickly the CO_2 gas may recondense to liquid.

3.1.3 Specimen coating

Although a range of coating techniques can be used for SEM specimens (48), diode sputter coating with gold in a low pressure environment of an inert gas, such as argon, remains the most popular. For conventional SEM the metal film coating the specimen needs to be 10–12 nm thick to eliminate electrical charging efficiently and to conduct away the heat which can rapidly build up in the specimen. A second aim of coating is to produce a layer in which the grains are smaller than the resolution of the microscopy being undertaken. Therefore, with high resolution field emission SEM it is necessary to coat the specimen with a layer of platinum or carbon (44).

3.1.4 Osmium etching methods

Conventional methods of specimen preparation for SEM in most instances preclude the observation of internal cell structure. High resolution cryo-SEM of rapidly frozen and fractured specimens is one way of overcoming this limitation (44), while an osmium maceration technique perfected by Tanaka (49) has been successfully applied to plant tissue (50). This latter technique is inexpensive and no specialist equipment is required other than a liquid nitrogen cooled metal block which can be placed in a small polystyrene container. Specimens are fixed in osmium tetroxide, cryoprotected in DMSO, fractured once frozen, and macerated over an extended period in weak osmium tetroxide. This procedure extracts the cytoskeleton and cytoplasmic matrix from the cytoplasm and following further osmication leaves the endomembrane system well fixed. In this way all the membrane-bounded organelles can be observed with the SEM. For high resolution work a microscope fitted with a field emission gun operating at low accelerating voltages is recommended.

Protocol 14. Fracturing and osmium maceration

- dimethylsulphoxide (DMSO)
- 1% buffered osmium tetroxide
- 2% aqueous tannic acid

1. Fix material for 2–16 h in 1% osmium tetroxide in an appropriate buffer.
2. Wash in buffer.
3. Infiltrate with 15%, 30%, and 50% aqueous DMSO for 30 min in each solution[a].
4. Freeze specimens in liquid nitrogen.
5. Fracture specimens on a liquid nitrogen cooled metal block with a razor blade; alternatively, if suitable, specimens such as leaves can be snapped under liquid nitrogen.
6. Thaw fractured specimens in 50% DMSO and rinse in buffer.
7. Macerate for 2–14 days in 0.1% buffered osmium tetroxide.
8. Fix for 1 h in 1% osmium tetroxide followed by 16 h in 2% aqueous tannic acid and a further 1 h in 1% osmium tetroxide.
9. Dehydrate, critical-point dry (see *Protocol 13*), mount on stubs, and sputter coat with gold or carbon.

[a] DMSO must be handled in a fume cupboard.

3.1.5 Replica techniques

A very quick non-destructive method of preparing samples of plant surfaces for the SEM is to make a replica. Low viscosity silicone rubber solutions can be gently painted on to the surface of interest and then left to set (e.g. Silastic 3110RTV plus catalyst 4 from BDH or Kerr silicone rubber impression materials Extrude or Reflect). The replica can then be peeled off the plant surface and a secondary replica made from cellulose acetate or epoxy resin. This technique has the advantage that sequential replicas can be made of the same surface so that morphological changes with time can be studied (44, 51).

3.2 Low temperature SEM

Low temperature SEM affords the opportunity to study fully frozen hydrated specimens on a cold stage in the microscope without any pretreatment other than freezing and coating (44). As it circumvents the need for fixation, dehydration, and critical-point drying or freeze-drying the chances of inducing artefacts are minimized (52) and the time taken for specimen preparation is greatly reduced. As the required resolution when compared with TEM is relatively poor, in most cases freezing is simply carried out by plunging a specimen into super-cooled nitrogen slush. However, if high resolution work is to be carried out with a field emission SEM then specimens will have to be preserved by freeze-fixation as detailed in Section 2.2. In the majority of cases cryopreparation for SEM is carried out in a commercially available apparatus either separate from the microscope with a cooled vacuum transfer system or bolted onto the specimen chamber of the SEM. The major steps in the procedure are as follows:

(a) Mount specimen on cryo-stub.
(b) Plunge-freeze in nitrogen slush.
    i. Specimen can be fractured to reveal internal structure.
    ii. Partial freeze-drying can be carried out to remove surface water or internal water.
(c) Sputter coat with gold.
(d) Insert on to cold stage of the SEM. (The specimen can initially be observed uncoated and then removed from the specimen chamber and gold-coated after further drying or fracturing.)
(e) Carry out microscopy between −100 °C and −175 °C.
(f) If it is needed again, store the specimen on the stub in liquid nitrogen.

Acknowledgements

I wish to thank Mr Barry Martin for his invaluable help with various cryo-techniques and Dr Béatrice Satiat-Jeunemaitre for supplying *Figure 1* and for her helpful comments on the manuscript.

References

1. Hayat, M. A. (1986). *Basic techniques for transmission electron microscopy*. Academic Press, Orlando, FL.

2. Glauert, A. M. (1975). *Practical methods in electron microscopy*, Vol. 3, Part 1. Elsevier, Amsterdam.
3. Hayat, M. A. (1989). *Principles and techniques of electron microscopy: biological applications*. Macmillan Press, Basingstoke.
4. Harris, J. R. (1991). *Electron microscopy in biology: a practical approach*. IRL Press, Oxford.
5. Hall, J. L. and Hawes, C. (1991). *Electron microscopy of plant cells*. Academic Press, London.
6. Coetzee, J. (1985). In *Botanical Microscopy 1985* (ed. A. W. Robards), pp. 17–38. Oxford University Press.
7. Roland, J. C. and Vian, B. (1991). In *Electron microscopy of plant cells* (Ed. J. L. Hall and C. Hawes), pp. 1–66. Academic Press, London.
8. Smith, M. and Croft, S. (1991). In *Electron microscopy in biology: a practical approach* (ed. J. R. Harris), pp. 17–37, IRL Press, Oxford.
9. Reid, N. (1975). *Practical methods in electron microscopy* (ed. A. M. Glauert), Vol. 3, Part II. Elsevier, Amsterdam.
10. Lewis, P. R. and Knight, D. P. (1992). *Practical methods in electron microscopy* (ed. A. M. Glauert). Vol. 5, Part 1. Elsevier, Amsterdam.
11. Reynolds, E. S. (1963). *J. Cell Biol.*, **17**, 208.
12. Hawes, C. R. and Horne, J. C. (1983). *Biol. Cell*, **48**, 207.
13. Hawes, C. R. (1991). In *Electron microscopy of plant cells* (ed. J. L. Hall and C. Hawes), pp. 67–84. Academic Press, London.
14. Harris, N. (1979). *Planta*, **146**, 63.
15. Hepler, P. K. (1980). *J. Cell Biol.*, **86**, 490.
16. Thiery, J. P. (1967). *J. Microsc.*, **6**, 987.
17. Spiess, E., Zimmermann, H.-P., and Lünsdorf, H. (1987). In *Electron microscopy in molecular biology: a practical approach* (ed. J. Sommerville and U. Scheer), pp. 147–66. IRL Press, Oxford.
18. Harris, R. and Horne, R. (1991). In *Electron microscopy in biology: a practical approach* (ed. J. R. Harris), pp. 203–28. IRL Press, Oxford.
19. Michel, M., Hillman, T., and Müller, M. (1991). *J. Microsc.*, **163**, 3.
20. Michel, M., Gnägi, H., and Müller, M. (1992). *J. Microsc.*, **166**, 43.
21. Robards, A. W. and Sleytr, U. B. (1985). *Practical methods in electron microscopy* (ed. A. M. Glauert), Vol. 10. Elsevier, Amsterdam.
22. Menco, B. P. M. (1986). *J. Electron Microsc. Techn.*, **4**, 177.
23. Sitte, H., Edelmonn, L., and Neumann, K. (1987). In *Cryotechniques in biological electron microscopy* (ed. R. A. Steinbrecht and K. Zierold), pp. 87–113. Springer-Verlag, Heidelberg.
24. Robards, A. W. (1991). In *Electron microscopy of plant cells* (ed. J. L. Hall and C. Hawes), pp. 257–312. Academic Press, London.
25. Lancelle, S. A., Callaham, D. A., and Hepler, P. K. (1986). *Protoplasma*, **131**, 153.
26. Ding, B., Turgeon, R., and Parthasarathy, M. V. (1991). *Protoplasma*, **165**, 96.
27. Heuser, J. E., Reese, T. S., Dennis, M. J., Jan, Y., and Evans, L. (1979). *J. Cell Biol.*, **81**, 275.
28. Heuser, J. (1989). *J. Electron Microsc. Technique*, **13**, 244.
29. Hirokawa, N. and Heuser, J. E. (1981). *J. Cell Biol.*, **91**, 399.
30. Hawes, C. and Martin, B. (1986). *Cell Biol. Int. Rep.*, **10**, 985.

31. McLean, B. and Juniper, B. E. (1986). *Planta*, **169**, 153.
32. Moor, H. (1987). In *Cryotechniques in biological electron microscopy* (ed. R. A. Steinbrecht and K. Zierold), pp. 175–91. Springer-Verlag, Berlin.
33. Dahl, R. and Staehelin, L. A. (1989). *J. Electron Microsc. Techn.*, **13**, 165.
34. Craig, S. and Staehelin, L. A. (1988). *Eur. J. Cell Biol.*, **46**, 80.
35. Staehelin, L. A., Giddings, T. H., Kiss, J. Z., and Sack, F. D. (1990). *Protoplasma*, **157**, 75.
36. Kiss, J. Z., Giddings, T. H., Jr, Staehelin, L. A., and Sack, F. D. (1990). *Protoplasma*, **157**, 64.
37. Willison, J. H. M. (1975). *Planta*, **126**, 93.
38. Fineran, B. A. (1978). In *Electron microscopy and cytochemistry of plant cells* (ed. J. L. Hall), pp. 280–341. Elsevier/North Holland, Amsterdam.
39. Fujikawa, S. (1991). In *Electron microscopy in biology: a practical approach* (ed. J. R. Harris), pp. 173–201. IRL Press, Oxford.
40. Heuser, J. (1980). *J. Cell Biol.*, **85**, 560.
41. Coleman, J., Evans, D., Hawes, C., Horsley, D., and Cole, L. (1987). *J. Cell Sci.*, **88**, 35.
42. Satiat-Jeunemaitre, B., Martin, B., and Hawes, C. (1992). *Protoplasma*, **167**, 33.
43. Glenney, J. R., Jr (1987). In *Electron microscopy in molecular biology: a practical approach* (ed. J. Sommerville and U. Scheer), pp. 167–78. IRL Press, Oxford.
44. Jeffree, C. E. and Read, N. D. (1991). In *Electron microscopy of plant cells* (ed. J. L. Hall and C. Hawes), pp. 313–414. Academic Press, London.
45. Jeffree, C. D., Read, N. D., Smith, J. A. C., and Dale, J. E. (1987). *Planta,* **172**, 20.
46. Livesey, S. A., del Campo, A. A., McDowall, A. W., and Stansy, J. T. (1991). *J. Microsc.*, **161**, 205.
47. Robards, A. W. (1978). In *Electron microscopy and cytochemistry of plant cells* (ed. J. L. Hall), pp. 343–415. Elsevier/North Holland, Amsterdam.
48. Newbury, D. E., Joy, D. C., Echlin, P., Fiori, C. E., and Goldstein, J. I. (1986). *Advanced scanning electron microscopy and analysis*. Plenum Press, New York.
49. Tanaka, K. (1980). *Int. Rev. Cytol.*, **68**, 97.
50. Blackmore, S. and Barnes, S. H. (1988). *Ann. Bot.*, **62**, 605.
51. Williams, M. H. and Green, P. B. (1988). *Protoplasma*, **147**, 77.
52. Beckett, A., Read, N. D., and Porter, R. (1984). *J. Microsc.*, **136**, 87.

5

In situ hybridization of RNA

G. I.McFADDEN

1. Introduction

In situ hybridization is a cytochemical technique for localizing specific nucleic acid sequences. It is based on the phenomenon of Watson–Crick base pairing (commonly termed hybridization) between two polynucleotides with complementary sequences. A labelled polynucleotide (the probe) is allowed to hybridize with complementary sequences (the targets) in the tissue. The probe–target hybrids are then made visible with some form of marker (*Figure 1*). *In situ* hybridization is analogous to Southern or Northern blots except that the target sequence is not extracted from the tissue for electrophoresis and immobilization on a membrane; instead, it is left *in situ*.

Resolution levels for the localization of nucleic acid species range from macroscopic to ultrastructural, depending on the system used (1). By far the most widespread application of *in situ* hybridization is the localization of mRNA at the light microscopic level. This approach is most useful for

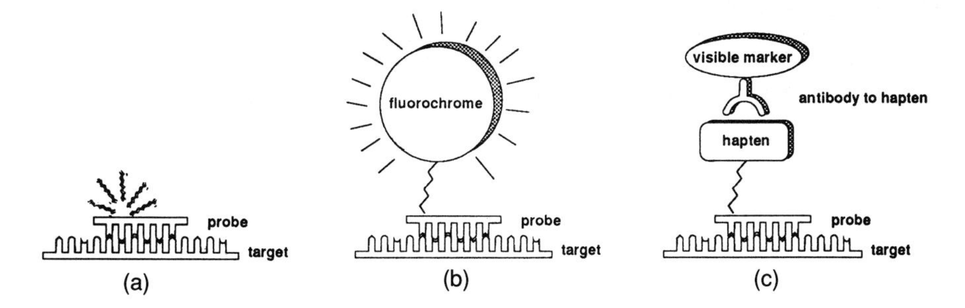

Figure 1. Three main types of *in situ* hybridization method. Isotopic labelling (a) incorporates a radionuclide into the probe. Emitted β particles are detected by autoradiography. Direct non-isotopic labelling (b) incorporates a reporter molecule (typically a fluorochrome) into the probe. Sites of probe binding are visualized by fluorescence microscopy. Indirect non-isotopic labelling (c) involves incorporating a hapten such a biotin or DIG^{TM} into the probe. The hapten is then located with a visible marker introduced via an affinity or immuno intermediary detection system.

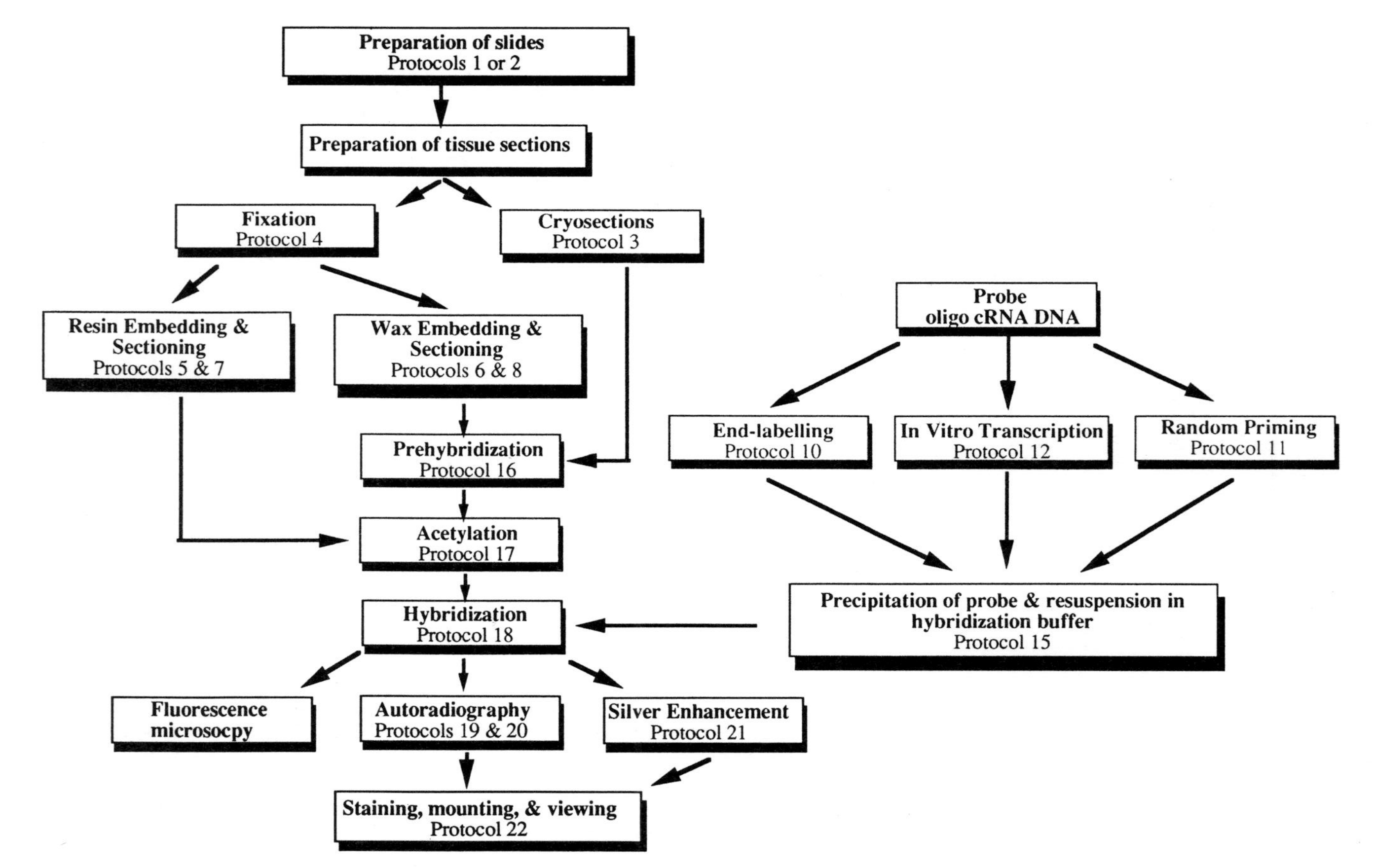

Figure 2. Flow chart for various protocols provided in this chapter.

demonstrating gene expression in a particular organ, tissue, or even cell layer within the organism of interest. Emerging technologies such as localization of DNA elements on isolated chromosomes, and localization of genes or transcripts at the subcellular levels by confocal or electron microscopy (2), are providing our first glimpses of the cell's inner workings. This chapter examines the different methodologies used to localize mRNAs at a light microscopic level. Factors affecting the resolution and sensitivity of *in situ* hybridization techniques are examined. Because of the wide range of applications, there is unfortunately no single universal protocol for localizing nucleic acid species. Each worker will have to adopt a protocol suited to their system. Some guidelines for selecting a suitable approach are provided here (*Figure 2*).

2. Specimen preparation, labels, detection systems, and probes

The usefulness of any *in situ* hybridization technique is dependent on the resolution and sensitivity achieved. Three factors govern successful localization: the efficacy of the probe, the efficiency of the detection system, and the method of specimen preparation.

2.1 Specimen preparation

For the majority of applications it will be necessary to prepare sections of the tissue. Three types of sections can be used:

- cryosections
- wax sections
- resin sections

Cryosections are the most straightforward to prepare and give excellent preservation of nucleic acids. Production of quality cryosections is greatly facilitated by the use of a Cryostat Frozen Sectioning Aid (Instrumedics Inc.). This device allows persons with minimal experience to produce high quality cryosections. Relatively thin sections (1–2 μm) can be produced and even serial sectioning is readily undertaken. The major advantage of the Cryostat Frozen Sectioning Aid is the section collecting system. Sections are collected while still frozen using adhesive-coated slides. The adhesive is polymerized by a UV flash, and the section, which remains frozen throughout the entire procedure, is irreversibly bonded to the slide. These sections are never lost during the hybridization and washing procedures, thus avoiding a well known bugbear of *in situ* hybridization. The adhesive does not interfere with *in situ* hybridization steps. Protocols for cryosections (with and without the Cryostat Frozen Sectioning Aid) are provided in *Protocol 3*.

Embedding of tissue in wax or plastic requires fixation and dehydration prior to infiltration with the embedding medium. Aldehyde fixation is optimal and a combination of 4% paraformaldehyde and 0.5% glutaraldehyde suits most systems. The fixative should be prepared fresh in a buffer isotonic to the growing conditions. Do not use Tris in conjunction with aldehyde fixatives; phosphate or 'Good' buffers such as Hepes or Pipes are best. Fixation of plant tissues with extensive waxy cuticles can be improved by use of the phase-partition fixation technique (see *Protocol 4*).

Protocol 1. Coating of slides for wax and resin sections

- weak solution of Decon 90 detergent (Rhone Poulenc)
- poly-L-lysine (0.1 mg/ml)
- distilled water

1. Clean slides in detergent[a], rinse well with water, and air dry.
2. Dip slides in poly-L-lysine (0.1 mg/ml) for 5 min.
3. Drain and air dry away from dust.

[a] Plastic slide-staining boxes holding 25 slides from Vit-Labs/Vitri are useful for these procedures.

Protocol 2. Coating of slides of cryosections

- weak solution of Decon 90 detergent
- 1% gelatin
- 0.25% paraformaldehyde
- distilled water

1. Clean slides in detergent, rinse well with water, and air dry.
2. Dip slides in 1% gelatin.
3. Drain.
4. Transfer to 0.25% fresh paraformaldehyde, drain, and allow to dry.

Protocol 3. Preparation of cryosections

- Freon 22 (DuPont)
- liquid nitrogen
- OCT compound and embedding moulds (Tissue Tek)
- coated slides (see *Protocol 2*)
- three flat blocks of dry ice
- 2% glutaraldehyde, 33% ethylene glycol (in 0.15 M PO_4 buffer, pH 7.2)

1. Prepare Freon 22 (DuPont) slush in a Freon liquifier and liquid nitrogen dewar vessel
2. Fill an embedding mould with chilled OCT compound.
3. Immerse fresh tissue in mould and immediately transfer to the Freon slush until frozen.
4. Store frozen block at −20 °C overnight or at −70 °C for longer periods.
5. If using the Cryostat Frozen Sectioning Aid, prepare sections following the manufacturer's recommendations then proceed to step 8.
6. When preparing cryosections without a Cryostat Frozen Sectioning Aid, collect cryosection by thawing on to a coated slide (see *Protocol 2*) and immediately transfer to flat block of dry ice.
7. Leave slide on dry ice for 30 min.
8. Mark slide with diamond pencil to record details and circle the section on the back of slide.
9. Put slides in rack and dip in 2% glutaraldehyde, 33% ethylene glycol (in 0.15 M PO_4 buffer, pH 7.2) at 4 °C.
10. Fix for 5–10 min maximum.
11. Proceed immediately to *Protocol 16.*

Protocol 4. Phase partition fixation

This is a procedure originally developed for *in situ* hybridization of *Drosophila* embryos and now adapted to plant tissue (3). Use this fixation prior to embedding in either wax or resin.

- 50 mM Pipes (pH 7.3)
- paraformaldehyde

Protocol 4. *Continued*

- glutaraldehyde
- *n*-heptane

1. Prepare fresh paraformaldehyde (4% w/v) and glutaraldehyde (0.5% v/v) in 50 mM Pipes (pH 7.3).
2. Combine fixative with an equal volume of *n*-heptane and shake for 1 min.
3. Allow phases to separate then decant heptane (upper) phase.
4. Immerse small pieces of tissue (about 2 mm × 2 mm × 2 mm) in fixative-loaded heptane for 10 min at room temperature.
5. Transfer tissue pieces to paraformaldehyde (4%) and glutaraldehyde (0.5%) in 50 mM Pipes (pH 7.3) and fix for a further 2 h at room temperature.
6. Wash in buffer three times and proceed to either resin embedding (*Protocol 5*) or wax embedding (*Protocol 6*).

Protocol 5. Resin embedding

- 10%, 30%, 50%, 70%, and 100% ethanol
- LR Gold (London Resin Co.)
- benzoyl peroxide paste (London Resin Co.)
- gelatin capsules

1. Partially dehydrate tissue in a graded ethanol series (10%, 30%, 50%, and 70% for 15 min each).
2. After the 70% ethanol step, transfer specimens to a 3:1 mixture of ethanol/LR Gold and rotate overnight at room temperature.
3. Transfer specimens to a 1:1 mixture of ethanol/LR Gold for 3 h.
4. Transfer specimens to a 1:3 mixture of ethanol/LR Gold for a further 3 h.
5. Transfer specimens to 100% LR Gold and leave overnight.
6. Polymerize LR Gold with benzoyl peroxide paste[a].

[a] Addition of benzoyl peroxide paste to 1% (w/v) will usually effect polymerization within a few hours. It is advisable to run a test polymerization to establish the optimal amount of benzoyl peroxide paste. Make up fresh resin with benzoyl peroxide and transfer specimens into this resin for polymerization. Oxygen must be excluded during polymerization. Best results are achieved by polymerizing in gelatin capsules in darkness. Store blocks at room temperature.

Protocol 6. Wax embedding

- 10%, 20%, 30%, 40%, 50%, 60%, 70%, 90%, and 100% ethanol
- Clearene (Surgipath Medical Industries Inc.)
- Paraplast embedding wax chips (BDH Chemicals)
- embedding moulds (Tissue Tek no. 4566)

1. Fix tissue as described in *Protocol 4.*
2. Dehydrate through ethanol series (10%, 20%, 30%, 40%, 50%, 60%, 70%, 90%, and 100%) allowing at least 15 min for each step.
3. Transfer tissue into a glass tube containing a mixture of three parts ethanol and one part Clearene for 60 min.
4. Transfer tissue to a mixture of one part ethanol and one part Clearene for 60 min.
5. Transfer tissue to a mixture of one part ethanol and three parts Clearene for 60 min.
6. Transfer tissue to 100% Clearene for 60 min.
7. Add several chips of Paraplast embedding wax to each sample in Clearene and leave overnight at room temperature.
8. Add several more chips of Paraplast and transfer tube to 42 °C waterbath for 3 h.
9. Place several Paraplast chips in an embedding mould and melt wax in an oven at 60 °C.
10. Using warmed forceps, transfer tissue pieces from tubes to embedding moulds.
11. Incubate for 24 h at 62 °C. Remove and store at room temperature.

Protocol 7. Preparation of resin sections

1. Cut sections of desired thickness on suitable microtome. Sections from 0.2 μm up to 2 μm are best. Thicker sections swell when wetted and are easily lost during subsequent labelling steps.
2. Collect sections from boat with loop or blunt glass rod and transfer to small drop of distilled water on coated slides (see *Protocol 1*).
3. Slowly dry off water on 40 °C hotplate. It is important that the water does not boil under the section and produce wrinkles. Only perfectly flat sections can be used.
4. Bake sections onto slides for 2 h at 60 °C then store at room temperature.

Protocol 8. Preparation of wax sections

- poly-L-lysine coated slides (see *Protocol 1*)
- Clearene
- 30%, 60%, 80%, and 100% ethanol
- distilled water
- pronase (0.1 mg/ml in Tris–HCl pH 7.6)
- glycine and histidine (2 mg/ml in PBS)

1. Cut sections of desired thickness (3–6 μm) on suitable microtome. A cryostat adjusted to room temperature with a disposable knife (Feather, Japan) produces suitable sections.
2. Transfer sections to droplet of water on poly-L-lysine coated slides (see *Protocol 1*). Dry down on 42 °C hotplate for 2 h.
3. Dewax sections by soaking slides in Clearene for 10 min followed by two washes in ethanol for 5 min each.
4. Rehydrate sections by transferring through graded ethanol series (95%, 80%, 60%, and 30%) allowing at least 5 min in each step.
5. Transfer to distilled water. Air dry and store at 4 °C.
6. Digest sections with pronase (0.1 mg/ml in Tris–HCl pH 7.6) for 10 min at room temperature. The amount of digestion may need to varied with different tissue types. Monitor nucleic acid retention by Acridine Orange staining (see *Protocol 9*).
7. Deactivate pronase by dunking slides in glycine and histidine (2 mg/ml in PBS).
8. Wash twice in water and air dry.

Protocol 9. Acridine Orange staining to check retention of nucleic acids in wax, resin, and cryo-sections

- 0.2 M glycine (pH 2)
- Acridine Orange (0.5 mg/ml in glycine buffer)
- phenylenediamine (0.1% in PBS:glycerol 2:1 (v/v))

1. Preincubate slides in 0.2 M glycine (pH 2) for 5 min at room temperature.
2. Stain in Acridine Orange (0.5 mg/ml in glycine buffer[a]) for 20 min.

3. Rinse extensively with glycine buffer.
4. Drain sections of excess buffer, add 50 μl of phenylenediamine (0.1% in PBS:glycerol, 2:1 (v/v)), apply coverslip.
5. View sections on UV fluorescence microscope. The nuclei should be yellow and the cytoplasm orange. Orange fluorescence of RNA in the nucleolus may be visible.

[a] Prepare a 10 mg/ml stock of Acridine Orange in distilled water and store at 4 °C.

2.2 Labels and detection systems

There are two major categories of labels and detection systems for *in situ* hybridization: isotopic and non-isotopic. Both have relative merits, depending on the system used. Isotopic labelling is familar to most molecular biologists and standard probe production techniques can be used. Mastery of the delicate procedures for microautoradiography requires some investment of time, however. Non-isotopic systems provide better user safety, a much longer probe life, and potentially higher resolution and sensitivity.

2.2.1 Isotopic labels

Isotopic labels are detected by autoradiography. Latent silver grains produced by β particles colliding with silver halide in an emulsion are chemically developed (4). Macroscopic autoradiography can be performed by apposing fine grain, single-coated X-ray film directly to the labelled sections. Microautoradiography involves coating the sections with a very thin layer of melted emulsion. After a suitable exposure period, silver grains in the emulsion are chemically developed (see *Protocol 10*).

Three isotopes emitting β particles are commonly used for *in situ* hybridization, namely ^{32}P, ^{35}S, and ^{3}H. The β particles from ^{32}P have the highest energy and spread of the particles results in low resolution. However, ^{32}P has the highest specific activity and this isotope gives maximum sensitivity in low resolution systems. Though seldom used, ^{33}P is potentially well suited to *in situ* hybridization applications because it combines a high specific activity with a relatively low β particle energy (5). ^{35}S β particles have a lower emission energy and provide better resolution. The longer half-life of ^{35}S extends probe longevity slightly. ^{35}S-labelling is most useful for applications localizing mRNAs to tissues or cell layers in sections. Very fine resolution is achieved with ^{3}H, which has the lowest emission energy of the three. Subcellular localization can be achieved using ^{3}H-labelled probes for electron microscopy (6). The specific activity of ^{3}H-labelled probes is relatively low and prolonged exposures may be necessary, but the probes do have the advantage of being extremely long-lived.

2.2.2 Non-isotopic labels

A wide variety of non-isotopic labels is now in use and a burgeoning selection of visible markers is available to complement these different labels. Newly developed systems involve attaching fluorochromes directly to the probe (*Figure 1*); sites of probe attachment are then visualized by fluorescence microscopy. More traditional approaches rely on attaching a tag (known as a hapten) to the probe, and subsequent detection of the tag using immunocytochemical techniques, or affinity labelling techniques in the case of biotin tagging. The visible marker is introduced via an antibody or affinity intermediary (see *Figure 1*) and so these methods are referred to as indirect. Markers can include fluorochromes, insoluble coloured precipitates, or metal deposits.

The most widely used system for non-isotopic labelling uses enzymic incorporation of haptenized nucleotides into a probe molecule production from a cloned template (7). The hapten is attached at the C5 position on a pyrimidine (*Figure 3*). The C5 position is not involved in base pairing during hybridization, and modifications do not adversely affect recognition of the nucleotide substrate by commonly used nucelic acid modifying enzymes. The hapten is attached to the base by an allylamine spacer arm which keeps the hapten accessible to the detection reagents, even though the probe is duplexed with the target. An 11 or 14 atom spacer arm is most effective. Labelling of RNA probes is normally done by incorporating haptenized rUTP. Oligonucleotides and double-stranded DNA probes are labelled using haptenized dUTP as a dTTP analogue.

Biotin (vitamin H) was the first hapten employed for *in situ* hybridization. It was originally chosen because of the extraordinary affinity of the egg white protein avidin for biotin (8). Avidin and the bacterial cell wall protein,

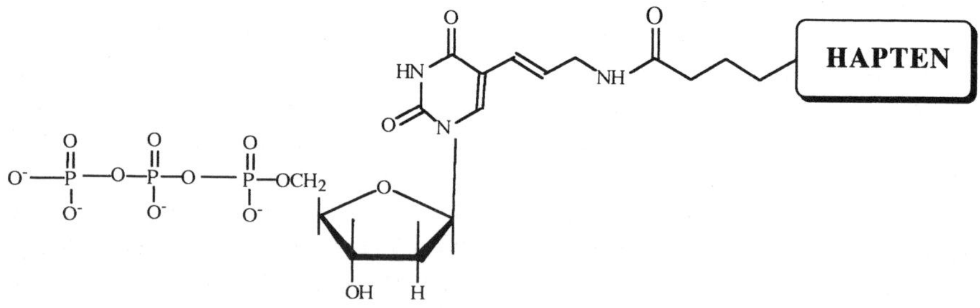

Figure 3. Generalized structure of a hapten-labelled nucleotide analogue. The hapten, which can include tags such as biotin or DIG™, or reporter molecules such as fluorochromes, is attached to the pyrimidine at C5 via a spacer arm. The labelled analogues are substrates for commonly used nucleic acid modifying enzymes and are readily incorporated into probes in various *in vitro* labelling procedures.

streptavidin, bind to biotin with dissociation constants (K_{dis}) in the order of 10^{-15}, and, when these proteins are coupled to indicator molecules, an excellent affinity detection system can be devised.

Virtually any moiety can be used as a hapten providing that a suitable means for detecting the hapten is available and that the hapten's presence does not disrupt hybridization. Digoxigenin, the aglycone of the plant cardiac glycoside digoxin, has recently been introduced as an alternative to biotin, which was sometimes found to be unsuitable as a tag in animal tissues with high levels of endogenous biotin. Digoxigenin (DIG™) is incorporated into probes in the same ways as biotin. Detection is via an anti-digoxigenin antibody. The only other hapten commonly used for *in situ* hybridization is dinitrophenol (9). Whether any particular hapten is superior in plant systems is not yet clear, although plant scientists studying foxgloves (the source of digoxin) would of course be unwise to employ DIG™.

2.3 Probe types

There are three major types of *in situ* hybridization probe in common use:

- synthetic oligodeoxyribonucleotides (oligos)
- double-stranded DNA probes (i.e. cDNAs or genomic clones)
- single-stranded RNA probes (i.e. antisense/sense RNAs, cRNAs, or 'riboprobes')

Each of these probe types has advantage and disadvantages.

2.3.1 Oligos

Oligos (which are usually between 15 and 50 nucleotides in length) are now readily available to most workers and can be produced relatively quickly. Working with oligos requires no cloning expertise and probes can be produced by any worker from published sequence data. A notable advantage of oligos is that they can be directed to particular portions of the target species. Several different oligos directed at the coding region, introns, or untranslated regions can be used to localize the same target and provide information about target processing in different tissues.

Various labelling and detection strategies are possible with oligos. Isotopic labelling can be achieved by kinasing with T4 polynucleotide kinase using [γ-^{32}P]NTP, [γ-^{3}H]NTP, or [γ-^{35}S]NTP (see *Protocol 10*) or by tailing with terminal transferase to add hapten-dNTPs at the 3′ end (10).

Non-enzymic attachment of haptens to oligos is also possible using the 'amino-link' system (11). An amine group is incorporated at the 5′ end of the oligo during synthesis. When the amine-bearing oligo is mixed with an *N*-hydroxysuccinimide (NHS) ester of a fluorochrome, biotin, or DIG™ at alkaline pH, the unprotonated amine attacks the ester linkage in the NHS-hapten resulting in an amide bond between the oligo and the hapten molecule. The NHS hapten esters have allylamine linker arms. Very recently biotin (12) or fluorescein amidites (13) have become available, allowing biotin or fluorescein to be incorporated directly into the oligonucleotide during synthesis. The amidites allow incorporation of the hapten at any chosen site in the oligo.

2.3.2 Double-stranded DNA probes

These probes are produced from cDNA or genomic clones that have been amplified in a cloning vector supported by a suitable bacterial host. Double-stranded probes can be labelled with isotopes or non-isotopic labels by either nick translation or, preferably, random priming (see *Protocol 11*). Double-stranded refers to the fact that both the coding strand and the non-coding strand are included in the probe. While such probes are reasonably satisfactory for localization of DNA, double-stranded probes suffer from some major disadvantages when localizing RNA targets. Only the coding strand hybridizes with RNA targets and the labelled non-coding strand is not only superfluous, it actually hybridizes back to the coding strand thereby competing with the target. Moreover, because the unlabelled template remains with the probe, it competes with the labelled probe to hybridize with targets, severely reducing the signal. Additionally, the low yield of probe produced (typically less than 100 ng) is only sufficient for a small number of *in situ* hybridization experiments.

2.3.3 Single-stranded RNA probes

RNA probes complementary to cellular RNA can be produced by *in vitro* transcription. The antisense RNA or cRNA (complementary RNA) is produced from a vector in which the non-coding strand is sited downstream of a bacteriophage RNA polymerase promoter. Discrete length 'run off' transcripts incorporating either isotopically or non-isotopically labelled nucleotides are made from a linearized plasmid template using phage RNA polymerase (see *Protocol 12*). Large quantities of probe can be produced in this manner. The unlabelled DNA template is relatively insignificant as a competitor to the probe but can be removed by DNase digestion. Another advantage of RNA probes is the stronger hybrids formed between two strands of RNA allowing a higher stringency (see Section 2.3.4). A drawback with RNA probes is a tendency to bind non-specifically to rRNAs if used at too high concentrations.

2.3.4 Hybridization conditions

Hybridization of the chosen probe to the target is governed by a range of physical parameters. The hybridization occurs through base pairing to create hydrogen bonds linking the probe to the target. The number of hydrogen bonds determines the strength of the hybrid. Shorter probes form less stable hybrids. Probes that do not have a perfectly complementary sequence (often known as heterologous probes) also form less stable hybrids. For a successful *in situ* hybridization experiment, conditions maximizing specific annealment of the probe to the target must be determined. The conditions under which hybridization is attempted are known as the 'stringency', where low stringency allows weaker hybrids to remain as hybrids while high stringency is less permissive. The stability of the hybrid is represented by the melting temperature (T_m), the temperature at which half the population of hybrids become dissociated or 'melted'. A guide to stringency conditions can be obtained from the following relationship which predicts T_m for two RNA molecules:

$$T_m = 79.8\ ^\circ\mathrm{C} + 18.5(\log M) + 0.584(\%G/C) - 300 + 200M/L - 0.35(\%F) - 1.4(\%\text{mismatch}).$$

where M is the concentration of NaCl ionic strength (mol/litre), %G/C is the mole percentage of guanine/cytosine pairs in the probe target hybrid; L is the length of the probe in base pairs; %F is the percentage (v/v) of formamide in the hybridization buffer; and %mismatch is the percentage of non-complementary base pairs between the hybridizing strands. Thus, by increasing the salt concentration, or by decreasing either the formamide concentration or the hybridization temperature, the stringency can be lowered allowing short probes or probes with a small degree of sequence mismatch to be used. Conversely, higher stringency can be attained by reversing these parameters. Formamide increases the solubility of the bases and is a duplex-destabilizing agent added to allow hybridization at lower temperatures, thus minimizing damage to the tissue sample. Standard hybridization buffer includes 50% formamide and 0.15 M NaCl (see *Protocol 15*, footnote *b*). The simplest way to optimize an *in situ* hybridization system is to run a series of experiments using a standard buffer at different temperatures and monitor the signal-to-noise ratio. As a guide, DNA–DNA hybridization should be done at 37–42 °C, DNA–RNA hybridization at 42 °C, and RNA–RNA hybridization at 50–70 °C. When working with oligos, decrease the formamide concentration to 40%.

Post-hybridization washes are done at a temperature equivalent to or higher than the hybridization temperature. Salt concentration can be reduced in the wash buffer to increase stringency, but best results are obtained by optimizing the hybridization conditions rather than trying to reduce noise with stringent post-hybridization washing; once probe has annealed it is hard to dislodge.

Protocol 10. Isotopic labelling of oligos

- oligo for labelling
- [γ-^{32}P]ATP (Amersham International, U.K)
- 10 × PNK buffer (0.5 M Tris–HCl pH 7.6, 0.1 M $MgCl_2$, 50 mM dithiothreitol, 1 mM spermidine–HCl, 1 mM EDTA pH 8.0)
- T4 polynucleotide kinase (20 U) (Promega Corp.)
- Sephadex G50 spin column (Pharmacia)
- glycogen (0.2 μg/μl in water)
- 4 M LiCl
- ethanol (ice-cold)
- hybridization buffer (see *Protocol 15*, footnote [b])

1. Combine the following reagents
   - 1 μl of oligo (10 pmol/μl) in water
   - 5 μl [γ-^{32}P]ATP (10 mCi/ml)
   - 20 U T4 polynucleotide kinase
   - 5 μl 10 × PNK buffer
   - water to give a final volume of 50 μl
2. Incubate at 37 °C for 1 h.
3. Heat inactivate kinase by incubating at 68 °C for 10 min.
4. Remove unincorporated nucleotides by passing reaction mix through Sephadex G-50 spin column.
5. Add 0.2 μg of glycogen to the eluate and precipitate by adding 0.1 vol. of 4 M LiCl and 2.5 vol. of ice-cold ethanol.
6. Leave for 30 min at −70 °C or 2 h at −20 °C.
7. Centrifuge at 12 000 *g* for 30 min.
8. Drain off ethanol and resuspend in appropriate volume of hybridization buffer.

Protocol 11. Biotin, digoxigenin, or fluorochrome labelling of DNA probe by random priming

This is an adaptation of the method of Feinberg and Vogelstein (15) which should be used for isotopic labelling. This method is suited for labelling of linear, double-stranded DNA template. The template duplex is denatured and hexanucleotides (random primers) are annealed to the template strands at random sites. The primers are then extended with DNA polymerase incorporating labelled nucleotides. A typical reaction will produce about 100 ng of labelled DNA.

- template DNA
- 100 ng random primers (Amersham)
- 10 × Messing buffer (50 mM Tris–HCl pH 7.6, 100 mM $MgCl_2$, 100 mM dithiothreitol)
- 1 mM dATP, dCTP and dGTP
- 1 mM hapten-labelled dUTP (bio-11-dUTP (Sigma) or DIG-11-dUTP (Boehringer Mannheim) or fluorochrome-dUTP (Boehringer Mannheim)
- 1 mM dTTP
- Klenow subunit of DNA polymerase 1 (20 U, Promega Biotech)
- distilled water
- BioSpin 30 column (Bio-Rad)

1. Isolate the cloned insert from the vector and determine the DNA concentration (15).
2. Combine 100 ng of insert with 100 ng of random primers and make up the volume to 10 μl with water.
3. Boil this mix for 3 min then chill on ice for 5 min.
4. Now add 5 μl of 10× Messing buffer.)
5. Add 5 μl of a mixture of 1 mM dATP, dCTP, and dGTP.
6. Add 5 μl of 1 mM bio-11-dUTP or 1 mM DIG-11-dUTP or 5 μl of 1 mM fluorochrome-dUTP.
7. Add 1 μl of dTTP (1 mM).
8. Add 2 μl (20 U) of DNA polymerase 1 (Klenow subunit).
9. Add distilled water to give a final volume of 50 μl.
10. Incubate at 20 °C overnight.
11. Remove unincorporated nucleotides with a (BioSpin 30) spin column.

In vitro transcription of biotinylated or digoxigenylated RNA probes (*Protocol 12* produces single-stranded RNA probes complementary to the target (see reference 16 for more information). The cloned sequence must be subcloned into an *in vitro* transcription vector such that the non-coding strand is downstream of a bacteriophage RNA polymerase promoter engineered into the vector. The polymerase then transcribes a complementary or antisense RNA from the template. Most commercially available vectors have a different RNA polymerase promoter on either side of the cloning site so that the sense RNA can be transcribed from the same construct simply by using a different polymerase. By cutting the vector with a restriction endonuclease downstream of the insert it is possible to limit the transcription to discrete length run-offs and produce labelled probe molecules of defined size. The pBluescript KS II vector (Stratagene) has restriction sites for *Bss*HII flanking the RNA polymerase promoters allowing a transcription cassette, suitable for both sense and antisense transcriptions, to be released with a single enzyme digestion.

A typical reaction will produce several micrograms of labelled RNA probe. RNA is easily destroyed by ribonucleases. These enzymes are extremely resilient and are easily transferred from the skin to reaction tubes and pipette tips. Always wear gloves while preparing RNA probes, autoclave all tubes and tips, and treat water for all buffers with diethylpyrocarbonate (17).

Protocol 12. *In vitro* transcription of biotinylated or digoxigenylated RNA probes

- template DNA
- 5× transcription buffer (200 mM Tris pH 7.5, 30 mM $MgCl_2$, 10 mM spermidine, 50 mM NaCl)
- 100 mM dithiothreitol
- RNasin (Promega)
- 10 mM ATP, 10 M M GTP, 10 mM CTP
- 6 mM UTP
- 2 mM bio-11-UTP (Sigma) or 10 mM DIG™-11-UTP (Boehringer Mannheim)
- RNA polymerase (20 U)
- DEPC-treated water
- 20% (w/v) SDS
- BioSpin 30 column (Bio-Rad)
- 2% (/v) blue dextran 2000 (Pharmacia)

1. Linearize template and determine concentration (16, 17)
2. At room temperature add these reagents in this order:
   (a) 4 µl of 5× transcription buffer
   (b) 2 µl of 100 mM dithiothreitol
   (c) 1 µl of RNasin
   (d) 1 µl of 10 mM ATP

(e) 1 μl of 10 mM GTP

(f) 1 μl of 10 mM CTP

(g) 1 μl of 6 mM UTP

(h) 5 μl of 2 mM bio-12-UTP or 1 μl of 10 mM DIG™-11-UTP

(i) 1 μl of template DNA (0.5–2.0 μg)

(j) 1 μl RNA polymerase (20 U; T7, T3 or SP6 may be used depending on which promoter is upstream of the insert)

(k) sufficient DEPC-treated water to make a final reaction volume of 20 μl

3. Incubate at 30 °C for 3–4 h.
4. Stop transcription by adding 1 μl of 20% SDS
5. Set aside 3 μl of the reaction mix for agarose gel electrophoresis (see below).
6. Remove unincorporated nucleotides using a spin column (BioSpin 30). When processing non-isotopic probes, it can be useful to add 0.1 vol. of 2% blue dextran 2000 to the reaction mix prior to using the spin column. The dye should co-elute with the probe and will be precipitated also. It does not interfere with hybridization.

Removal of the DNA template is not done. The full length probe is used without size reduction by alkaline hydrolysis as is usually recommended (18).

Protocol 13. Dot blot of probe

This procedure serves to test that the labelling has worked and that the anti-hapten antibody and gold markers are working.

- NA45 nitrocellulose paper (Schleicher & Schuell) or Hybond N (Amersham)
- PBS
- probe
- 3% gelatin (w/v) in PBS
- anti-hapten antibody (5 μg/ml in PBS)
- colloidal gold–secondary antibody conjugate
- IntenSE™ silver enhancement kit (optional) (Amersham)
- distilled water

1. Cut a 1 cm × 5 cm piece of NA45 nitrocellulose paper or Hybond N.
2. Wet the paper in PBS and air dry.

Protocol 13. *Continued*

3. Take a 1 μl aliquot of probe (eluate from spin column used in *Protocol 11* or *12*) and spot on to paper. Add an equivalent spot of unlabelled nucleic acid as a negative control
4. Dry for 1 h.
5. Wet the paper in PBS.
6. Block the paper in 3% gelatin in PBS at 25 °C for 1 h.
7. Incubate the paper in anti-hapten antibody (goat anti-biotin (Sigma) at 5 μg/ml or sheep anti-digoxigenin (Boehringer) at 5 μg/ml) for 1 h at room temperature on a shaker.
8. Rinse paper in PBS three times.
9. Incubate paper at room temperature in colloidal gold conjugated to suitable secondary antibody[a] until some of the dots turn pink.
10. If necessary, it is possible to intensify the gold dots using the IntenSE™ silver enhancement system (Amersham).
11. Wash the paper in distilled water three times.
12. Perform silver enhancement as described in the manufacturer's protocol if necessary.

[a] 10 nm particles conjugated with goat anti-rabbit antibody (Sigma) or 15 nm particles with a goat anti-rabbit antibody (Amersham) are suitable for use with a rabbit anti-biotin primary antibody (Enzo). 15 nm particles with rabbit anti-sheep (EY Labs are suitable for use with a sheep anti-DIG™ (Boehringer Mannheim). 15 nm particles with protein G (19) can also be used in conjunction with either of these primary antibodies. Protein A–gold should not be used with the sheep-anti-DIG™ $F(ab)_2$ fragment from Boehringer Mannheim.

Protocol 14. Gel electrophoresis of labelled probe

When preparing RNA probes it is advisable to check the transcripts by electrophoresis. A simple non-denaturing gel is adequate for this purpose.

- 1 litre TBE (0.1 M Tris–borate, 2 mM EDTA, 0.1 M boric acid)
- agarose
- loading dye
- distilled water
- transcription mix (aliquot reserved from *Protocol 12*)
- RNA ladder (Bethesda Research Labs)

1. Dissolve 0.5 g of agarose in 50 ml of TBE in a microwave oven.
2. Add 2 μl of 1% (w/v) ethidium bromide (ethidium bromide is toxic).

3. Pour agarose solution itno 'mini-gel' casting mould and insert comb (see reference 17 for more detail).
4. Allow gel to set before removing comb.
5. Add 1 μl of loading dye (0.25% (w/v) bromophenol blue, 40% glycerol) and 7 μl of water to the the 3 μl set aside from the transcription reaction.
6. A useful standard is 3 μg of 'RNA ladder' 9.5–0.24 kbp (Bethesda Research Labs/Gibco). Add the dye and water as above.
7. Load the samples and electrophorese at 5 V/cm for 1 h.
8. View the gel using a 300 nm UV transilluminator.

The apparent molecular weight of the labelled transcripts will not be correct due to the non-denaturing conditions and the incorporation of hapten moieties. An approximate estimate of the yield can be made by comparing the fluorescent intensity of the transcript band with that of one of the bands in the RNA ladder standard. It is also possible (but not optimal) to run a gel using a portion of the probe after it has been precipitated and resuspended in hybridization buffer (see *Protocol 15*, footnote *b*).

Protocol 15. Precipitation of probe

Once the unincorporated nucleotides have been removed using the spin column it is necessary to precipitate the probe and resuspend it in hybridization buffer. This protocol works for either DNA or RNA probes.

- 4 M LiCl (RNA probes) or 3 M NaOAc (DNA probes)
- *Escherichia coli* RNA (40 μg/ml in water) (DNA probes only) (Sigma)
- ethanol
- hybridization buffer[b]

1. Estimate the volume of the spin column eluate.
2. Add 0.1 vol. of 4 M LiCl for RNA probes or 0.1 vol. of 3 M NaOAc (pH 4.8) for DNA probes. When precipitating random primed DNA probes, add 1 μl of *E coli* tRNA solution (40 μg/ml in water; store frozen). This acts as a carrier to facilitate precipitation of the small quantity of DNA.
3. Add 2.5 vol. of ethanol.
4. Stand the mixture in a freezer for at least 1 hour.
5. Spin in microfuge at 4 °C for 15 min.
6. Pour off supernatant and stand tube upside down to drain for 15 min.

Protocol 15. *Continued*

7. Resuspend the pellet in suitable volume[a] of hybridization buffer[b].

[a] The volume of buffer used to resuspend the probe is dependent on the amount of probe synthesized and the concentration at which the probe is to be used. Probe concentrations in the order of 1 μg/ml are usually suitable. Resuspend in a minimal volume and test a dilution series to establish which dilution produces the optimal signal-to-noise ratio in the your system. Remember, excess probe will increase the non-specific hybridization and create noise. Store probe in hybridization buffer at −20 °C or use immediately.

[b] Hybridization buffer: 50 mM piperazine ethane sulphonic acid (pH 7.2), 0.75 M NaCl, 5 mM EDTA, 100 μg/ml powdered herring sperm DNA (type D-3159, Sigma), 0.1% Ficoll 400 (Pharmacia, 0.1% polyvinyl pyrrolidine 40, 0.1% BSA, 50% deionized formamide. Aliquots of buffer should be stored at −20 °C.

Protocol 16. Prehybridization of wax or cryosections

- hybridization buffer (see *Protocol 15*)
- 1 × SSC[a]
- ethanol (technical grade)
- ethanol (AR grade)

1. Dip slides in hybridization buffer at room temperature for 5 min.
2. Transfer to hybridization buffer at 38–42 °C for 1 h.
3. Rinse in 1 × SSC.
4. Dehydrate in ethanol (technical grade) followed by ethanol (AR grade).
5. Drain and store for up to 4 weeks at 4 °C in a closed staining box containing about 4 ml of ethanol.

[a] Prepare a stock of 20 × SSC (3 M NaCl, 0.3 M sodium citrate) (store at room temperature).

Protocol 17. Acetylation of slides with wax or resin sections

Slides coated with poly-L-lysine should be acetylated to prevent probe adhesion to the coating.

- 0.1 M triethanolamine–HCl pH 8.0.
- acetic anhydride
- 2 × SSC (see *Protocol 16*, footnote *a*)

1. Dip slides in 0.1 M triethanolamine–HCl pH 8.0.

2. While stirring vigorously, add acetic anhydride to a concentration of 5 ml/l.
3. Wash slides in 2 × SSC for 5 min.
4. Drain and air dry.

Protocol 18. Hybridization of sections (wax, resin, or cryosections) with isotopic or non-isotopic probes

- circular coverslips (12 mm diameter)
- chloroform
- probe in hybridization buffer
- hybridization buffer (see *Protocol 15*)
- 4 × SSC, 2 × SSC, and 1 × SSC
- ethanol (technical grade), ethanol (AR grade)
- 200 mM ammonium acetate (resin sections only)

1. Wash several circular coverslips (12 mm diameter) in chloroform and dry.
2. Place coverslips on dark surface[a] and dispense 3–5 µl of probe on to each slip.
3. Invert slide on to coverslip. The circle scratched around the section with the diamond pencil acts as a useful guide. The section should be covered by a droplet of probe and no bubbles should be under the coverslip. If the section is thick, you may have to add a few extra microlitres of probe at one edge of the coverslip with a Gilson-type pipette.
4. Lay slides in hybridization box[b] that contains about 50 ml of hybridization buffer beneath the support rack. If the lid seals well, a vapour of buffer will prevent the probe under the coverslips drying out.
5. Hybridize at selected temperature overnight.
6. To remove coverslips, stand slides in a beaker of 4 × SSC. Gently agitate slides until coverslips fall off. This may take 10–20 min. Never prise off the coverslips as this dislodges the sections.
7. Transfer slides to rack and wash in 2 × SSC.
8. Wash in 1 × SSC at hybridization temperature.
9. Wash in <1 × SSC at selected temperature if higher stringency is required.

Protocol 18. *Continued*

10. Dehydrate in ethanol (technical grade) and then in ethanol (AR grade). If using resin sections, do not dehydrate in ethanol, but simply rinse slides in 200 mM ammonium acetate and air dry.

[a] The lid from a Vitri staining box is suitable.
[b] A plastic meat defrosting box with a draining rack is ideal.

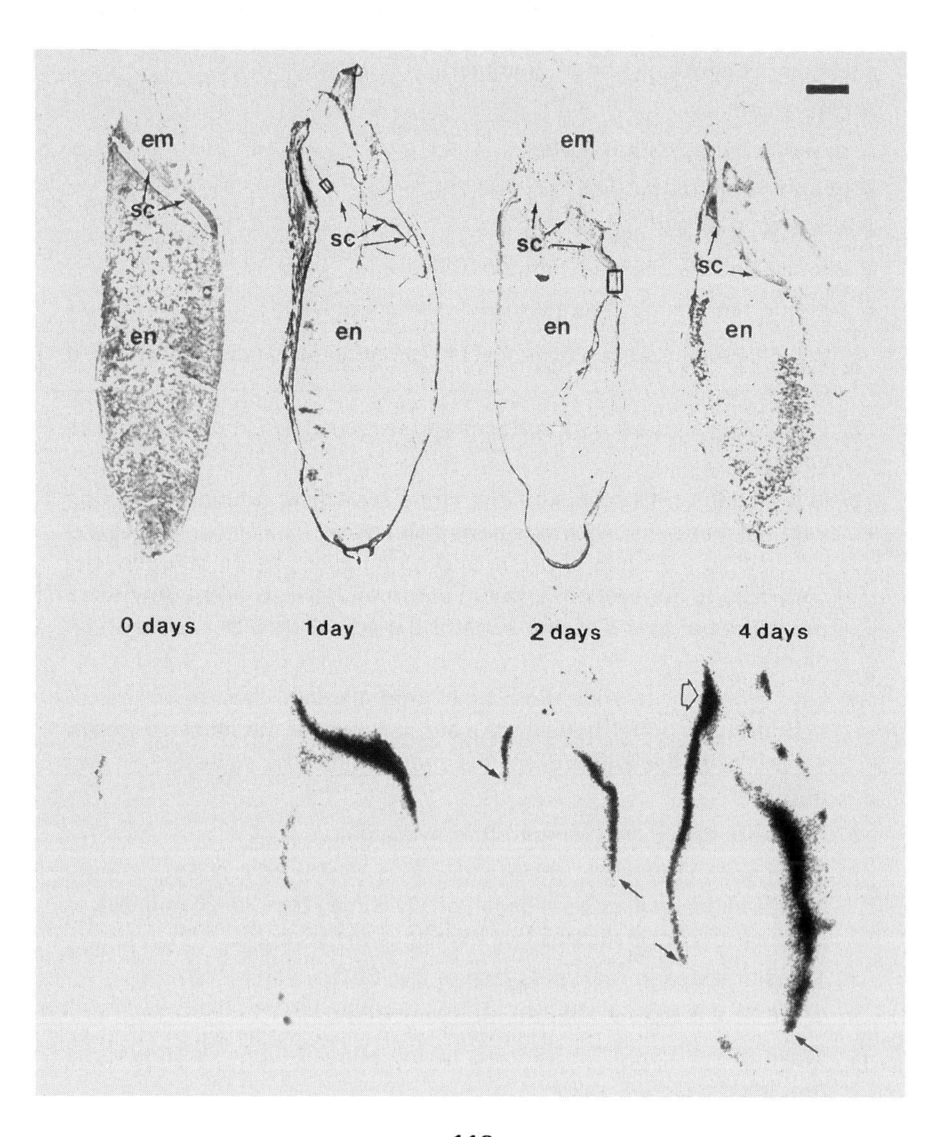

Protocol 19. X-ray film autoradiography(^{32}P and ^{35}S probes only)

1. Tape slides on to sheet of stiff paper in X-ray film cassette.
2. Tape additional new slides around the probed slides to form a flat surface.
3. In the darkroom under a red safelight (Kodak Wratten no. 2), place a sheet of Kodak XAR X-ray film over the slides.
4. Expose at room temperature without an intensifying screen for 4–12 h (^{32}P) or 24–72 h (^{35}S).
5. Develop film in Kodak Liquid X-ray Film Developer Type 2 at 20 °C, wash, fix, and rinse under running water.
6. Finer resolution can be achieved with DuPont Cromex MRF 32 single-coated film. Expose for four times the suitable exposure determined with XAR film in step 4.

Protocol 20. Liquid emulsion autoradiography

Liquid emulsion can only be handled under the prescribed safelight (Kodak Wratten no. 2). Since long drying periods are necessary, it is convenient to have a darkroom with double doors or a large lightproof box in which to store slides while they dry. The emulsion should be stored in 4 ml aliquots in scintillation vials away from extraneous radiation. Prolonged storage of emulsion increases background. Blank slides should always be included to check the status of the emulsion. Good resolution requires a thin emulsion layer.

- distilled water
- Kodak NTB-2 emulsion
- Kodak D19 developer
- Kodak Unifix

Figure 4. Localization of mRNA encoding a glucan hydrolase in barley grains (21). Cryosections of grains germinated for 0, 1, 2, and 4 days (top panel) were probed with a ^{32}P-labelled cDNA (15) and sites of message accumulation were visualized (bottom panel) by X-ray film autoradioagraphy (*Protocol 19*). The changing pattern of gene expression in the grains during germination is readily identified at a macroscopic level. Expression is detected first in the scutellum (sc) at 1 day, and after 2–4 days in the aleurone layer (arrows) surrounding the endosperm (en).

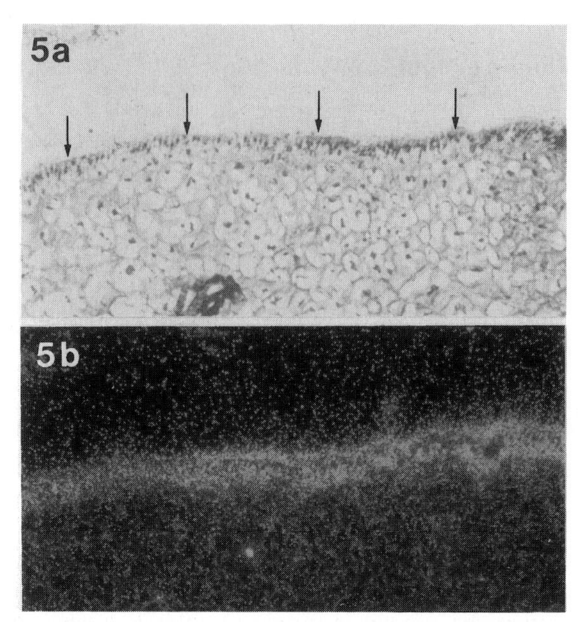

Figure 5. Light microscopic localization of glucan hydrolase mRNA in a cryosection of a barley grain using a ^{32}P-labelled cDNA detected by liquid emulsion autoradiography (*Protocol 20*). This procedure allows high resolution and the region illustrated in this figure is shown by a rectangle in the scutellum in the 1 day grain of *Figure 4*. Dark field optics (b) reveals a row of brilliant white silver grains which, when compared with the bright field image (a), are seen to correlate with a single cell layer (arrows) known as the scutellar epithelium.

Protocol 20. *Continued*

1. Add 2 vol. of distilled water to aliquot of Kodak NTB 2 emulsion.
2. Stand the mixture in 40 °C water bath for 1 h.
3. Warm a Pasteur pipette by gently squirting emulsion in and out several times.
4. Hold slide at a slight angle and slowly squirt a small stream of emulsion over the section.
5. Stand the slide on a piece of absorbent paper and allow to dry for 1 h. We find this method more reliable than dipping for obtaining a thin layer.
6. Repeat for all slides.
7. Transfer slides to tray and place in lightproof, airtight container[a] containing silica gel.
8. Expose for suitable period[b].
9. Develop in Kodak D19 for 2 min at 15 °C.

10. Wash gently in water for 2 min at 15 °C.

11. Fix in Kodak Unifix at 15 °C for 2 min.

12. Wash at 15 °C for 30 min in gently running water. (Maintenance of constant temperature minimizes wrinkling and cracking of the emulsion.)

13. Air dry at room temperature.

[a] A large instant coffee tin is ideal.
[b] Exposure periods range from about 1 day up to several weeks, depending on the isotope used and the abundance of the target. It is best to have several duplicate slides that can be taken out and developed at intervals to check for adequate exposure.

Protocol 21. Silver-enhanced colloidal gold marker

This marker has been used in conjunction with biotin probes and could probably be used in conjunction with other haptens by employing an anti-hapten antibody conjugated to colloidal gold. The silver enhancement procedure can be repeated several times in order to obtain a stronger signal. Simply monitor the silver deposition after each round and repeat if desired. Excessive noise will be evident beyond about four rounds.

- 1% gelatin (Bio-Rad)
- PBS
- streptavidin–gold conjugate (either 10 nm or 15 nm particles; Amersham)
- 0.1% Tween 20 in PBS (Bio-Rad)
- 1% glutaraldehyde in PBS
- distilled water
- 3% gelatin (Amersham) or 5% gum arabic
- IntenSE M™ kit (Amersham)

1. Block slides in 1% gelatin in PBS in room temperature for 30 min.
2. Incubate in streptavidin–gold conjugate diluted 1:20 (v/v) in 1% gelatin in PBS for 1 h at room temperature.
3. Wash three times in 0.1% Tween 20 in PBS.
4. Wash in PBS.
5. Fix in 1% glutaraldehyde in PBS.
6. Wash in water three times.
7. Air dry.

Protocol 21. *Continued*

8. Incubate sections in 3% gelatin or 5% gum arabic at room temperature for 15–30 min.
9. Perform silver enhancement using the IntenSE M™ kit as per manufacturer's protocol.
10. Stain and mount sections as per *Protocol 22*.

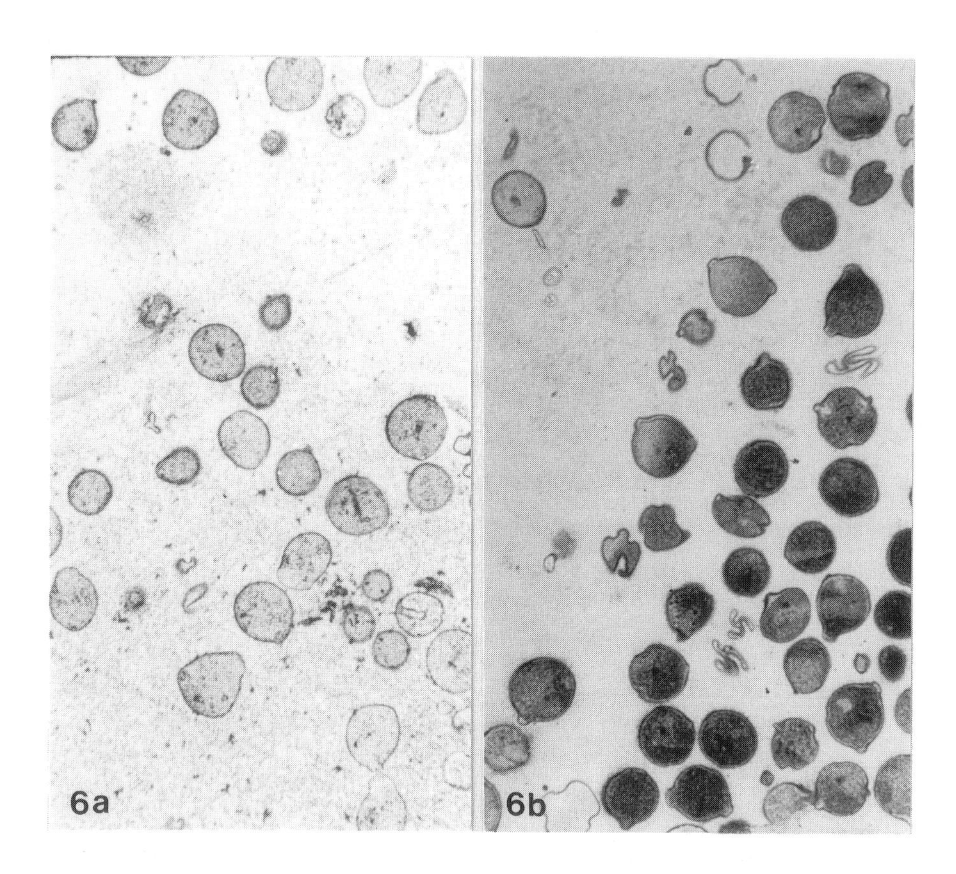

Figure 6. Localization of mRNA encoding a protein translation elongation factor (EF1α provided by Dr Marilyn Anderson and Dr Angela Atkinson) in germinating ornamental tobacco pollen grains. Resin sections (*Protocols 5* and 7) of the pollen grains were hybridized with sense (a) and antisense (b) RNA probes labelled with biotin (*Protocol 12*). Probe binding sites were then visualized with streptavidin–colloidal gold followed by silver enhancement (*Protocol 21*). Viewed in bright field microscopy, the signal is visible as a dark brown stain over the pollen grains hybridized with the antisense probe. Some silver deposit on the negative control using the sense probe is interpreted as noise.

Protocol 22. Counterstaining, mounting, and viewing of sections

- 0.25% aqueous Toluidine Blue
- distilled water
- 'Merckoglass' mounting medium (Merck)
- 5 g lead weights

1. Sections marked with flurochromes should be viewed immediately on a fluorescent microscope using the appropriate filter system and fluorescent micrographs made prior to staining and mounting.
2. Stain with a drop of 0.25% aqueous Toluidine Blue for 1 min.
3. Rinse with water and allow to dry.
4. Make a permanent mount by placing about 20 μl of Merckoglass on the section, cover with coverslip, and weigh down with small piece of lead (5 g)[a].
5. Allow to harden overnight.
6. Sections marked with silver-enhanced colloidal gold or liquid emulsion can be viewed by standard bright field microscopy. The silver particles can also be viewed by dark field optics[b] in which they appear as brilliant white points (see *Figure 5*).

[a] Fishing sinkers are ideal.
[b] Consult reference 20 for setting up dark field optics.

2.4 Controls

A number of different control experiments can be used to verify the authenticity of the signal. Some very convincing false positives can occur with *in situ* hybridization, so good controls are essential. When localizing RNAs, it is useful to predigest some sections with RNase which should remove most of the signal (digest with 1 mg/ml pancreatic RNase (Boehringer Mannheim) in 2 mM $MgCl_2$, 0.1 M Tris–HCl pH 7.5 at 37 °C for 1 h then wash extensively with 5 mM EDTA, 0.1 M Tris–HCl pH 7.0 before air drying). Similarly, DNase digestion can be used to eliminate DNA targets. An 'irrelevant' probe (i.e. one for which there is no target) can be used in parallel to detect non-specific binding. When using indirect non-isotopic systems, it is recommended that several sections are processed in which one of the detection reagents is omitted in order to demonstrate that no spurious signal is being generated. A positive control is very useful for optimizing a new *in situ* hybridization experiment. Using a probe for an abundant target such as rRNA or a highly expressed housekeeping gene can help in optimizing parameters. Checking retention of nucleic acids in preparations by Acridine Orange staining (*Protocol 9*) is also recommended.

Acknowledgements

Ms Ingrid Bonig provided excellent assistance in the development of the protocols presented here. I thank Professor Adrienne Clarke for supporting this work and making the facilities of the Plant Cell Biology Research Centre available. A Senior Research Fellowship from the Australian Research Council is gratefully acknowledged.

References

1. McFadden, G. I. (1989). *Cell Biol. Int. Rep.*, **13**, 3.
2. McFadden, G. I. (1991). In *Electron microscopy of plant cells* (ed. J. L. Hall and C. R. Hawes), pp. 219–55. Academic Press, London.
3. McFadden, G. I., Bonig, I., Cornish, E., and Clarke, A. E. (1988). *Histochem. J.*, **20**, 575.
4. Pardue, M. L. (1986). In *Drosophila. A practical approach* (ed. R. B. Robertson), pp. 111–37. IRL Press, Oxford.
5. Evans, M. R. and Read, C. A. (1992). *Nature*, **358**, 520.
6. Penschow, J. D., Haralambidis, J., Coghlan, J. P. (1991). *J. Histochem. Cytochem*, **39**, 835.
7. Langer, P. R., Waldrop, A. A., and Ward, D. C. (1981). *Proc. Natl Acad. Sci. USA*, **78**, 6633.
8. Wilchek, M. and Bayer, E. A. (1989). *Trends Biochem. Sci.*, **14**, 408.
9. Narayanswami, S. and Hamkalo, B. A. (1991). *Genetic analysis: Techniques and Applications*, **8**, 14.
10. *Nonradioactive in situ hybridization: application manual* (1992). Boehringer Mannheim, Mannheim, Germany.
11. User bulletin no. 49 (August 1988). Applied Biosystems, Foster City, CA.
12. Misiura, C., Durrant, I., Evans, M. R., and Gait, M. J. (1990). *Nucleic Acids Res.*, **18**, 4345.
13. FAM Amidite Product Bulletin (1992). Applied Biosytems, Foster City, CA.
14. Tescott, L. H., Eberwine, J. H., Barchas, J. D., and Valentino, K. L. (1987). In *In situ hybridization: applications to neurobiology* (ed. K. L. Valentine, J. B. Eberwine, and J. D. Barchas), pp. 3–24. Oxford University Press, New York.
15. Feinberg, A. P. and Vogelstein, B. (1983). *Anal. Biochem.*, **132**, 6.
16. Protocols and Applications Guide (1991). Promega Corp., Madison, WI.
17. Ausubel, F. M., Brent, R., Kingston, R.E., Moore, D. D., Seidman, J. G., Smith, J. A., and Struhl, K. (1988). *Current Protocols in Molecular Biology*. Wiley Interscience, New York.
18. Angerer, L. M. and Angerer, R. C. (1989). *DuPont Biotechnol. Update*, **4** (5), 1.

19. Bendayan, M. (1990). In *Colloidal gold*, Vol. 1 (ed. M. A. Hayat) pp. 34–94, Academic Press, San Diego, CA.
20. O'Brien, T. P. and McCully, M. E. (1986). *The study of plant structure. Principles and selected methods*. Termacaphi Pty, Melbourne.
21. McFadden, G. I., Ahluwalia, B., Clarke, A. E., and Fincher, G. B. (1988). *Planta*, **173**, 500.

6

DNA–DNA *in situ* hybridization—methods for light microscopy

T. SCHWARZACHER, A. R. LEITCH,
and J. S. HESLOP-HARRISON

1. Introduction

The great potential of DNA–DNA *in situ* hybridization arises from its unique ability to combine cytological information about chromosome and nuclear morphology with molecular information about DNA sequence structure. *In situ* hybridization enables the localization of any particular DNA sequence along metaphase chromosomes (*Figure 1*) and hence is valuable for the physical mapping of sequences and investigation of their long-range organization in the genome. Examples of the types of sequence that have been mapped by *in situ* hybridization in plants range from single copy and repeated genes to tandem and dispersed repetitive, non-transcribed sequences. Indeed, for many highly repetitive sequences, *in situ* hybridization is the method of choice to map them to a linkage group. The rDNA sequences are repeated hundreds or even thousands of times in plant genomes and have only been mapped reliably by *in situ* hybridization (e.g. in *Arabidopsis thaliana* (1), wheat (2), and barley (3)).

Data about the chromosomal location of sequences from *in situ* hybridization are often used to complement those from analysis by Southern hybridization. For example, a highly repeated retrotransposon-like element from barley, BIS1, was shown to be dispersed over most of the chromosomes, but excluded from the centromeric regions (4). Similarly, a repetitive sequence in *Arabidopsis thaliana*, pAL1, was isolated and characterized by Southern hybridization (5) but only shown to be located at all 10 centromeres by *in situ* hybridization. Comparative studies of sequence distribution and location in related species enables karyotype evolution and genome relation to be studied.

In situ hybridization can also locate sequences within interphase nuclei, and hence enables structural studies to investigate such diverse areas as nuclear architecture, chromosome packing, condensation and decondensation, gene expression, and chromosome interactions. Probe locations can be detected in

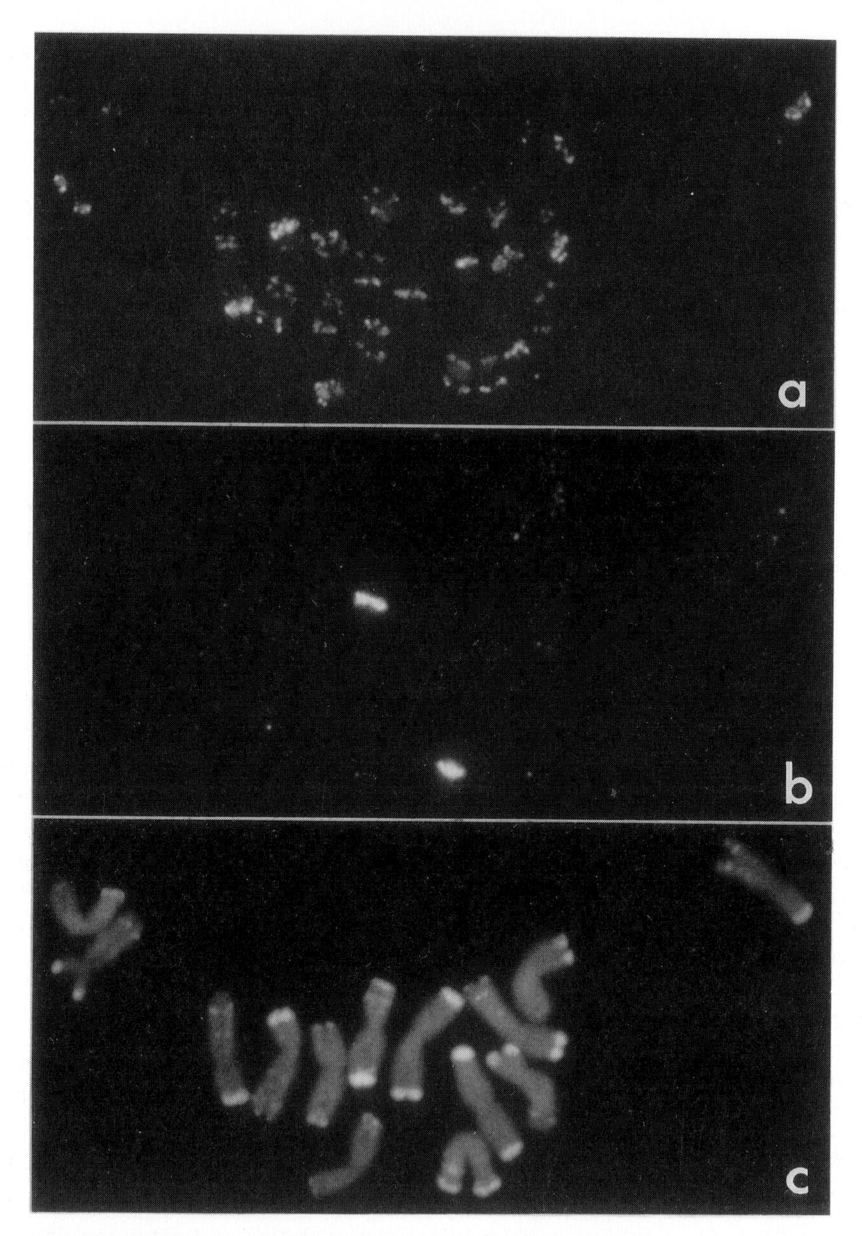

Figure 1. Fluorescent *in situ* hybridization to root tip metaphase chromosomes of rye (*Secale cereale* L., $2n = 14$) using two different probes simultaneously. A tandemly repeated, heterochromatin-specific clone (pSc119.2) has been labelled with rhodamine-dUTP and several major and minor hybridization sites fluoresce brightly under green excitation (a). The clone pTa71, a 9 kb rDNA repeat isolated from wheat, was labelled with digoxigenin-dUTP and was detected by fluorescein conjugated anti-digoxigenin. Under blue excitation one pair of strongly fluorescing rDNA sites is visible on the short arm of

two dimensions or in three-dimensional reconstructions of physical and optical sections. We have followed the behaviour of whole genomes and individual chromosomes throughout the cell cycle and shown that the interphase nucleus is highly ordered (6, 7). Using high resolution *in situ* hybridization to sections which were analysed in the electron microscope, Leitch *et al.* (8) studied the organization of expressed and unexpressed rDNA genes in different cereal nuclei. Finally, the reconstruction of serial mechanical sections (6) and optical sections (9, 10) allows three-dimensional analysis of the nucleus as *in vivo*.

2. Probes

DNA cloned in recombinant DNA libraries is the most common source of probes used for *in situ* hybridization, but some applications use total genomic DNA, flow sorted chromosomes, synthetic oligomers, bands cut from size-fractionated restriction enzyme digests, or satellite bands from centrifuge separations.

2.1 Testing of probes

Some probes used for *in situ* hybridization may be extensively characterized, including sequencing or transformation into host plants. Others may be relatively untested. However, in our experience, it is almost invariably worthwhile testing a potential probe by Southern hybridization to several size-separated restriction enzymes digests of genomic DNA before or simultaneously with an *in situ* hybridization experiment. This procedure will not only give a standard to compare with the strength of hybridization *in situ*, but also give extensive diagnostic information should results be negative. Can the probe be labelled successfully? Was the probe homologous to the species under test? Does the clone give the expected Southern hybridization pattern?

2.2 Low- and single-copy cloned sequences

There are increasing numbers of reports of detection of hybridization sites using probes with lengths of homologous DNA at the target sites in the chromosome ranging from under 1 kb to 10 kb (11–16). In plants, the techniques required are not yet as routine as those in human. Optimal chromosome preparation, denaturation, and hybridization conditions are

---

chromosome 1R (b). The preparation was simultaneously stained with DAPI to show the chromosome morphology. Major heterochromatin bands fluoresce brightly under UV excitation (c). Compare the *in situ* hybridization signal of a and b with the DAPI picture. Photographed using Kodak Tmax 400. Magnification: 1600×.

prerequisites for success—an increase of 10^6 over the typical, single copy detection sensitivity on a Southern blot is required (5 μg/track loading of a plant such as barley with a 1*C* genome size of 5 pg).

2.3 Repetitive DNA sequences

Highly repeated sequences are widely used for *in situ* hybridization and may show tandem or dispersed distributions along chromosomes. *In situ* hybridization enables their distribution to be mapped (1, 3) and comparative studies made. Repetitive sequences may be essentially specific to one chromosome or one species, or more widespread among many species. Some sequences may be valuable for chromosome identification: chromosomes with similar morphologies may be identified by the repeat hybridization pattern. Double target hybridization experiments, using two different probes, can thus localize one sequence on a chromosome identified by the second sequence (see *Figure 1*).

2.4 Synthetic oligomers

Synthetic oligomers have been labelled and used directly as probes for *in situ* hybridization experiments to plant and mammalian chromosome preparations (17–20). The binding specificity may be higher than that of cloned DNA sequences in mammals (21) and in plants (T. Schwarzacher, unpublished observations). Now that sequence data and oligonucleotide synthesis facilities are so widely available, such probes are likely to become more widely used in the future. In particular, we see applications to the detection of locations of gene families on chromosomes, and of sequences common to a number of genes, in promoters, introns, or coding regions. Highly sensitive techniques, and in particular primed *in situ* hybridization (PRINS) (22, 23) now being developed mean that the use of short, consensus sequences as probes is likely to be increasingly valuable. Ultimately, we expect that sites of hybridization of any sequence from 20 bp upwards will be detectable, making the use of cloned probes redundant.

2.5 Genomic DNA as a probe

Total genomic DNA has been used as a probe for *in situ* hybridization to detect and characterize parental genomes in hybrids (6, 7, 24) and alien chromosomes or chromosome segments incorporated into recombinant wheat lines (25–27). The differentiation of closely related species can be tested on Southern hybridization blots (28). To reduce cross-hybridization between very closely related species used for hybrids or recombinant lines, unlabelled total genomic DNA of the other parental or recipient species has been added in high amounts (20–100 times more than the labelled probe) to the probe mixture.

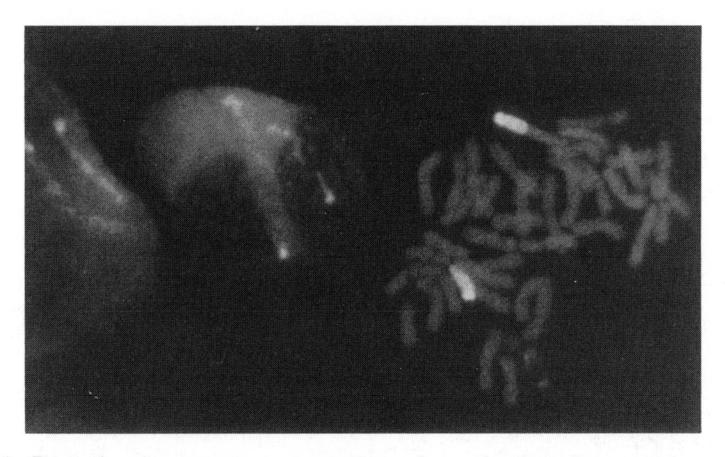

Figure 2. Root tip chromosome preparation of a wheat variety carrying a 1B/1R translocation after *in situ* hybridization using total genomic DNA from rye as a probe. In the partial metaphase, the rye arm fluoresces strongly showing the translocation breakpoint with the weakly stained wheat arm clearly. At interphase the two chromosome arms form distinct elongated domains. Photographed using Kodak Ektar 1000. Magnification 900×.

At metaphase, total genomic DNA hybridizes to the chromosomes of its origin along most of their lengths (*Figure 2*). Hence visualization of the parental genomes or alien chromatin is possible at all stages of the cell cycle. At interphase and prophase, individual alien chromosomes and chromosome segments are seen as distinct domains: one domain is present in monosomic lines and two in disomic lines. Human chromosome domains have also been identified and found to form domains using total genomic human DNA as a probe to human–rodent cell fusion hybrids which had eliminated more than 90% of the human chromosomes (29, 30).

The genomic *in situ* hybridization method is of value to plant breeding programmes for screening of recombinant lines. The method is quick, and informative and can be adapted easily to many species. On metaphase chromosomes, the size and genomic location of alien chromatin can be assessed, while interphase analysis shows the presence and number of alien segments.

2.6 Flow-sorted chromosomes

Chromosomes sorted by flow cytometry are widely used to generate probes for human *in situ* hybridization. Complete DNA libraries from individual sorted chromosomes enable 'chromosome painting', where particular chromosomes or subchromosomal regions are coloured following *in situ* hybridization (31, 32). Chromosome painting is informative for the analysis of translocations and trisomies in cancers and in the diagnosis of genetic disorders. Using human chromosome paints, chromosomal rearrangements between different primates have been characterized (33).

DNA can be directly amplified from a few sorted chromosomes using PCR (polymerase chain reaction) and used as a probe *in situ*. If DNA from a deleted chromosome identified on a flow histogram can be sorted, *in situ* hybridization to normal human chromosomes can define the deleted region: the rest of the chromosome shows hybridization signal while the deleted region shows only counterstain (34, 35).

Recent advances in the flow cytogenetics of plants (36, 37) indicate that similar techniques will be directly applicable. We expect that probes derived from sorted chromosomes will enable extensions of the painting methods beyond alien chromosomes in the next few years, at least for important crop species such as wheat.

3. Targets

3.1 Chromosome spread preparations

High-quality preparations are a prerequisite for successful *in situ* hybridization. Chromosomes and nuclei must not be overlapping and should be free of cytoplasm, cell walls, debris, and dirt. Since most *in situ* hybridization experiments aim to localize a probe on metaphase chromosomes, preparations should have as many complete metaphases as possible. Methods described below relate to the standard protocols used in our laboratory; most other groups use similar techniques.

We pre-treat our material to accumulate metaphases, and fix it in alcohol:acetic acid prior to digestion with cellulase and pectinase. Most workers partially digest cells walls with enzymes to free chromosomes (11, 38). As an alternative, living root tips or suspension cell cultures can be treated wth enzyme to make a protoplast suspension which is then fixed (39, 40). Protoplasts freed by digestion can be either squashed or dropped on to slides which are pre-cleaned to limit selective loss of metaphases (see *Protocol 1*) (11, 24, 38).

We routinely use both the dropping and spreading techniques. Dropped preparations are of uniform density and many similar slides can be made from the same suspension. Chromosomes spread apart well, are free of cytoplasm, and are usually undistorted. However, the enzyme digestion time requires careful adjustment and dropping suspensions usually require cells from many roots with a good metaphase index.

After preparation, slides must be checked and rigorously selected using phase contrast microscopy (or after staining small chromosomes with DAPI; see Section 5.6, *Protocol 9*). We routinely discard half of our slides from good batches of fixations, and half of all batches of fixations.

Protocol 1. Chromosome preparations on glass slides

- enzyme buffer: solution A, 0.1 M citric acid; solution B, 0.1 M tri-sodium citrate
- cellulase (from *Aspergillus niger*; Calbiochem 21947); optionally cellulase Onozuka RS (Yakult Pharmaceutical)
- pectinase (from *Aspergillus niger*; solution in glycerol, Sigma P-9932)
- 100% ethanol or methanol
- glacial acetic acid and 45% aqueous acetic acid
- chromium trioxide solution in 80% (w/v) sulphuric acid (Merck)
- optional: colchicine or 8-hydroxyquinoline
- microscope slides and coverslips (no. 1, 18 × 18 mm)
- dissecting instruments: fine forceps (no. 5), needles, razor blade
- dry ice or liquid nitrogen

A. *Cleaning glass microscope slides*

1. Place slides into chromium trioxide solution for at least 3 h at room temperature.
2. Wash slides in running tap water for 5 min, rinse them thoroughly in distilled water, and air dry.
3. Place slides in 100% ethanol, remove, and dry with a tissue immediately prior to use.

B. *Accumulation of metaphases and fixation*

1. Prepare one of the following metaphase arresting agents:
   (a) ice water (for cereals and grasses): put distilled water in a clean bottle, shake vigorously to aerate, keep at −20 °C until the water starts to freeze, and shake again
   (b) 0.05% (w/v) colchicine in water (for most plant tissues)
   (c) 2 mM 8-hydroxyquinoline in water (for dicotyledonous plants, particularly those with small chromosomes such as *Arabidopsis thaliana*).
2. Collect plant material. Any tissue containing dividing cells can be used. Root tips (10–20 mm) from young seedlings, newly grown roots at the edge of plant pots, or roots grown in hydroponic systems are all suitable. Alternatively, flower buds, anthers, carpels, apical meristems, or tissue culture material can be used.

Protocol 1. *Continued*

3. Place plant material in metaphase arresting agent and incubate as follows:
   (a) 24 h at 0 °C for ice water or
   (b) 3–6 h at room temperature or 10–20 h at 4 °C for colchicine or
   (c) 0.5–2 h at room temperature followed by 0.5–2 h at 4 °C for 8-hydroxyquinoline.
4. Prepare fixative by mixing three parts 100% ethanol or methanol and one part glacial acetic acid.
5. Briefly blot material dry and plunge into an excess of fixative. For immediate use, leave for at least 10 h at room temperature; otherwise leave for 2 h at room temperature before storage at 4 °C for 1–4 weeks or at −20 °C for up to 3 months.

C. *Enzyme digestion*

Enzyme treatment weakens the cell walls so that cells separate and chromosomes are freed from cytoplasm. For the dropping method, cells should just fall apart, while for squashing, material should remain intact or dividing cells will be lost into the medium. The time of digestion, strength of enzyme, and sometimes the ratio of cellulase to pectinase need to be adjusted depending on the method, species used, source of cells, and fixation.

1. Make a 10× stock solution of enzyme buffer (pH 4.6) by mixing 60 ml solution B and 40 ml solution A. Dilute 1:10 with distilled water for use.
2. Wash material from *Protocol 1*B twice in 5 ml enzyme buffer for 10 min each at room temperature to remove fixative.
3. Make 10–20 ml 2× enzyme solution by mixing 2% cellulase (and optionally add Onozuka RS 0.2–0.5%) and 20% pectinase in enzyme buffer. Ready made enzyme solution can be stored in aliquots at −20 °C.
4. Transfer plant material to 1–2 ml 2× enzyme solution and incubate at 37 °C for 0.5–2 h.
5. Follow D or E.

D. *Squashing method*

1. Follow *Protocol 1*C with shorter digestion times.
2. Transfer to enzyme buffer, leaving for at least 15 min.
3. Place enough material for one preparation (typically one root tip or anther) in 45% acetic acid in an embryo dish or small Petri dish for a few minutes.

4. Make chromosome spreads. Under a stereo microscope tease the material to fragments with a fine needle in one or two drops of 45% acetic acid on a clean slide (*Protocol 1*A), isolate the meristem, and discard other tissues, particularly the root cap which is tough and prevents squashing. Apply a coverslip. Carefully tap the coverslip with a needle and then gently squash the material between glass slide and coverslip. Check the preparation with a phase contrast microscope.

5. Put slide on dry ice for 5–10 min (preferred method) or immerse into liquid nitrogen, then flick off the coverslip with a razor blade, and air dry.

7. Screen slides under phase contrast. Small chromosomes can be stained with DAPI (see *Protocol 9*) and analysed under the fluorescence microscope.

8. Slides can be stored desiccated at 4 °C or −20 °C for several weeks.

E. *Dropping method*

Digestion and washing steps are carried out in 1.5 ml microcentrifuge tubes at room temperature. If plenty of material is available, 10–15 ml polypropylene or glass centrifuge tubes can be used. A squash preparation is normally made to check the metaphase index and state of the material.

1. Follow *Protocol 1*C. Check material from time to time and try to disperse cells with a glass rod or pipette tip to make a suspension.
2. When cells fall apart, centrifuge the tube for 3 min at 450 *g*, discard the supernatant, and shake up the pellet.
3. Resuspend the pellet in fresh enzyme buffer and repeat step 2.
4. Repeat step 3 twice.
5. Resuspend the pellet in fresh fixative.
6. Centrifuge for 3 min at 1000 *g*, discard the supernatant, and shake up the pellet.
7. Repeat steps 5 and 6 at least twice.
8. Resuspend in fixative to a final volume of 50–100 μl.
9. Drop 10–20 μl of the suspension into a clean glass slide (*Protocol 1*A), shake the slide a little, blow gently on it, and then allow to air dry.
10. Check the cell density and spreading quality with a phase contrast microscope.
11. Slides can be stored desiccated at 4 °C or −20 °C for several weeks.

3.2 Sectioned preparations for high resolution microscopy

High resolution *in situ* hybridization can be conducted on thin sectioned biological material (<0.5 μm) which can be examined in the light or electron microscope (see also Chapter 5). Fresh material is fixed using cross-linking fixatives (e.g. glutaraldehyde and formaldehyde) which give good morphological preservation of cell ultrastructure and retention of target DNA. Acrylic resins, rather than epoxy resins, are used to embed the material because they are hydrophilic and give better access to the probe and detection reagents. Sections (typically between 0.25 and 0.1 μm thick) are cut from the embedded material using an ultramicrotome and mounted on to a glass slide or electron microscope grid. DNA accessibility to probe and detection reagents needs to be improved by treatment with proteases. A protocol for preparing material for sectioning prior to *in situ* hybridization is given in *Protocol 2* (see also *Protocol 4* of Chapter 7 and *Protocols 4* and *5* of Chapter 5). Some workers also use paraffin embedding (41; see also Chapter 5).

3.3 Whole cells

In situ hybridization can be conducted on whole cells by fixing a piece of tissue with cross-linking (e.g. glutaraldehyde) or precipitating (e.g. ethanol: acetic acid) fixatives and sectioning the tissue to 5–10 cells' thickness by hand or using a Vibratome. Permeability of individual plant cells to probes and detection reagents can be improved by pretreating the material with wall-degrading enzymes (as for making plant nuclei spreads) prior to the prehybridization steps in *Protocol 4*. Finally, the material is soaked in 10% glycerol in 2 × SSC and placed in the freezer for 1 h before *in situ* hybridization. Enzyme or fluorochrome detection of probe hybridization signal is possible in whole cells. This method keeps whole cells intact so that they can be imaged using confocal microscopy which can be used to make three-dimensional reconstructions.

Protocol 2. Fixation and resin embedding for sectioning

- 25% (v/v) glutaraldehyde (electron microscopy (EM) grade, Taab Laboratories) in water
- saturated aqueous picric acid (Sigma) (**warning**: this compound is explosive if allowed to become dry)
- phosphate buffer (pH 6.9: three parts 0.1 M Na_2HPO_4, two parts 0.1 M KH_2PO_4)
- 100% ethanol

- LR White resin (medium grade; London Resin Company)
- poly-L-lysine (mol. wt 300 000)
- benzyl alcohol (**care**: benzyl alcohol can dissolve the cement of a diamond knife)

1. Make 10 ml glutaraldehyde fixative by mixing 0.8 ml 25% glutaraldehyde with 9.2 ml phosphate buffer and then add 20 μl picric acid.
2. Fix material for 2 h at room temperature.
3. Wash material (2 × 5 min) in phosphate buffer.
4. Dehydrate material in water/ethanol series: 10 min each in 10%, 20%, 30%, 50%, 70%, 90%, 100%, and 100% ethanol.
5. Embed material in LR White resin by replacing ethanol with LR White in the following ratios: ethanol:LR White 3:1 (30 min), 1:1 (30 min) then 100% LR White for 3 days changing the solution six times.
6. Polymerize resin at 65 °C for 15 h.
7. Make poly-L-lysine-coated slides by immersing slides in 1 mg/ml poly-L-lysine in deionized water and air dry.
8. Cut sections (0.1–0.25 μm thickness) with a glass or diamond knife. Float sections on to a solution of 1% (v/v) benzyl alcohol in water to allow them to re-expand after the compression caused by sectioning.
9. Dry sections on to a glass slide that has been coated with poly-L-lysine.

4. Probe labelling: types and methods

In situ hybridization was first developed in the late 1960s (42, 43) using probe DNA labelled with radioactive atoms. Since then, radioactively labelled probes have been used extensively for mapping of genes (44, 45). The development of non-radioactive labels has revolutionized *in situ* hybridization in the light microscope because of their improved speed, resolution, safety, and convenience. The use of several different labels for simultaneous multicolour detection of hybridization of several probes is another major advantage (46–48). Early problems with reduced sensitivity compared with radioactive probes have largely been overcome by improvements in immunochemistry, microscopy, photographic films, low-light video cameras, and image processing (3, 49, 50).

4.1 Non-radioactive indirect labels

The two naturally occurring haptens, biotin (a vitamin) and digoxigenin (a steroid from *Digitalis purpurea*, foxglove) are widely used for non-radioactive *in situ* hybridization. The molecules are available commercially linked to nucleotides, and are incorporated into the DNA probe by standard labelling techniques (see Section 4.4). Hybridization sites are detected by avidin, its derivatives or antibodies (for biotin), or anti-digoxigenin (for digoxigenin) linked to fluorescent molecules or to enzymes which precipitate chromogenic substrata (see Section 2.2.2 of Chapter 5). Total DNA is usually stained separately to show chromosome morphology. We prefer fluorescent detection systems because the locations of hybridization signals are defined, characteristic, and easy to distinguish from dirt or background signal (see *Figure 1*). Several probes can be detected simultaneously using detection reagents with different fluorescent tags—typically fluorescing blue, green, or red (see Section 5.5, *Table 1*). We have also found that chromosome preparations can be re-probed after fluorescent *in situ* hybridization in a second *in situ* hybridization experiment (51).

4.2 Direct fluorochrome labels

Nucleotides directly conjugated to red, blue, and yellow fluorochromes have been available commercially since the early 1990s. Since no detection steps are needed after the post-hybridization washes, they are quicker and simpler to use and no background is introduced during detection steps. Sensitivity is more limited than other systems since there is no signal amplification. Hence direct fluorochrome labels are used mostly to detect hybridization sites of middle- to high-copy number probes or sets of pooled probes, although anti-fluorochrome antibodies could amplify signal and digital imaging systems (see below) may overcome problems of low brightness. Fluorochrome-labelled probes can be hybridized together with biotin- or digoxigenin-labelled probes, increasing the number of simultaneously detected probes (47, 48). We now routinely use the combination of a direct rhodamine-labelled probe with fluorescein isothiocyanate (FITC) detection of a digoxigenin-labelled probe (*Figure 1*).

4.3 Radioactively labelled nucleotides

Radioactively labelled nucleotides are now rarely used for DNA–DNA *in situ* hybridization in plants at the light microscope level; in mammalian studies they are occasionally used. However, the long exposure times involved, lack of spatial resolution of signal, and complex protocol mean the technique cannot be recommended for *in situ* localization of DNA sequences to chromosomes. When detecting RNA or DNA in tissues where spatial resolution is less critical and non-radioactive probe detection by immunochemicals can be difficult, radioactive labels are often used; protocols are described by McFadden (Chapter 5).

4.4 Labelling protocols

All the standard protocols for incorporation of labelled nucleotides into nucleic acids can be used for probe labelling with biotin-, digoxigenin-, or fluorochrome-labelled nucleotides. These include nick translation, randomly primed labelling, PCR-based amplification, and label incorporation and end-tailing (the latter particularly for short oligomeric probes). Many companies sell kits for labelling with reliable and easy-to-follow protocols; most companies offer useful advice by telephone and in newsletters.

For randomly primed labelling and nick translation we have developed modified protocols which are listed in *Protocol 3* (see also Chapter 5, *Protocol 11*). We start with high quality DNA—that suitable for digestion with low concentrations of *Eco*RI is normally satisfactory for labelling. We generally use random priming for small quantities of short (<1 kb) cloned inserts which we cut out of the plasmid vector, run on an agarose gel, and purify. Nick translation is more convenient for total genomic DNA or large cloned inserts. PCR-based amplification in the presence of labelled nucleotides is possible and yields large amounts of labelled probe. However, we find that it requires optimization for each probe, a costly procedure in time, nucleotide, and enzyme. Some sequences seem to label poorly by one method, so we test alternatives. Although we have not analysed the reasons in detail, hapten-conjugated nucleotides are not as readily incorporated by polymerases as are unconjugated nucleotides, and presumably result in incomplete strands, while some repetitive sequences have unusual secondary structures which may slow down label incorporation.

After labelling, we routinely check the incorporation of biotin or digoxigenin by a small test blot (either dot blotted or size-separated on a gel and transferred to a nylon membrane) following standard protocols for non-radioactive Southern hybridization with the same haptens. Direct fluorochrome-labelled probes are checked under a fluorescent microscope by placing a drop of labelled DNA on a glass side and exciting it with the appropriate wavelength. A drop including labelled probe will fluoresce uniformly in the expected colour.

Protocol 3. Labelling DNA by random priming or nick translation

All solutions are purchased sterile, filter-sterilized, or autoclaved

- labelled nucleotides: one of the following
  - (a) digoxigenin-11-dUTP (1 mM solution, Boehringer Mannheim)
  - (b) biotin-11-dUPT (powder, Sigma) or biotin-14-dATP (0.4 mM solution GIBCO-BRL)
  - (c) fluorescein-12-dUTP (1 mM solution, Boehringer Mannheim) or fluorescein-11-dUTP (FluoroGreen, 1 mM solution, Amersham)
  - (d) rhodamine-4-dUTP (FluoroRed, 1 mM solution, Amersham)
  - (e) coumarin-4-dUTP (FluoroBlue, 1 mM solution, Amersham)

Protocol 3. *Continued*

- unlabelled nucleotides: dATP, dCTP, dGTP, TTP (lithium salt, 100 mM solution in Tris–HCl pH 7.5, Boehringer Mannheim)
- modifying enzyme: for random priming use Klenow enzyme (DNA polymerase I, large fragment, labelling grade, 6 U/μl, Boehringer Mannheim); for nick translation use DNA polymerase I/DNase I mixture (0.4 U/μl, Gibco BRL)
- for random priming only: hexanucleotide reaction mixture (in 10× buffer, Boehringer Mannheim) or 'random' hexanucleotides or primers
- 3 M sodium acetate or 4 M lithium chloride
- 0.3 M EDTA pH 8
- ice-cold 100% ethanol
- ice-cold 70% ethanol
- dithiothreitol (DTT)

Optional reagents, if buffers or nucleotide mixtures have to be prepared:

- 1 M Tris–HCl, pH 8, pH 7.8, pH 7.5, pH 7.2
- 1 M $MgCl_2$
- 20 mg/ml bovine serum albumin (BSA, aqueous solution, nuclease-free, Sigma)
- dithiothreitol
- 'random' hexanucleotides (Boehringer Mannheim, GIBCO-BRL)

The quantities of stock solutions are for approximately 20 labelling reactions.

1. Prepare DNA for labelling. Total genomic DNA should be clean (test by cutting with *Eco*RI) and sheared or sonicated to 5–10 kb fragments. Cloned DNA should be either linearized whole plasmid or cut out and cleaned insert.
2. Prepare labelled nucleotide mixture and store at −20 °C:
   - for digoxigenin: make a 18.5 μl mM solution of TTP in Tris–HCl pH 7.5 and mix with 11.5 μl digoxigenin-11-dUTP
   - for biotin: if necessary make a 0.4 mM solution in Tris–HCl pH 7.5
   - for directly labelled nucleotides: make 10 μl 1 mM solution of TTP and mix with 20 μl of fluorochrome-conjugated dUTP
3. Prepare 75 μl unlabelled nucleotide mixture by combining the three nucleotides not present as labelled nucleotide. Make 0.5 mM solution

of each nucleotide in 100 mM Tris–HCl pH 7.5 and mix 1:1:1. Store at −20 °C.

4. Follow *Protocol 3*A or B.

A. *Random priming*

1. Prepare 50 μl hexanucleotide reaction buffer, if ready-made buffer is not available from manufacturer, by mixing 0.5 M Tris–HCl pH 7.2, 0.1 M $MgCl_2$, 1 mM dithioerythritol, 2 mg/ml BSA, and 62.5 A_{260} units/ml 'random' hexanucleotides; store at −20 °C.

2. Take 50–200 ng of DNA for labelling and mix with sterile distilled water to make 12.5 μl in a 1.5 ml microcentrifuge tube. (Firmly tape or clip the lid to prevent opening during step 3).

3. Denature the DNA in boiling water for 10 min, put on ice for 5 min, and centrifuge briefly.

4. Add 3 μl unlabelled nucleotide mixture, 1.5 μl of labelled nucleotide mixture, and 2 μl of hexanucleotide reaction buffer; briefly vortex mixture.

5. Add 1 μl of Klenow enzyme and mix gently.

6. Incubate at 37 °C for 6–8 h or overnight.

7. Add 2 μl 0.3 M EDTA to stop the reaction and continue with *Protocol 3*C.

B. *Nick translation*

1. Prepare 150 μl 10× nick translation buffer by mixing 0.5 M Tris–HCl pH 7.8, 0.05 M $MgCl_2$, and 0.5 mg/ml BSA, stored at −20 °C. Prepare 20 μl of 100 mM DTT.

2. Prepare the reaction in a 1.5 ml microcentrifuge tube by adding

- 5 μl 10 nick translation buffer
- 5 μl unlabelled nucleotide mixture
- 2–3 μl of labelled nucleotide mixture
- 1 μl dithiothreitol
- 1 μg of DNA

adjust to the final volume of 45 μl with distilled water and briefly vortex.

3. Add 5 μl of DNA polymerase I/DNase I mixture and mix gently.

4. Incubate at 15 °C for 1–2 h.

5. Add 5 μl 0.3 M EDTA to stop the reaction and continue with *Protocol 3*C.

Protocol 3. *Continued*

C. *Ethanol precipitation to remove unincorporated nucleotides*

1. Add 0.1 vol. sodium acetate or lithium chloride.
2. Add 0.66–0.75 vol. cold 100% ethanol and mix gently.
3. Incubate at −20 °C overnight or −80 °C for 1–2 h.
4. Prepare 10 ml 100× TE buffer by mixing 1 M Tris–HCl pH 8 and 0.1 M EDTA. For use, dilute 100 μl with 9.9 ml distilled water and filter sterilize.
5. Spin the tubes containing precipitated DNA at −10 °C for 30 min, 12 000 *g*.
6. Discard the supernatant, add 0.5 ml of cold 70% ethanol, and spin at −10 °C for 5 min, 12 000 *g*.
7. Discard supernatant and leave pellet to dry.
8. Resuspend the DNA in 10–30 μl 1 × TE.

5. Hybridization and denaturation

Chromosome preparations are pretreated to reduce non-specific probe and detection binding, increase permeability to probe and detection reagent, and stabilize the target DNA sequences (*Protocol 4*). Hybridization mix containing the labelled probe, blocking DNA, and low-salt buffers is made up as described in *Protocol 5*. The hybridization mix and the DNA in the chromosome preparations are denatured to make them single-stranded (*Protocol 6*). The probe is then hybridized to the chromosomes *in situ*. Weakly bound probe is removed in a series of washing steps. The stringent wash, described in *Protocol 7*, allows sequences with more than about 85% homology to remain hybridized but can be altered if desirable. Note too that some repetitive sequences have extreme AT:GC nucleotide ratios, which changes the stringency of any wash, and that probe length alters stringency.

5.1 Pretreatments

RNase A is used to remove endogenous RNA which might hybridize to the probe giving a high 'background' signal. Treatment of slides with pepsin can increase probe and detection reagent accessibility by digesting proteins over the surface of chromosomes. It is particularly effective if cytoplasm is associated with the chromosome preparations, and can be omitted if the chromosome preparations appear free of cytoplasm. Paraformaldehyde fixation minimizes the loss of DNA from chromosomes during the subsequent denaturation steps.

Protocol 4. Pretreatment of slide preparations

- 20× SSC stock: 3 M NaCl, 0.3 M sodium citrate adjusted to pH 7
- DNase-free RNase A
- pepsin (porcine stomach mucosa, activity: 3200–4500 U/mg protein, Sigma)
- paraformaldehyde (EM grade)
- 1 M Tris–HCl pH 7.5
- 1 M NaCl
- 0.1 M NaOH
- 0.01 M HCl
- 70%, 90%, and 100% ethanol
- plastic coverslips: pieces of appropriate size cut off from autoclavable waste disposal bags

A. *For spread preparations*

1. Dry chromosome spreads (from *Protocol 1*) in an incubator at 37 °C overnight.
2. Make up 1 litre of 2 × SSC by diluting 20 × SSC stock 1:10.
3. Make up 10 ml RNase A stock by dissolving 10 mg/ml of RNase in 10 mM Tris–HCl, pH 7.5, and 15 mM NaCl. Boil for 15 min and allow to cool. Store at −20 °C in aliquots.
4. Dilute RNase stock 1:100 with 2 × SSC to give a final concentration of 0.1 mg/ml.
5. Add 200 μl of RNase A to each preparation, cover with a plastic coverslip and incubate for 1 h at 37 °C in a humid chamber.
6. Start preparing paraformaldehyde (for steps A.12 and B.7). In the fume hood, add 4 g of paraformaldehyde to 80 ml water, heat to 60 °C for about 10 min, clear the solution with about 20 ml 0.1 M NaOH, allow to cool down, and adjust the final volume to 100 ml with water. (The paraformaldehyde solution must be freshly prepared.)
7. Wash slides in 2 × SSC for 3 × 5 min. This step and all further washing steps are carried out in Coplin jars using enough liquid to immerse the whole slide fully.
8. Make up 5 μg/ml pepsin in 0.01 M HCl. (A stock of 500 μg/ml in 0.01 M HCl can be stored at −20 °C in aliquots and diluted prior to use.)
9. Place slides in 0.01 M HCl for 2 min.
10. Add 200 μl pepsin on to each preparation, cover with a plastic cover slip, and incubate for 10 min at 37 °C.

Protocol 4. *Continued*

11. Stop pepsin reaction by placing the slides in water for 2 min and then wash in 2 × SSC for 2 × 5 min.
12. Place slides into paraformaldehyde in a Coplin jar and incubate for 10 min at room temperature.
13. Wash slides in 2 × SSC for 3 × 5 min.
14. Incubate slides for 3 min each in 70%, 90%, and 100% ethanol and then air dry. Slides can be kept for up to 24 h before denaturation and hybridization (*Protocols 5* and *6*).

B. *For sections*

1. Take preparations from *Protocol 2* and follow *Protocol 3*A steps 2–7.
2. Make up 200 ml proteinase K buffer consisting of 20 mM Tris–HCl pH 8 and 2 mM $CaCl_2$, and 300 ml proteinase K buffer containing 50 mM $MgCl_2$.
3. Make up 1 μg/ml proteinase K in proteinase K buffer (can be stored in aliquots at −20 °C).
4. Incubate slides in proteinase K buffer for 2 × 5 min.
5. Add 200 μl proteinase K, cover with a plastic coverslip and incubate for 10 min at 37 °C.
6. Stop reaction by placing slides in proteinase K buffer containing $MgCl_2$ and wash 3 × 5 min.
7. Place slides into paraformaldehyde in a Coplin jar and incubate for 10 min at room temperature.
8. Wash slides in 2 × SSC for 3 × 5 min.
9. Slides can be kept for a few hours before continuing with denaturation and hybridization (*Protocols 4* and *5*).

5.2 Hybridization mixture

The conditions used during hybridization and later in the post-hybridization washes determine the stringency of hybridization. The stringency (expressed in percent) is calculated by subtracting the difference between the melting temperature and the hybridization temperature from 100 and gives the amount of homology required for probe and target DNA to form stable hybrids (see Chapter 5). Stringency depends on the length and base pair composition of the probe DNA, temperature, and concentrations of salt and helix-destabilizing agents (e.g. formamide) in the solution. Temperature and concentrations can be varied to control hybridization stringency. Usually we hybridize with 50% formamide in 2 × SSC at 37 °C (75–80% stringency, *Protocols 5* and *6*) and wash in 20% formamide in 0.1 × SSC at 42 °C (80–85% stringency, *Protocol 7*).

Protocol 5. Preparing the hybridization mixture

- formamide: good but not the highest grade (e.g. Sigma catalogue number F 7508); store in aliquots at −20 °C
- dextran sulphate
- 10% SDS in water
- salmon sperm DNA (5 μg/μl)
- 20 × SSC stock: 3 M NaCl, 0.3 M sodium citrate adjusted to pH 7; filter (0.22 μm) sterilize 5 ml
- labelled probe DNA (from *Protocol 3*)
- plastic coverslips: pieces of appropriate size cut from autoclavable waste disposal bags

1. Make 50% (w/v) dextran sulphate in water and filter (0.22 μm) sterilize. Aliquots can be stored at −20 °C.
2. Autoclave salmon sperm DNA to fragments of 100–300 bp (typically 5 min at 100 kg/m). Can be stored in aliquots at −20 °C.
3. Prepare the hybridization mixture in a 1.5 ml microcentrifuge tube by adding:

• formamide	20 μl	
• 50% dextran sulphate	8 μl	
• 20 × SCC	4 μl	
• 10% SDS	0.5–1 μl	
• probe DNA	1–5 μl	(final concentration 0.5–2 μg/ml)
• salmon sperm DNA	1–5 μl	(final concentration 5–50 μg/ml, i.e. 10–200× of probe concentration)

   adjust to a final volume of 40 μl with distilled water and briefly vortex the mixture.

4. Denature the hybridization mixture at 70 °C for 10 min and then transfer to ice for 5 min.
5. Take slides from *Protocol 3*. Spread preparations are dry. Slides with sections are taken out of 2 × SSC. Rapidly blow off excess liquid and proceed before slides are desiccated.

Protocol 5. *Continued*

6. Add 40 μl denatured hybridization mix to each slide and cover with a plastic coverslip. Ensure no bubbles are trapped.
7. Immediately proceed to *Protocol 6*.

5.3 Denaturation and hybridization

Denaturation of the chromosomal DNA is a critical step for the *in situ* hybridization procedure. For successful hybridization, the chromosomal DNA needs to be single-stranded but without loss of DNA by over-denaturation. Widely used methods involve acid, alkali, or formamide (52) at various temperatures, concentrations, and times. Most commonly, the slide preparations are dipped into 50–100 ml of preheated denaturation solution (e.g. 70% formamide at about 70 °C for 2 min). As the temperature drops substantially when the slide is added, it is advisable to denature slides separately or in pairs. After denaturation, slides are dehydrated through an ice-cold ethanol series. The high amounts of formamide may expose the operator to fumes, and the turbulent solution may wash away preparations, especially sections. The method is not suitable for acrylic resin sections which should not be treated with alcohols.

We have adopted a method to reduce exposure to toxic formamide fumes and to ensure reproducibility for both sectioned and spread material. Slide preparations are denatured in 30–50 μl of hybridization mixture (which can be denatured in advance; see *Protocol 5*) by placing them in a modified programmable thermal cycler (53). A thermocouple is mounted next to the slides to measure and control the temperature. We have found that the denaturation time and temperature can be easily optimized and are reproducible for each type of material and species used. Alternatively, slides can be incubated in a humid chamber floating in a water bath with the external probe of a digital thermometer placed in the chamber for temperature control by opening and closing the water bath lid as needed.

Protocol 6. Denaturation of slide preparations and hybridization

- modified programmable thermal cycler: instead of the block containing holes for microcentrifuge tubes, make a small humid chamber with shelves for slides to rest on; mount a thermocouple probe within a glass capillary next to the slides, and cover with an insulating lid

or

- metal tin with a convex lid to prevent drips on to slides

Follow either method A or B.

A. *Modified thermal cycler*

1. Program the thermal cycler for one cycle at 75–80 °C for 5–10 min using the mode which starts the timing after the target temperature is reached. Then program to cool down slowly to 37 °C (if necessary program several temperature steps of a few minutes each) and finally hold the temperature at 37 °C. Place some filter paper in the chamber and moisten it with distilled water.
2. Place the slides with hybridization mixture (from *Protocol 5*) in the moist chamber and start the machine.
3. After the program has reached the hybridization temperature of 37 °C, slides can be left in the machine or can be transferred to an incubator or water bath at 37 °C. In all cases make sure slides are in a moist chamber to prevent drying out. Incubate overnight.

B. *Floating dish method*

1. Prepare a humid chamber by placing filter paper in the bottom of a metal tin with a lid and soak with distilled water. Place in a water bath to equilibrate the internal temperature to 85–90 °C. This might take about 30 min. Temperature is best monitored by placing the external probe of a digital thermometer in the chamber.
2. Quickly place the slides from (*Protocol 5*) in the preheated humid chamber and incubate for 5–10 min. Monitor temperature carefully. It will take a few minutes for the temperature to rise again: make sure at least two-thirds of the incubation time is at the target temperature (e.g. 85 °C).
3. Transfer humid chamber to 37 °C incubator or water bath and leave slides to hybridize overnight.

5.4 Post-hybridization washes

After hybridization overnight the slides are taken through several washing steps to remove non-specifically and weakly bound probe. They include a stringent formamide wash, generally a few percent more stringent than during hybridization, after which the slides are treated with 2 × SCC to remove the formamide. *Preparations must not dry out at any stage.*

Protocol 7. Post-hybridization washes

- formamide: good but not the highest grade (e.g. Sigma catalogue number F 7508); store in aliquots at −20 °C

Protocol 7. *Continued*

- 20 × SSC stock: 3 M NaCl, 0.3 M sodium citrate adjusted to pH 7

1. Prepare 500 ml 2 × SCC, 200 ml 0.1 × SSC, and 200 ml 20% formamide in 0.1 × SSC (stringent formamide wash) and heat to 42 °C.
2. Remove slides from the moist chamber (*Protocol 6*), carefully remove coverslip, and place slides in a Coplin jar containing 2 × SSC at 42 °C.
3. Pour off 2 × SSC and replace with stringent formamide wash. Incubate for 2 × 5 min at 42 °C shaking gently.
4. Wash slides in 0.1 × SSC for 2 × 5 min at 42 °C and then in 2 ×SSC for 2 × 5 min at 42 °C.
5. Allow to cool down to room temperature.

5.5 Fluorescent detection of probe hybridization sites

Table 1 lists the different fluorochromes used as conjugates to nucleotides or antibodies and as DNA stains for chromosomes. Fluorochromes are molecules that are excited by light at a certain wavelength and then emit light

Table 1. Properties of fluorochromes used in signal-generating systems for *in situ* hybridization and as stains for chromosomal DNA

Fluorochrome	Excitation (max. nm)	Emission (max. nm)	Fluorescence colour
(a) Conjugated to detecting and antibody molecules			
Coumarin (AMCA)[a]	350	450	blue
Fluorescein (FITC)[b]	495	515	green
R-phycoerythrin	525	575	orange-red
Rhodamine	550	575	orange-red
$Rhodamine_{600}$ (TRITC)[c]	575	600	red
Texas Red	595	615	red
Ultralite 680	625	680	infra-red
(b) Conjugated directly to nucleotides			
Coumarin (AMCA)[a]	350	450	blue
Fluorescein (FITC)[b]	495	515	green
$Rhodamine_{600}$ (TRITC)[c]	575	600	red
(c) DNA stains			
DAPI[d]	355	450	blue
Hoechst 33258	356	465	blue
Chromomycin A3	430	570	yellow
Quinacrine	455	495	yellow-green
Propidium iodide	340, 530	615	red

[a] 7-amino-4-methyl-coumarin-3-acetic acid
[b] fluorescein isothiocyanate
[c] tetramethyl rhodamine isothiocyanate
[d] 4′,6-diamidino-2-phenylindole

of longer wavelength (of lower energy). The most common types have characteristics of UV excitation (blue fluorescence), blue excitation (yellow/green fluorescence), and green excitation (red fluorescence); using suitable imaging systems, red-excited infra-red emitting dyes such as Ultralite 680 are now being used. Epifluorescence microscopes use lamps and filter blocks to select light of the correct wavelength for excitation and emission. Where several fluorochromes are used simultaneously (*Figure 1*), each needs to have a different pair of excitation and emission maxima. Digoxigenin- or biotin-labelled probes need to be detected by immunocytochemistry to attach marker molecules for visualization (*Protocol 8*). For direct fluorochrome-labelled probes this step is not needed and preparations can be counterstained immediately (Section 5.6, *Protocol 9*).

Protocol 8. Detection of probe hybridization sites

- 20 × SSC stock: 3 M NaCl, 0.3 M sodium citrate adjusted to pH 7
- Tween 20
- bovine serum albumin (BSA; e.g. Sigma Fraction V, globulin-free)
- for digoxigenin-labelled probes: anti-digoxigenin conjugated to fluorescein or rhodamine raised in sheep (Boehringer Mannheim)
- for biotin-labelled probes: avidin conjugated to fluorescein, Texas Red, rhodamine, or coumarin (e.g. Vector Laboratories)
- plastic coverslips: pieces of appropriate size cut from autoclavable waste disposal bags

1. Make up 1 litre detection buffer by diluting 20 × SSC to 4 ×SSC and adding 0.2% (v/v) Tween 20; heat to 37 °C.
2. Take slides from *Protocol 7* and incubate in detection buffer for 5 min.
3. Make up 5% BSA block in detection buffer.
4. Add 200 μl of BSA block to each slide, cover with a plastic coverslip, and incubate for 5 min at room temperature.
5. Make up 10–20 μg/ml anti-digoxigenin and/or avidin in BSA block.
6. Remove coverslip, drain slide, and add 30 μl of the anti-digoxigenin and/or avidin solution. Cover and incubate slides in a humid chamber for 1 h at 37 °C.
7. Wash slides in detection buffer for 3 × 8 min at 37 °C.

5.6 Counterstaining of chromosomal DNA

All steps are carried out at room temperature. Depending on the fluorochromes attached to the probe (or the detection reagents), select appropriate counterstains (see *Table 1*). They enable easy screening and identification of chromosomes and nuclei. Several DNA stains can also be applied together to produce specific bands (54). Use DAPI for green (fluorescein) and red (rhodamine or Texas red) fluorescing probes, but not for blue (coumarin) fluorescing probes. Fluorescein-detected *in situ* hybridization signal can be counterstained with propidium iodide (PI), which fluoresces red under many excitation wavelengths; the fluorescein signal will appear yellow as a result of overlapping red and green fluorescence. It is important to avoid overstaining with counterstain, since PI in particular can obscure weak hybridization sites.

Protocol 9. Counterstaining chromosomal DNA

- 0.2 M Na_2HPO_4
- 0.1 M citric acid
- PBS (phosphate buffered saline): 0.13 M NaCl, 0.007 M Na_2HPO_4, 0.003 M NaH_2PO_4, adjust to pH 7.4 (can be bought ready-made)
- DAPI (4′,6-diamidino-2-phenylindole)
- PI (propidium iodide)
- antifade solution: AF1 (Citifluor Ltd) or Vectashield mounting medium (Vector Laboratories)
- glass coverslips: no. 0, 24 mm × 40 mm or 24 × 50 mm

1. Make McIlvaine's buffer (pH 7.0) by mixing 18 ml of Na_2HPO_4 and 82 ml citric acid.
2. Prepare DAPI stock solution of 100 μg/ml in water. DAPI is a potential carcinogen. To avoid weighing out the powder, order small quantities and use the whole vial to make the stock solution. Aliquot and store at −20 °C (it is stable for years). Prepare a working solution of 2 μg/ml by dilution in McIlvaine's buffer, aliquot, and store at −20 °C.
3. Add 100 μl of DAPI per slide, cover with a plastic coverslip, and incubate for 10 min. Wash briefly in PBS and drain.
4. Prepare PI stock solution of 100 μg/ml in water. PI is a potential carcinogen. To avoid weighing out the powder, order small quantities and use the whole vial to make the stock solution. Aliquot 50 μl or 100 μl in 1.5 ml microcentrifuge tubes and store at −20 °C. Dilute with PBS to 2–5 μg/ml prior to use. PI does not keep in diluted form.
5. Add 100 μl of PI, cover with a plastic coverslip and incubate for 10 min. Wash briefly in PBS and drain the slide.
6. Apply one or two drops of anti-fade on to the wet slide and cover with a glass coverslip. Firmly squeeze excess anti-fade from the slide with filter paper.

7. Slides can be viewed immediately, but the signal stabilizes after storing for a few days in the dark at 4 °C. Slides can be kept at 4 °C for up to a year.

6. Visualization of hybridization sites

6.1 Epifluorescence light microscopy

A good epifluorescence microscope is required to view fluorescent hybridization sites. For descriptions of the fluorescence microscope see Chapters 1 and 10. We use a Leitz Aristoplan microscope with a 100 W HBO mercury lamp and appropriate filter blocks for the fluorochromes being used. It is important to have UV-transparent, high numerical aperture objectives: we use a 25× oil objective for scanning the slides and a 100× objective for photography. The confocal microscope is used routinely by several laboratories for imaging two-dimensional material such as chromosome spreads which carry *in situ* hybridization signal (55, 56).

Fluorescent *in situ* hybridization signals fade easily under high energy excitation, particular UV. To prevent rapid fading, the preparations are mounted in anti-fade (see *Protocol 9*) and should be stored in the fridge before viewing. It is also important to view a particular field only briefly before taking a picture, and to photograph with the longer (green or blue) excitation wavelengths first (normally, the *in situ* hybridization signal before DAPI counterstain). One or two exposures of several seconds each are possible without significant degradation of the image.

We normally record our *in situ* hybridization results on 400 or 1000 ASA, 35 mm colour print film. Colour print films are easy to use, since their exposure latitude is wide and adjustment of contrast and colour is possible during printing. We recommend Fujicolor 400 and Fujichrome 100 films. Other suitable films include Kodak TMAX 400 for black and white photography (*Figure 1*) and Agfa and Fuji high speed slide films. Of course, the resolution and bandwidth of 35 mm film is far in excess of that realistically obtainable by digital imaging systems.

6.2 Quantification and image processing

In situ hybridization, like most Southern hybridization, is not a truly quantitative technique, since many unknown and uncontrollable factors influence hybridization including probe and detection reagent penetration, uniformity of chromosomal DNA denaturation, and specificity of antibody labelling. Nevertheless, the technique is valuable for determining relative copy numbers of repeat units where the repeat occurs on several chromosomes (3, 16).

With the development of sensitive and affordable video camera systems, digital imaging techniques are increasingly used to supplement conventional analysis and photography (57). Advantages of digital analysis are that

- very weak fluorescent signal can be documented more readily
- the potential of image processing greatly facilitates the analysis of signal, as optical filtering techniques, such as thresholding and background subtractions, enhance the signal-to-noise ratio
- quantitative data on signal intensities or measurements of distances between signals and other markers can be easily made.

It is important to employ image processing techniques cautiously. Information and resolution can be easily lost, the wrong signal can be amplified, and the ratio of signals can be distorted. Unfortunately, reproduction of original photographs in journals is not always of high quality and has led to the publication of enhanced images which are no nearer the raw data than drawings.

6.3 Reconstructions

Nuclei can be reconstructed from both physical and optical sections. The most straightforward method for reconstruction is the use of acetate films. Photomicrographs can be taken of physical sections of nuclei with *in situ* hybridization signal and the photographs used to trace specific outlines on to an acetate film. The acetate films of all nuclear sections are then aligned and stacked one on top of the other. This enables fast examination of the three-dimensional relationship of nuclear components and *in situ* hybridization signal and compensation for overlapped, distorted, and damaged sections is straightforward. Alternatively, each section of the nucleus can be separately digitized on a microcomputer; the nuclei can then be reconstructed and rotated to give different viewing angles to best illustrate particular structural features (6).

Serial optical sections (e.g. from a data set acquired using a confocal microscope) can also be used to reconstruct nuclei with *in situ* hybridization signal (see Chapter 1). Each optical section can be superimposed on a computer monitor and the three-dimensional coordinates of all pixels rotated to give a three-dimensional perspective of the nuclei.

Acknowledgements

We acknowledge Drs I. J. Leitch, K. Anamthawat-Jónsson, J. Maluszynska, and W. Mosgöller for sharing their experience and protocols, and Mrs M. Shi for technical assistance. We thank BP Venture Research for support of this work.

References

1. Maluszynska, J. and Heslop-Harrison, J. S. (1991). *Plant J.*, **1**, 159.
2. Mukai, Y., Endo, T. R., and Gill, B. S. (1991). *Chromosoma*, **100**, 71.
3. Leitch, I. J. and Heslop-Harrison, J. S. (1992). *Genome*, **35**, 1013.
4. Moore, G., Cheung, W., Schwarzacher, T., and Flavell, R. (1991). *Genomics*, **10**, 469.
5. Martinez-Zapater, J. M., Estelle, M. A., and Somerville, C. R. (1986). *Mol. Gen. Genet.*, **204**, 417.
6. Leitch, A. R., Schwarzacher, T., Mosgöller, W., Bennett, M. D., and Heslop-Harrison, J. S. (1991). *Chromosoma*, **101**, 206.
7. Anamthawat-Jónsson, K., Schwarzacher, T., and Heslop-Harrison, J. S. (1993). *J. Hered.*, **84**, 78.
8. Leitch, A. R., Mosgöller, W., Shi, M., and Heslop-Harrison, J. S. (1992). *J. Cell Sci.*, **101**, 751.
9. Cremer, T., Remm, B., Kharboush, I., Jauch, A., Wienberg, J., Stelzer, E., and Cremer, C. (1991). *Rev. Eur. Technol. Biomed.*, **13**, 50.
10. Dekken, H. and Bauman, J. G. J. (1990). *Cytometry*, **11**, 579.
11. Ambros, P. F., Matzke, M. A., and Matzke, A. J. M. (1986). *Chromosoma*, **94**, 11.
12. de Pace, C., Delvre, V., Scarascia Mugnozza, G. T., Maggini, F., Cremonini, R., Frediani, M. *et al.* (1991). *Theor. Appl. Genet.*, **83**, 17.
13. Huang, P., Hahlbrock, K., and Somssich, I. E. (1988). *Mol. Gen. Genet.*, **211**, 143.
14. Mouras, A., Negrutiu, I., Horth, M., and Jacobs, M. (1989). *Plant Physiol. Biochem.*, **27**, 161.
15. Gustafson, J. P., Butler, E., and McIntyre, C. L. (1990). *Proc. Natl Acad. Sci. USA*, **87**, 1899.
16. Leitch, I. J. and Heslop-Harrison, J. S. (1993). *Genome*, **36**, 517.
17. Zaki, S. R., Austin, G. E., Chan, W. C., Conaty, A. L., Trusler, S., Trappier, S., Lindsey, R. B., and Swan, D. C. (1990). *Genes Chromosomes Cancer*, **2**, 266.
18. Koch, J. E., Kolvraa, S., Petersen, K. B., Gregersen, N., and Bolund, L. (1989). *Chromosoma*, **98**, 259.
19. Schwarzacher, T. and Heslop-Harrison, J. S. (1991). *Genome*, **34**, 317.
20. Moyzis, R. K., Buckingham, J. M., Cram, L. S., Dani, M., Deaven, L. L., Jones, M. D., Meyne, J., Ratliff, R. L., and Wu, J.-R. A. (1988). *Proc. Natl Acad. Sci. USA*, **85**, 6622.
21. Meyne, J. and Moyzis, R. K. (1989). *Genomics*, **4**, 472.
22. Koch, J., Mogensen, J., Pedersen, S., Fischer, H., Hindkjær, J., Kolvraa, S., and Bolund, L. (1992). *Cytogenet. Cell Genet.*, **60**, 1.
23. Koch, J., Hindkjaer, J., Mogensen, J., Kolvraa, S., and Bolund, L. (1991). *Genet. Anal. Tech. Appl.*, **8**, 171.
24. Schwarzacher, T., Leitch, A. R., Bennett, M. D., and Heslop-Harrison, J. S. (1989). *Ann. Bot.*, **64**, 315.
25. Heslop-Harrison, J. S., Leitch, A. R., Schwarzacher, T., and Anamthawat-Jónsson, K. (1990). *Heredity*, **65**, 385.
26. Mukai, Y. and Gill, B. S. (1991). *Genome*, **34**, 448.

27. Schwarzacher, T., Anamthawat-Jónsson, K., Harrison, G. E., Islam, A. K. M. R., Jia, J. Z., King, I. P., *et al.* (1992). *Theor. Appl. Genet.*, **84**, 778–786.
28. Anamthawat-Jónsson, K., Schwarzacher, T., Leitch, A. R., Bennett, M. D., and Heslop-Harrison, J. S. (1990). *Theor. Appl. Genet.*, **79**, 721.
29. Schardin, M., Cremer, T., Hager, H. D., and Lang, M. (1985). *Hum. Genet.*, **71**, 281.
30. Pinkel, D., Straume, T., and Gray, J. W. (1986). *Proc. Natl Acad. Sci. USA*, **83**, 2934.
31. Lichter, P., Cremer, T., Borden, J., Manuelidis, L., and Ward, D. C. (1988). *Hum. Genet.*, **80**, 224.
32. Pinkel, D., Landegent, J., Collins, C., Fuscoe, J., Segraves, R., Lucas, J., and Gray, J. (1988). *Proc. Natl Acad. Sci. USA*, **85**, 9138.
33. Wienberg, J., Jauch, A., Stanyon, R., and Cremer, T. (1990). *Genomics*, **8**, 341.
34. Carter, N. P., Ferguson-Smith, M. A., Perryman, M. T., Telenius, H., Pelmear, A. H., Leversha, M. A., *et al.*, (1992). *J. Med Genet.*, **29**, 299.
35. Cotter, F. E., Das, S., Douek, E., Carter, N. P., and Young, B. D. (1991). *Genomics*, **9**, 473.
36. Wang, M. L., Leitch, A. R., Schwarzacher, T., Heslop-Harrison, J. S., and Moore, G. (1992). *Nucleic Acids Res.*, **20**, 1897.
37. Schubert, I., Dolezel, J., Houben, A., Scherthan, H., and Wanner, G. (1993). *Chromosoma*, **102**, 96.
38. Schwarzacher, T., Ambros, P., and Schweizer, D. (1980). *Plant Syst. Evol.*, **134**, 293.
39. Mouras, A., Saul, M. W., Essad, S., and Potrykus, I. (1987). *Mol. Gen. Genet.*, **207**, 204.
40. Murata, M. (1983). *Stain Techn.*, **58**, 101.
41. Emmerich, P., Jauch, A., Hofmann, M.-C., Cremer, T., and Walt, H. (1989). *Lab. Invest.*, **61**, 235.
42. Pardue, M. L. and Gall, J. G. (1969). *Proc. Natl Acad. Sci. USA*, **64**, 600.
43. John, H. A., Birnstiel, M. L., and Jones, K. W. (1969). *Nature*, **223**, 582.
44. Ferguson-Smith, M. A. (1991). *Am. J. Hum. Genet.*, **48**, 179.
45. Harper, M. E. and Saunders, G. F. (1981). *Chromosoma*, **83**, 431.
46. Leitch, I. J., Leitch, A. R., and Heslop-Harrison, J. S. (1991). *Genome*, **34**, 329.
47. Ried, T., Baldini, A., Rand, T. C., and Ward, D. C. (1992). *Proc. Natl Acad. Sci. USA*, **89**, 1388.
48. Dauwerse, J. G., Wiegant, J., Raap, A. K., Breuning, M. H., and van Ommen, G. J. B. (1992). *Hum. Mol. Genet.*, **8**, 593.
49. Lichter, P. and Ward, D. C. (1990). *Nature*, **345**, 93.
50. Trask, B., Feritta, A., Christensen, M., Youngblom, J., Bergmann, A., Copeland, A., *et al.* (1993). *Genomics*, **15**, 133.
51. Heslop-Harrison, J. S., Harrison, G. E., and Leitch, I. J. (1992). *Trends Genet.*, **8**, 372.
52. Raap, A. K., Marijnen, J. G. J., and van der Ploeg, M. (1986). *Cytometry*, **7**, 235.
53. Heslop-Harrison, J. S., Schwarzacher, T., Anamthawat-Jónsson, K., Leitch, A. R., Shi, M., and Leitch, I. J. (1991). *Technique*, **3**, 109.
54. Schweizer, D. (1981). *Hum. Genet.*, **57**, 1.
55. Albertson, D. G., Sherrington, P., and Vaudin, M. (1991). *Genomics*, **10**, 143.

56. Boyle, A. L., Gwyn Ballard, S., and Ward, D. C. (1990). *Proc. Natl Acad. Sci. USA*, **87**, 7757.
57. Lichter, P., Boyle, A. L., Cremer, T., and Ward, D. C. (1991). *Genet. Anal. Tech. Appl.*, **8**, 24.

7

Immunocytochemistry for light and electron microscopy

N. HARRIS

1. Introduction

Immunohistochemical and immunocytochemical techniques became increasingly popular during the 1980s with the wider availability of, firstly, specific polyclonal and, later, monoclonal antibodies. The development and increasing range of experimental options, and details of their use, have been described elsewhere (1–3) and a more exhaustive range of protocols is given in another volume in this series (4). The technique is particularly valuable in association with biochemical and physiological studies since it illustrates the temporal and spatial patterns of accumulation of specific proteins during cellular differentiation and tissue development. In conjunction with localization of mRNAs it can be used to determine the patterns of regulation of gene expression and, with molecular analyses, makes a valuable contribution to our understanding of the regulation of gene expression. Used at an ultrastructural level the technique also gives detail of the distribution of antigens and the steps involved in their processing and intracellular transport; volumes dedicated to electron microscopic immunolocalization include reference 5.

The general procedure for immunolocalization involves: preparation of suitable antibodies, specimen preparation, incubation of primary antibodies with specimen, removal of non-specifically bound antibodies and localization of specifically bound antibodies. An outline of the various options and strategies for immunolocalization by light and electron microscopy is given in *Figure 1*.

2. Antibodies

2.1 Immunoglobulins

Antibodies are immunoglobulin proteins, of which there are five classes: IgA, IgD, IgE, IgG, and IgM. Each immunoglobulin is formed from two identical

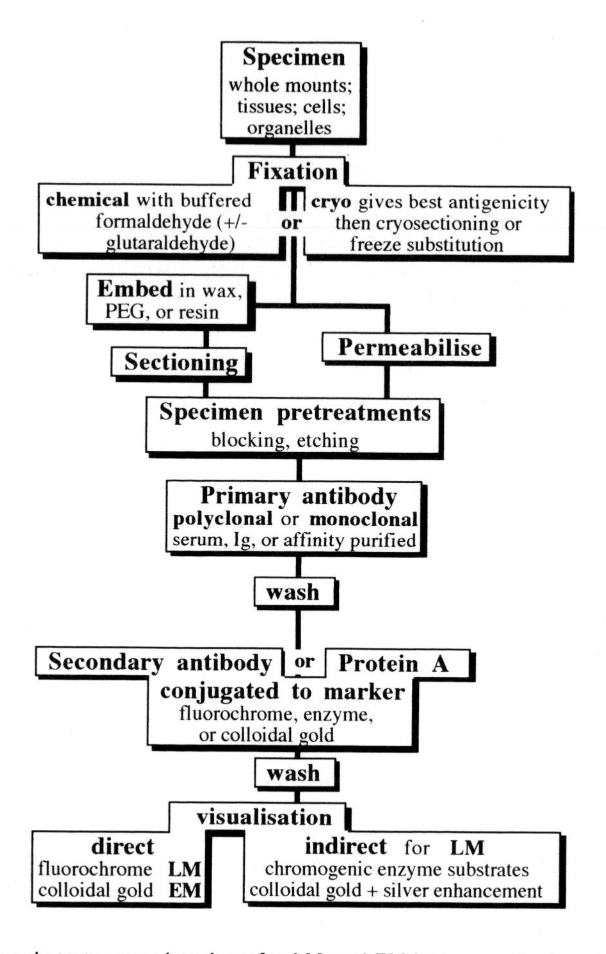

Figure 1. General strategy and options for LM and EM immunocytochemistry.

heavy chains and two identical light chains, with disulphide bridges linking within and between chains. Within the chains there are constant, variable, and hypervariable domains. The molecule can be cleaved to smaller components. Cleavage with papain gives two monovalent, heterogeneous antigen binding fragments (Fab fragments) and two homogeneous crystallizable fragments (Fc). Cleavage with pepsin gives bivalent F(ab′)2 fragments. Although IgG antibodies are generally the most common within serum, the first antibodies to be formed following immunization are IgMs; these contain five subunits, each with two light and two heavy chains, which can be cleaved to Fab and F(ab′)2 fragments. IgGs and IgMs and their fragments are most commonly used for immunolocalization. Immunoglobulins account for about

10% of serum proteins, and between 3 and 10% of the immunoglobulins result from the injected antigen.

2.2 Polyclonal antibodies

Polyclonal antibodies for immunolocalization are produced following immunization of animals such as rabbit, mouse, sheep, goat, etc. Between 10 and 200 μg antigen are required to generate an immune response, with boosters required when the titre drops (after 4–6 weeks). Antibodies are obtained in serum, after removal of blood cells, or may be fractionated by various methods. The mixed population of antibodies generated by numerous different cells (polyclonal antibodies) reacts to a variety of determinants, which may include several on each antigenic molecule.

Protocol 1. Fractionation of IgGs from serum[a]

- primary serum
- ammonium sulphate (1.75 M and solid)
- 17.5 mM sodium dihydrogen phosphate, pH 6.3
- column of DEAE–cellulose
- sodium azide
- bench centrifuge
- dialysis tubing
- spectrophotometer

1. Collected blood may be cooled to 4 °C and treated immediately, or allowed to clot: if a clot is formed this is broken up with wooden applicator sticks prior to centrifugation.
2. Centrifuge at 5000 *g* and collect the yellow to pale pink supernatant (to remove whole cells): a cloudy layer with fat may be seen at the top of the supernatant; this may be avoided by starving animals for 24 h prior to bleeding, or not collected with the rest of the supernatant.
3. Complement (the non-immunoglobulin component of serum) is treated by heating at 56 °C for 10–20 min.
4. Decomplemented serum may be stored for later immuno studies, or used for IgG fractionation.
5. Ammonium sulphate fractionation (all steps at 4 °C unless stated otherwise): add ammonium sulphate to serum to 45% saturation; stir for 30 min; centrifuge at 30 000 *g* for 15 min; discard supernatant; wash pellet with 1.75 M ammonium sulphate (at 0 °C) until pellet is white; dissolve in 17.5 mM NaH_2PO_4 pH 6.3 and dialyse against same (optional: centrifuge at 10 000 *g* to remove lipoproteins).

Protocol 1. *Continued*

6. DEAE–cellulose chromatography: load antibody solution onto a column of DEAE–cellulose previously equilibrated with the buffer (1 vol. antibody solution to 4 vol. DEAE–cellulose); wash with 2 × column volume of buffer and collect fractions; eluted IgG in fractions is monitored by absorbance at 280 nm; pool appropriate fractions and treat as in 5.
7. Store samples after freezing or lyophilizing: for short-term storage add sodium azide to 0.02%.

OR

- protein A–Sepharose (Sigma) column
- 10 mM sodium phosphate buffer pH 8.0, 0.15 M sodium chloride, 1 mM EDTA
- 150 mM sodium chloride, 0.58% (v/v) acetic acid, pH 2.7
- saturated aqueous Tris
- sodium azide

1–4. As above.

5. Prepare column (1 column volume to 2.5 volumes of serum): equilibrate protein A–Sepharose in 10 mM sodium phosphate buffer pH 8.0, 0.15 M NaCl, 1 mM EDTA.
6. Add unpurified serum to column and rinse with several volumes of buffer (until A_{280} of eluant is close to zero).
7. Elute IgG with 150 mM NaCl, 0.58% (v/v) acetic acid pH 2.7 and collect fractions in tubes containing saturated Tris to a final pH of 7.0.
8. Store samples after freezing or lyophilizing: for short-term storage add sodium azide to 0.02%.

[a] Fractionation of polyclonal IgGs may not always give the anticipated advantages for immunocytochemistry since the non-specific IgGs, which contribute to background, are also increasing in titre.

Various other methods for antibody purification, including the use of commercially prepared small columns and different forms of affinity columns, may be used to clean polyclonal samples. Practical and theoretical aspects of such methods are detailed elsewhere within the *Practical Approach* series (6, 7).

2.3 Monoclonal antibodies

Monoclonal antibodies are produced by immortalized cell lines generated by the fusion of antibody-secreting cells with non-secreting myeloma cells. Mice or rats are widely used for the primary immunization, with B lymphocytes from the spleen or lymph nodes used as the secreting cells. The hybridomas are usually maintained in culture and the antibodies obtained in the supernatant fluid. Each hybridoma produces a specific, monoclonal antibody. Although there are some advantages in using monoclonal antibodies, such as absence of non-specific antibodies and no variation within or between batches, there are some associated disadvantages. A monoclonal antibody should recognize only one form of epitope and it may be important that screening of monoclonal antibodies, for immunolocalization studies, is carried out using not the native antigen used to raise the monoclonal antibody but antigen that has been treated (e.g. fixed) as in tissue preparation, since fixation may alter the conformation of the target epitope. Details of the production and screening of monoclonal antibodies are available elsewhere (8, 9).

2.4 Storage and use of antibodies

Diluted antibodies are usually stored at 4 °C. It is preferable to store concentrated antibodies at −20 °C; use aliquots to prevent damage from repeated cycles of thawing and refreezing. Antibodies should be stored in containers with low protein absorption (borosilicate glass, polypropylene, or polycarbonate); bovine serum albumin (at 0.1–1.0%) may be added to reduce losses due to polymerization and absorption.

2.4.1 Titre and dilution

An antibody titre is the highest dilution which gives optimum specific staining. Fractionation of IgGs to give higher titres may not be of particular value with all polyclonal samples since non-specific antibodies may also be concentrated.

Typical working dilutions for immunocytochemistry are 1:100 to 1:2000 for polyclonal antibodies and 1:10 to 1:100 for monoclonal antibodies from cell culture supernatants. Monoclonal antibodies in ascites fluid may be used in dilutions up to 1:1 000 000.

To determine the optimum dilution for an antibody and the reagent(s) used in its localization a simple 'grid' experiment is carried out in which three or four dilutions of the primary antibody (e.g. 1:50; 1:100, and 1:250 for a polyclonal) and used with each of a series of dilutions of the localizing reagents (e.g. complex of secondary antibody with fluorochrome or enzyme/substrate combination).

2.4.2 Incubation conditions

As well as being dependent upon antibody titre and dilution, the period and temperature of incubation are also critical to the level of immunostaining. There is an inverse relationship between the incubation time required and the titre of the antibody. High titres may be used for incubations of only a few minutes whilst low titres may require incubation up to 24 h; antigen saturation in the presence of excess antibody takes approximately 15–20 min, and in practice titres requiring an incubation of 30–60 min at room temperature, or overnight at 4 °C, are often used. The lower titres used for overnight incubations may be valuable where a polyclonal antibody is only available in limited amounts.

3. Specimen preparation

3.1 For embedding in wax or polyethylene glycol

Immunolocalization studies may be carried out with permeabilized whole mounts, with isolated organelles, or with sections of tissues. The localization of cytoskeletal components in whole mounts is described in detail in Chapter 10, and the immunolocalization of hapten probes used for non-isotopic *in situ* hybridization is described in Chapters 5 and 6. This chapter is concerned with immunolocalization in tissue sections, at the histological, cytological, and ultrastructural levels.

Tissue may be fixed chemically or by freezing; the former is more common as cryofixation can often lead to problems with retention of structural integrity in highly vacuolate plant cells. There are, however, some examples where cryofixation has been used very successfully at the light microscopy (LM) level (see Chapter 5) and at the EM level (10). Embedding and sectioning are, however, generally preceded by chemical fixation. Buffered aldehydes act as a suitable primary fixative, although with a higher proportion (2.5–4%) of the monovalent aldehyde formaldehyde being included in comparison with fixatives used for conventional structural studies (Chapters 3 and 4). The bivalent fixative glutaraldehyde may be used to stabilize structures but at lower concentrations (0.5–1.5%) than in conventional fixatives, or it may be omitted completely. A variety of buffers have been used (4) although phosphate or Tris buffers are used most commonly, including TBS (Tris buffered saline).

Wax is a classical embedding medium, although more defined forms such as Paraplast are now used routinely; polyethylene glycol (PEG) 1000 has also been used very successfully as an embedding medium, its major advantage being a greater retention of cytological integrity. PEG is water-soluble but tissues are often dehydrated prior to infiltration and embedding since this may give better sectioning qualities.

Sections for immunohistochemistry are cut from wax or PEG-embedded tissue at 8–12 μm and mounted on to coated slides. Wax sections may be floated, as a ribbon, on warm water to remove wrinkles prior to transfer to slides; PEG sections are transferred directly to slides which have only a minimum film of water to aid attachment. PEG is water-soluble and sections lose their integrity if floated on water; PEG is suitable as a medium for intact tissues with good cell–cell contacts, but wax is preferable for sections that include several independent components within an ordered spatial arangement e.g. transverse section of a flower bud.

Slides may be coated with a conventional 'subbing' solution (1% gelatin, 0.1% chrome alum (chromium potassium sulphate dodecahydrate) dissolved in warm water), with poly-L-lysine (Chapter 5, *Protocol 1*), or with TESPA (Chapter 3, Section 1.1). TESPA-coated slides are used to collect sections for parallel *in situ* hybridization and immunocytochemical studies.

3.2 For embedding in resin

Both epoxy- and acrylic-based resins have been used in immunocytochemical studies. The more hydrophilic acrylic resins are now used most frequently. Widely available commercial resins include Lowicryl K4M (Agar, Polyscience), LR White, and LR Gold (London Resin Company). Lowicryl was developed for immunocytochemical use and can be polymerized by incorporating a UV-absorbing catalyst and illuminating with 360 nm light. The reaction is, however, exothermic and the polymerization should be carried out within a cold chamber. Although often suitable for animal tissues, Lowicryl has developed a reputation for being less reliable with plant tissues, although some laboratories have obtained good results. A frequently used convenient alternative is LR White; this can be cured in an oven at the relatively low temperature of 50 °C; polymerization initiated by catalyst is not recommended for immunological work.

4. Staining methods

A wide variety of staining methods is used to localize specifically bound primary antibodies, involving either direct or increasingly complex arrangements to couple a 'marker' to the primary antibody. A range of 'markers' is available, each with various advantages and disadvantages.

4.1 Direct method

The direct method involves use of a fluorochrome- or enzyme-labelled primary antibody. Since the method is relatively straightforward it can be completed quite quickly, and non-specific reactions are largely related to the quality of the primary antibody. The method requires labelling of each primary antibody, and does not allow for amplification of any signal; it is now used only rather rarely.

4.2 Indirect method

Indirect methods are commonly used; specifically bound but unlabelled primary antibody is localized by a labelled secondary directed against the primary, or by a protein A–label conjugate. This approach is generally more sensitive than direct labelling and more versatile, since a range of primary antibodies raised in the same species can be localized in different samples using the same secondary. The indirect method allows for amplification of the signal.

4.2.1 Two-step indirect method

The two-step indirect method utilizes a secondary antibody conjugated to an enzyme which, when incubated with appropriate substrates, will give a coloured reaction precipitate marking the distribution of the primary antigen. The method is versatile since the secondary conjugate and substrates can be used with a range of different primary antibodies, and the signal can be enhanced by use of longer enzyme/substrate incubations. Controls must examine non-specific binding of the primary, the secondary conjugate, and any endogenous activity against the substrates.

The *Staphylococcus* wall protein, protein A, binds to immunoglobulins from various species and may be used in place of the secondary antibody for immunocytochemistry in light and electron microscopy (11). Protein A binds strongly to rabbit immunoglobulins, but has a much lower affinity for those from goat, rat, and mouse; where these species provide the primary antibody, a secondary antibody or protein G is preferable to protein A.

4.2.2 Three-step indirect, and soluble complex, methods

Further amplification of the signal may be obtained by use of a three-step indirect method in which the incubation with the primary and enzyme-conjugated secondary antibodies are followed by incubation with an enzyme conjugated to an antibody to the secondary antibody. The same enzyme is used in both secondary and tertiary steps.

An alternative to the three-step method is the use of a soluble complex. The complex may be either a pre-formed enzyme–anti-enzyme or an avidin–biotin–enzyme complex. The primary antibody is raised in the same species as the antibody to the enzyme and an unlabelled secondary antibody is used as a link between the primary antibody and the complex. Methods involving soluble complexes are usually very sensitive and good for immunostaining fixed, wax-embedded samples.

Avidin–biotin methods are based on the high binding affinity between avidin, or streptavidin, and biotin. Biotin may be labelled with an enzyme and complexed with avidin, or avidin may be labelled directly. The complex is used after incubation with a biotinylated secondary, and the enzyme (peroxidase or alkaline phosphatase) is used to develop a stained reaction product.

The problem of endogenous biotin may be overcome by blocking (0.1% avidin for 20 min, 0.01% biotin for 20 min) prior to incubation with the primary antibody, or by use of digoxigenin-based methods. Avidin may bind to endogenous lectins; use of streptavidin, which is not glycosylated, avoids this problem.

4.3 'Markers'

The most widely used markers for immunolocalization include fluorochromes, enzymes, and colloidal gold. Fluorochromes are used in LM studies, enzymes are used predominantly in LM studies but have also been used in EM studies, and colloidal gold is used both for EM studies, where it has largely superceded ferritin as an electron-dense marker, and in LM studies where the small gold particles are enlarged by silver enhancement.

4.3.1 Fluorochromes

The range and some specific uses of fluorochromes for immunolocalization are detailed in Chapters 6 and 10. The protocols presented in these chapters are readily adaptable to other primary antibodies. Fluorochrome-based methods are quick and sensitive, but they require specialist microscopy facilities and the product is not permanent.

4.3.2 Soluble complexes

Commercially available enzyme–immune complexes include PAP (peroxidase–anti-peroxidase and APAAP (alkaline phosphate–anti-alkaline phosphatase); APAAP is used with tissues containing high levels of endogenous peroxidase. The avidin–biotin complex also include the same enzymes. Suppliers include Dako, Sigma, Vector Laboratories, and Boehringer.

The most widely used substrate for peroxidase is 3,3′-diaminobenzidene tetrahydrochloride (DAB); this produces a brown reaction precipitate which is insoluble in organic solvents. The brown colour can be intensified if cobalt ions are added to the medium. The reaction product can be osmicated, allowing visualization by EM. Cyanide and azide are reversible inhibitors used in controls.

Calf intestinal alkaline phosphatase used in immunohistochemistry may be localized with the substrate combinations NBT (indoxyl-tetranitro BT) and BCIP (5-bromo-4-chloro-3-indolyl phosphate), which give a blue reaction product, or naphthol AS-MX phosphate and Fast Red TR which give a red precipitate. The former blue product is not soluble in organic solvents, but the latter is and thus requires aqueous mounting or drying of sections before permanent mounting; the red colour is, however, more readily recorded photographically.

4.3.3 Colloidal gold

Colloidal gold (BioCell, Sigma) has proved extremely valuable in EM immunolocalization studies and is being used increasingly for LM studies after silver intensification/enhancement of the specifically bound colloid. Commercially available colloids range in size from 1 nm to 30 nm, and are available as stabilized colloid, for binding of the experimenter's own protein(s), or conjugated to a wide range of (secondary) antibodies, protein A, lectins, and streptavidin. Several colloidal gold particles may aggregate on a single primary immunoglobulin, giving a clustered appearance on EM sections; this may be avoided by use of protein A conjugates or by greater dilution of the secondary antibodies–gold conjugate.

Choice of colloid size is dependent upon the likely magnification at which the specimen is to be examined; for LM smaller (5 or 1 nm) colloid sizes are used as a base for silver enhancement, but for low magnification (<20 000×) EM studies larger (15 or 20 nm) colloids are used. As EM magnifications are increased so smaller colloids are more readily visible within specimen and micrographs. The density of colloid binding increases with decreasing size of colloid.

Monodispersive colloids, which have a very low variation in particle diameter and are essential to double labelling experiments, are quite expensive to purchase, but where a small range in diameter is acceptable some sizes can be made quite easily in the laboratory (see *Protocol 2*) Detailed protocols for preparation of different sizes of gold sols are given in references 12 and 13.

Protocol 2. Preparation of 12 nm colloidal gold, and antibody–gold (after 14)

- 1% $HAuCl_4$ (Sigma G4022)
- 0.1 M K_2CO_3
- 0.7% w/v sodium ascorbate
- ice
- 10% NaCl
- immunoglobulin at 1 mg/ml in TBS (0.01 M Tris, 0.15 M NaCl, pH 8.2)
- 10% aqueous bovine serum albumin (BSA) pH 8.2
- 20 mM Tris, 20 mM sodium azide, 1% BSA, 20% glycerol pH 8.2

1. Add 1 ml 1% $HAuCl_4$ and 1 ml 0.1 M K_2CO_3 to 25 ml distilled water.
2. Cool on ice[a] and, while stirring, quickly add 1 ml 0.7% sodium ascorbate (a purple-red sol should be obtained).
3. Adjust volume to 100 ml.
4. Heat until the colour of the solution changes to red.
5. Cool and add 0.7 ml 0.1 M K_2CO_3 (pH > 9).

6. Determine the amount of protein required to stabilize the sol: mix 0.25 ml sol with increasing amounts of protein, after 1 min add 0.25 ml 10% NaCl; the lowest concentration of protein that prevents a colour change to blue gives the amount required for stabilization.
7. To 30 ml gold sol add immunoglobulin to exceed the stabilization amount by 10%.
8. Add 0.3 ml 10% BSA to ensure complete stabilization.
9. Optional centrifugation steps which remove aggregates (13, 14).
10. Dialyse against 20 mM Tris, 20 mM sodium azide, 1% BSA, 20% glycerol pH 8.2.
11. Aliquot and store at 4 °C.

[a] Without cooling, particles with larger diameter are obtained.

5. Immunolocalization for light microscopy

The localization of specific proteins may be examined in permeabilized whole cells (Chapter 10, *Protocol 3*), in protoplasts (Chapter 10, *Protocol 2*), and in sections of tissue cut freshly, by cryotome, or after embedding in a removable matrix such as wax or PEG (see Chapter 3, *Protocol 1*) or in resin (see protocols in Chapters 3–6). *Protocol 3* outlines the procedures for localization of a single antigen type in sections of tissue without resin support matrix. Each combination of tissue type, tissue preparation, antigen, and primary antibody will require optimization of specimen preparation, dilution of primary antibody and other reagents, and the extent of any blocking required to prevent any non-specific staining; guide notes are given below, in the footnotes to the protocol, and in Sections 7 and 8 below on controls and troubleshooting.

Protocol 3. Outline procedure for immunolocalization in wax or PEG sections

- washing buffer: phosphate buffered saline (PBS) or TBS[a]
- BSA (Sigma)
- primary antiserum
- enzyme-conjugates seconary antibody (Sigma, Dako, or Boehringer)
- chromogenic substrates for conjugated enzyme (as conjugates)
- humid chamber[b]

Protocol 3. *Continued*

Sections must not dry out at any stage during the immunolocalization procedure.

1. Prepare wax sections (at 5–15 μm) as in Chapter 5, *Protocol 8*, or PEG sections as Chapter 3, *Protocol 1*.
2. Place slides in humid chamber[b] and incubate with blocking buffer[c] for 30 min.
3. Rinse briefly with washing buffer and remove excess[d].
4. Incubate with primary antiserum diluted in buffer, with blocking reagent when required[c]; for incubation conditions see Section 2.4.2 above.
5. Wash thoroughly with buffer, including blocking reagent[c] if necessary.
6. Incubate with diluted secondary antibody-fluorochrome, or secondary antibody-enzyme, or complex.
7. Wash thoroughly with washing buffer.
8. • **for fluorochrome markers**, add anti-fade mountant, coverslip, and examine (see Chapter 2, Section 5).
   • **for peroxidase-labelled secondary conjugates and complexes**, localize as in Chapter 3, *Protocol 15*.
   • **for alkaline phosphatase-labelled conjugates and complexes**, localize as in Chapter 3, *Protocol 9*.

[a] For PBS add to 100 ml distilled water 0.148 g Na_2HPO_4, 0.043 g KH_2PO_4, 0.72 g NaCl, 0.13 g NaN_3; for TBS add to 100 ml distilled water 0.242 g Tris, 0.13 g NaN_3, 0.9 g NaCl: adjust pH with 0.1 M HCl or 0.1 M NaOH.
[b] A Petri dish with buffer-moistened filter paper acts as a suitable humid chamber for incubation: glass slides should be separated from the moist filter paper using e.g. cocktail sticks or similar.
[c] Blocking reagent may be required to reduce background non-specific staining; e.g. add BSA or fish gelatin at 0.05–1% to buffer.
[d] Excess liquids may be removed with a deft 'flick of the wrist', or by laying a piece of Velin tissue (from suppliers of EM consumables) over the slide, allowing absorption of most liquid and then carefully peeling tissue away leaving a thin film of liquid still over the sections.

Protocol 4. LM immunogold localization of a single antigen type in resin sections, with silver enhancement[a]

- washing buffer (PBS or TBS)[b]
- BSA (Sigma)
- primary antiserum

- 1 or 5 nm colloidal gold conjugated to secondary antibody or protein A (BioCell, Sigma)
- silver enhancement reagents[c] (BioCell, Amersham)
- deionized or MilliQ water
- humid chamber[d]

Sections must not dry out at any stage during the immunoocalization procedure.

1. Cut semi-thin sections and transfer to drop of distilled water on glass slides; place slide on hot plate at 45–50 °C and allow water to dry: place slide in humid chamber.
2. If required, add blocking reagent[e] e.g. 0.05% BSA in buffer and incubate for 15 min at 37 °C.
3. Remove blocking reagent and add diluted primary antibody; incubate for 1–3 h at 37 °C or overnight at 4 °C.
4. Wash thoroughly with buffer and remove excess primary antibody[f].
5. Add diluted conjugate of 5 nm or 1 nm colloidal gold with either a suitable secondary antibody or protein A; incubate for 1 h at 37 °C.
6. Wash thoroughly with buffer and then with distilled water.
7. Wash with deionized water, remove excess, and apply silver enhancement reagents[c].
8. Wash thoroughly with distilled water, dry, and if required counterstain lightly (see Chapter 3, *Protocol 4*).
9. Examine by bright field or, for greater sensitivity, epipolarized microscopy (e.g. Nikon IGS system).

[a] Silver enhancement of immunogold can be used with non-resin-embedded sections, most successfully with meristematic tissues.
[b] See *Protocol 3*, footnote *a*.
[c] Several commercial silver enhancement kits are available; they can be used in low light so that the progress of the reaction may be monitored. Alternatively use 5.5 mM silver lactate and 75 mM hydroquinone in 20 mM citrate buffer pH 3.85; this solution should be prepared immediately before use and applied only in the dark or under photographic safe light. It is essential that reagents and equipment used are washed carefully with deionized water; this reduces accumulation of non-specific background. Several short periods (3–6 min) of enhancement should be used when required rather than a prolonged single incubation with the reagents; the latter results in accumulation of significant background precipitation.
[d] See *Protocol 3*, footnote, *b*.
[e] See *Protocol 3*, footnote *c*.
[f] See *Protocol 3*, footnote *d*.

6. Immunolocalization for electron microscopy

EM immunolocalization has been used particularly successfully with tissue cut from frozen (15) or resin-embedded blocks. Reviews (16, 17) indicate that when cryosectioning is not available cryofixation and freeze substitution, or chemical fixation (predominantly paraformadehyde-based), followed by dehydration and resin embedding can give specimens with good retention of antigenicity. The more hydrophilic acrylic resins (LR White, LR Gold, and Lowicryl K4M) are generally preferred to epoxy resin. A protocol for embedding in LR Gold is given in Chapter 5, and an alternative protocol for LR White is described in Chapter 6. See Chapter 4, Section 1 for suppliers of consumables for EM work.

Protocol 5. EM immunogold localization of a single antigen

- PBS or TBS washing buffer[a] with blocking reagent[b]
- primary antibody diluted in washing buffer
- gold colloid of appropriate size[c] conjugated to secondary antibody or protein A (BioCell)
- 2% aqueous uranyl acetate
- Formvar-coated nickel EM grids
- humid chamber[d]

Sections must not dry out at any stage during the immunolocalization procedure; incubations are carried out within a humid chamber[d]

1. Cut thin (0.07–0.1 μm) sections, pick up on to Formvar-coated nickel grids, and transfer to 40 μl drop of buffer on Parafilm within a humid chamber[d].
2. If required, transfer to 40 μl of buffer with blocking reagent[b] and incubate for 15 min at 37 °C.
3. Remove blocking reagent[e] and place grid on diluted primary antibody; incubate for 1–3 h at 37 °C or overnight at 4 °C.
4. Wash thoroughly with buffer and remove excess primary antibody[f].
5. Add diluted conjugate of colloidal gold[c] with either a suitable secondary antibody or protein A; incubate for 1 h at 37 °C.
6. Wash thoroughly with buffer and then with distilled water[f].
7. Stain with uranyl acetate for 15 min, wash with water, and examine by EM at 60 or 80 kV.

[a] For buffers see *Protocol 3*, footnote *a*.
[b] See *Protocol 3*, footnote *c*.
[c] Size of colloidal gold is determined by magnification used to examine specimen: 5 nm is used for EM studies employing <60 000×; 20, 15 or 10 nm colloid used for 5000–60 000× (the density of labelling is proportional to the amount of antigen but inversely proportional to the size of colloid).
[d] A Petri dish with buffer-moistened filter paper acts as a suitable humid chamber for incubation: EM specimen grids are incubated on drops of solution on a sheet of Parafilm or similar within the chamber.
[e] Excess liquid is drawn off using a piece of filter paper held against the edge of the specimen grid, ideally at the region where it is held by forceps; remove any liquid held by meniscus between tips of forceps.
[f] Wash EM grids by holding under a gentle stream of reagent from a Pasteur pipette, or by transferring grid along a series of 40 μl drops of washing buffer within the humid chamber.

Protocol 6. EM immunolocalization of two antigens on a single section

- PBS or TBS washing buffer[a] with blocking reagent[b]
- two primary antibodies (raised in different species) diluted in washing buffer
- two gold colloids of appropriate size[c] conjugated to appropriate secondary antibodies
- 2% aqueous uranyl acetate
- Formvar-coated nickel EM grids
- humid chamber[d]

Sections must not dry out at any stage during the immunolocalization procedure; incubations are carried out within a humid chamber. The two antibodies can be applied sequentially using *Protocol 5* steps 1–6 twice and then step 7, or they can be applied together as follows.

1–2. As *Protocol 5* steps 1 and 2.

3. Remove blocking reagent[e] and place grid on mixture of diluted primary antibodies (e.g. one from mouse and the other from rabbit); incubate for 1–3 h at 37 °C or overnight at 4 °C.

4. Wash thoroughly with buffer and remove excess primary antibody[f].

5. Add diluted conjugates of colloidal gold[c] with suitable secondary antibodies (e.g. 10 nm goat anti-mouse and 20 nm goat anti-rabbit); incubate for 1 h at 37 °C.

6. Wash thoroughly with buffer and then with distilled water[f].

Protocol 6. *Continued*

7. Stain with uranyl acetate for 15 min, wash with water, and examine by EM at 60 or 80 kV.
8. If quantifying result, repeat experiment but with secondary antibodies conjugated to the alternative gold colloid sizes (i.e. 20 nm goat anti-mouse and 10 nm goat anti-rabbit in comparison with example in step 5 above).

[a] For buffers see *Protocol 3*, footnote *a*.
[b] See *Protocol 3*, footnote *c*.
[c] See *Protocol 5*, footnote *c*; when labelling of two different antigens the two colloids should differ in size by 10 nm.
[d] See *Protocol 5*, footnote, [d]
[e] See *Protocol 5*, footnote, *e*.
[f] See *Protocol 5*, footnote *f*.

Neither *Protocol 5* nor *Protocol 6* includes pretreatment of the section to remove any of the embedding matrix. In some circumstances etching of the section, particularly for those embedded in epoxy resins, with periodic acid can lead to increased levels of immunolabelling (18). Sections for etching must be collected on gold grids. This approach is not always successful; embedding in acrylic resin usually reduces the requirement for etching.

7. Controls

Required controls for immunocytochemical investigations include reagent substitutions and, ideally, negative and positive tissue controls. The most critical of the reagents is the primary antibody: all experiments should include a control with non-immune serum from the same species, ideally a pre-immune serum from the same animal that produced the primary antibody. A poorer alternative is an antiserum of known different specificity; a 'no-primary' control is not particularly valuable, except to indicate non-specific binding of the secondary antibody or endogenous enzyme activity when enzyme–chromogenic substrates are used for localization.

8. Troubleshooting

Problems include: no immunostaining, a general background non-specific staining, and restricted non-specific staining; lignified cell walls frequently show the last.

8.1 No staining

This may be the result of:

(a) loss of antigenicity within the sample: try different specimen preparations changing
  i. primary fixative (e.g. reduce glutaraldehyde content)
  ii. embedding medium

(b) antigen only being present in very low amounts: increase primary antibody concentration or incubation periods, or try methods in which signal can be amplified

(c) primary antibody being poor; this may be the result of poor storage, use at wrong dilution, or using too many freeze–thaw cycles

(d) incorrect choice of secondary antibody or steps in its localization

8.2 General background staining

General background staining is usually the result of either ineffective 'washing' steps or use of excessive concentrations of the main reagents within the protocols. Background staining may be overcome by:

- increasing period and/or number of changes of washing buffer
- agitation of samples in wash buffer
- addition of detergent to wash buffer (e.g. 0.1% Tween)

8.3 Restricted background staining

Restricted non-specific staining is usually the result of hydrophobic and/or electrostatic (ionic) interactions between the reagents and specific components within the sample. Such non-specific staining may be reduced by:

(a) addition of, or increase in concentration of, serum or blocking protein to sample *prior to* incubation with primary; normal serum cannot be used for blocking in protocols utilizing protein A

(b) addition of serum or blocking protein to primary and secondary reagents (BSA or fish gelatin at 0.05–1%)

(c) increasing ionic strength of washing buffers (to 2.5% NaCl)

(d) diluting primary and/or secondary reagents

9. Tissue print immunoblots

For determining the distribution of specific protein in larger (>1 mm) tissue pieces the tissue print method (19) may be suitable, particularly if fine resolution is not required. Tissue sections are blotted on to nitrocellulose paper and the pattern of antigen distribution is demonstrated by treating the paper sequentially with the primary antibody, then with the secondary with conjugated enzyme, and then with substrates to give coloured deposit (see *Protocol 7*).

Protocol 7. Tissue print immunoblotting (after Cassab and Varner (19))

- Whatman 1 MM paper
- nictocellulose paper
- plain paper
- Kimwipes or similar tissues
- 0.2 M calcium chloride
- Toluidene Blue stain
- primary antibody diluted in TBS
- alkaline phosphatase-conjugated secondary antibody
- chromogenic substrates (see Chapter 3, *Protocol 9*)
- alcohol-washed and dried, single- or double-edged razor blades
- forceps
- small sealable polythene bags

1. Pretreat nitrocellulose paper with 0.2 M $CaCl_2$ for 30 min and air dry.
2. Put six layers of Whatman 1 MM paper onto a plastic plate, with one sheet of photocopy paper on top, and place the nitrocellulose paper on this.
3. Put sections from tissue using a new, washed single or double-edged razor blade; remove any excess surface moisture by blotting with a Kimwipe.
4. Carefully transfer the section, using forceps or a small paint brush, and lay it on the nitocellulose paper.
5. Place several layers of Kimwipes over the sections and then press gently but evenly for 15–20 sec.
6. Carefully remove the Kimwipes and transfer the sections to glass slides for anatomical examination.
7. Air dry the nitrocellulose paper.
8. Incubate paper in dilute primary antibody.
9. Wash with TBS.

10. Incubate with secondary antibody conjugated to alkaline phosphatase.
11. Wash with TBS.
12. Incubate with alkaline phosphatase substrates (see Chapter 3 *Protocol 9*, or footnote *a* below).
13. Wash and examine filter.

[a] Alternative substrates: 0.16 mg/ml BCIP (5-bromo-4-chloro-3-indolyl-phosphate) and 0.33 mg/ml NBT (nitro-blue tetrazolium salt) in 0.2 M Tris–HCl, 10 mM $MgCl_2$, pH 9.2.

10. Immunosorbent methods for viral identification

Immunotechniques have proved very valuable for the quick diagnosis of plant viruses (20). The 'Derrick' method (21), also referred to as I(mmuno) S(orbent)EM can be used to reduce the time required for identification of a virus from weeks to hours; it is particularly valuable for phloem viruses which are present in low numbers.

Protocol 8. Immunosorbent EM for virus identification (from 20).

- antiserum to test virus
- 0.06 M phosphate buffer pH 7
- negative stain (2% aqueous uranyl acetate or 2% sodium or potassium phosphotungstate or 2% ammonium molybdate)
- Formvar- or carbon-coated EM grids
- incubation chamber (see *Protocol 3*, footnote *b*)

1. Aliquot 20 μl drops of diluted antiserum on to Parafilm in incubation chamber.
2. Place coated EM grid on surface of drop and leave for 0.5–3 h at 37 °C or at room temperature[a].
3. Prepare sap extract: grind a small amount of infected material in phosphate buffer, centrifuge at 8000 *g* for 5–15[a] min, and collect supernatant.
4. Wash grid with phosphate buffer and drain excess liquid.
5. Place grid on 20 μl aliquot of sap supernatant and incubate in chamber for 24 h at 4 °C or 1–5 h at room temperature[a].

Protocol 8. *Continued*

6. Remove grid from sap, drain, and negative stain[b].

[a] Optimum times and temperatures should be determined empirically for each combination of primary antiserum and sap sample.
[b] If staining with uranyl acetate, wash thoroughly with distilled water as uranyl acetate is not compatible with phosphate and with pH values >5.

References

1. Bullock, G. R. and Petrusz, P. (ed.) (1982). *Techniques in immuno-cytochemistry*, Vol. 1. Academic Press, San Diego.
2. Polak, J. M. and Van Noorden, S. (1986). *Immunocytochemistry: modern methods and applications*. J. Wright and Sons, Bristol.
3. Polak, J. M. and Van Noorden, S. (1987). *An introduction to immuno-cytochemistry: current techniques and problems*. Royal Microscopical Society Handbook 11. Oxford University Press.
4. Beesley, B. (1993). *Immunocytochemistry: a practical approach*. IRL Press, Oxford.
5. Polak, J. M. and Varndell, I. M. (ed.) (1984). *Immunolabelling for electron microscopy*. Elsevier, Amsterdam.
6. Catty, D. (ed.) (1988). *Antibodies: a practical approach*, Vol. 1. IRL Press, Oxford.
7. Catty, D. (ed.) (1989). *Antibodies: a practical approach*, Vol. II. IRL Press, Oxford.
8. Campbell, A. M. (1984). *Monoclonal antibody technology: laboratory techniques in biochemistry and molecular biology*. Elsevier, Amsterdam.
9. Liddell, J. E. and Cryer, A. (1991). *A practical guide to monoclonal antibodies*. Wiley, Chichester.
10. Tokyasu, K. T. and Singer, S. J. (1976). *J. Cell Biol.*, **71**, 894.
11. Roth, J. (1984). In *Immunolabelling for electron microscopy* (ed. J. M. Polak and I. M. Varndell), pp. 113–21. Elsevier, Amsterdam.
12. Frens, G. (1973). *Nature Phys. Sci.*, **241**, 20.
13. Slot, J. W. and Gueze, H. J. (1985). *Eur. J. Cell Biol.*, **38**, 87.
14. Slot, J. W. and Gueze, H. J. (1984). In *Immunolabelling for electron microscopy* (ed. J. M. Polak and I. M. Varndell), pp. 129–42. Elsevier, Amsterdam.
15. Greenwood, J. S., Keller, G. A., and Chrispeels, M. J. (1984). *Planta*, **162**, 548–5.
16. Herman, E. (1988). *Annu. Rev. Plant Physiol. Plant Mol. Biol.*, **39**, 139.
17. VandenBosch, K. A. (1990). In *Electron microscopy of plant cells* (ed. J. L. Hall and C. Hawes), pp. 181–218. Academic Press, London.
18. Craig, S. and Goodchild, D. J. (1984). *Protoplasma*, **122**, 35.
19. Cassab, G. I. and Varner, J. E. (1987). *J. Cell Biol.*, **105**, 2581.
20. Wright, D. M. (1984). In *Immunolabelling for electron microscopy* (ed. J. M. Polak and I. M. Varndell), pp. 323–40. Elsevier, Amsterdam.
21. Derrick, K. S. (1973). *Virology*, **56**, 652.

8

Plant protoplast techniques

L. C. FOWKE and A. J. CUTLER

1. Introduction

Modern methods for isolating protoplasts (naked plant cells) by enzyme digestion were developed in the early 1960s. Mixtures of cellulases, hemicellulases, and pectinases are now routinely used to release spherical protoplasts from plant cells and tissues by enzyme digestion of cell walls. Large numbers of viable protoplasts are easily produced from a wide range of tissues representing many different plant species (e.g. 1, 2). Many protoplasts are totipotent and under appropriate conditions will regenerate cell walls, undergo sustained cell division, and ultimately develop into new plants.

Protoplasts have been used extensively to study the structure and function of plant cells and their organelles (3). They have been particularly important for studies of the plasma membrane which is normally enclosed within the cell wall. For example, the accessibility of the plasma membrane of protoplasts permits the application of electrical permeabilization (electroporation) to facilitate uptake of molecules into plant protoplasts (4, 5). This method has been particularly useful for genetic transformation of plants. Protoplasts can be induced to fuse and provide an excellent system to study plasma membrane interactions as well as the fate of organelles in developing hybrid cells (6). Rather interesting hybrid plants have been regenerated from such fusion products. Finally, large fragments of plasma membrane produced from freshly isolated protoplasts can be used to examine cell organelles associated with the inner surface of the membrane (7). The structure and distribution of both microtubules and coated vesicles have been examined by this technique.

This chapter provides methods for

- isolating protoplasts from angiosperm leaves and suspension cultures, gymnosperm somatic embryos, and a filamentous green alga
- permeabilization of protoplasts by electroporation
- fusion of protoplasts using polyethylene glycol (PEG)
- preparing large plasma membrane fragments from protoplasts

These methods are described for specific protoplast types but are applicable, with suitable modifications, to most other systems.

2. Protoplast isolation

2.1 Introduction

Plant cell walls vary markedly from tissue to tissue and even within a single tissue during growth and development. Thus it is not possible to identify a standard protoplast isolation method which can be applied to all plants. The establishment of new isolation protocols requires an empirical approach. To simplify the process of defining a protoplast isolation method, reliable protocols for four different tissues are presented in detail. The protocols include widely separated groups of plants and the starting material ranges from highly organized structures to suspensions of individual cells. For a complete discussion of protoplast isolation methods refer to Eriksson (1), Giles (2), and Wetter and Constabel (8).

2.2 Chemicals

The following chemicals are required for the four protoplast isolation protocols. Glass-distilled water should be used to prepare all solutions.

- sodium hypochlorite
- 70% ethanol
- Tween 80
- sorbitol
- mannitol
- Cellulysin (Calbiochem, 219466)
- Macerase (Calbiochem, 441201)
- cellulase R-10 (Kanematsu-Gosha (USA))
- cellulase RS (Kanematsu-Gosha (USA))
- Driselase (Sigma, D-8037)
- hemicellulase Rhozyme HP-150 (Genencor Inc.)
- pectinase (Sigma, P-4300)
- $CaCl_2 \cdot 2H_2O$
- acetone
- NaH_2PO_4
- Calcofluor White M2R (American Cyanamid Co.)
- fluorescein diacetate (Sigma, F-7378)
- NaCl
- $AgNO_3$

2.3 Apparatus

The following apparatus is required for the four protoplast isolation protocols:

- laminar flow cabinet
- benchtop centrifuge, swinging bucket
- conical 15 ml glass test tubes
- inverted microscope
- gyratory shaker
- freeze drier (for desalting enzymes only)
- pH meter
- nylon mesh for filters (50, 84 and 150 μm pore sizes) (Wilson Sieves)
- disposable syringes, 10 ml
- Nalgene syringe filter units, 0.45 μm pore size (Canlab, F3201–915)
- fine forceps
- razor blades
- pipettes with widened tips, 5 ml; glass Pasteur pipettes
- plastic Petri dishes, 100 mm × 15 mm
- Parafilm 'M' (Canlab, P1150-4)
- aluminum foil
- Miracloth (Calbiochem 475855)

2.4 Methods for protoplast isolation

Protoplasts should be isolated under aseptic conditions in a laminar flow cabinet. Release of protoplasts can be monitored with an inverted microscope. *Table 1* provides details of enzyme mixtures for the four protoplast isolation protocols.

Protocol 1. Isolation of protoplasts from suspension cultures of soybean (SB-1; *Glycine max* (L.) Merr.) (see reference 9)

- starting material: 2 day old subculture of SB-1

1. Allow 5 ml suspension to settle, remove supernatant, and resuspend the cells in 5 ml enzyme solution (*Table 1*).
2. Pour into a Petri dish, seal with Parafilm, wrap with aluminum foil, and incubate with slow agitation (25–50 r.p.m.) on a gyratory shaker for 4–8 h.

Protocol 1. *Continued*

3. Pass incubated material through a 50 μm filter.
4. Transfer protoplast suspension to a 15 ml conical test tube, centrifuge (100 *g*, 3 min), remove supernatant and resuspend in 0.55 M sorbitol. Repeat twice more. Purification of protoplasts may also be achieved by flotation on a sucrose or Ficoll density gradient (e.g. reference 10).

Table 1. Enzyme mixtures for protoplast isolation

	Source of protoplasts			
	SB-1	**Pea leaf**	**White spruce**	***Ulothrix***
Sorbitol	0.55 M	0.55 M		0.3 M
Mannitol			0.44 M	0.3 M
Cellulysin				2.0%
Cellulase R-10	0.6%			
Cellulase RS			1.0%	
Driselase	0.2%	0.5–1.0%	0.25%	
Hemicellulase	0.2%		0.25%	
Pectinase	0.2%	0.25–0.5%	0.25%	
Macerase				0.1%
NaH_2PO_4				2.0 mM
$CaCl_2 \cdot 2H_2O$			5.0 mM	2.0 mM
pH	5.8	6.8	5.8	5.8

Centrifuge enzyme solutions to remove large contaminants and sterilize by filtration through a Nalgene syringe filter unit, 0.45 μm pore size.

Protocol 2. Isolation of protoplasts from leaves of pea (*Pisum sativum* L. cv. Century) (see reference 8)

- starting material: shoot with young fully expanded leaves

1. Place shoot cuttings in water in the dark for 30 h before isolation of protoplasts.
2. Immerse leaves in 10% sodium hypochlorite or equivalent commercial bleach solution for 5 min or in 70% ethanol for 2 min. A drop of the wetting agent Tween 80 can be added to improve the sterilization by hypochlorite.
3. Wash leaves two or three times in distilled water or in a solution of the same osmolality and pH as the enzyme solution.

4. Remove the lower epidermis from the leaves using fine forceps. Cut leaves into sections and transfer 0.5–0.8 g material into 5 ml enzyme solution (*Table 1*) in a Petri dish.
5. Seal dishes with Parafilm, wrap in aluminum foil, and incubate with slow agitation (25–50 r.p.m.) for 4–6 h.
6. Pass incubated material through a 70 μm filter.
7. Wash protoplasts with 0.55 M sorbitol or by flotation on Ficoll (see *Protocol 1*, step 4).

Protocol 3. Isolation of protoplasts from somatic embryos of white spruce (*Picea glauca* (Moench) Voss) (see reference 11)

- starting material: 4–6 day old embryogenic suspension culture

1. Collect cells on Miracloth and transfer to 0.44 M mannitol, pH 5.8 for 1 h to preplasmolyse cells.
2. Collect cells on Miracloth and transfer 2 g tissue to 10 ml enzyme mixture (*Table 1*) in a Petri dish. Seal with Parafilm, wrap with aluminum foil, and incubate with slow agitation (25–50 r.p.m.) for 3–4 h.
3. Pass incubated material sequentially through 150 μm and 84 μm mesh filters.
4. Wash protoplasts with 0.44 M sorbitol or by flotation on Ficoll (see *Protocol 1*, step 4).

Protocol 4. Isolation of protoplasts from the filamentous green alga *Ulothrix fimbriata* (see reference 12)

- starting material: rapidly growing algae preferably entering mitosis

1. Transfer about 20 mg wet weight of algae to 20 ml enzyme solution (*Table 1*) in a Petri dish. Seal with Parafilm, wrap with aluminum foil, and incubate with slow agitation (25–50 r.p.m.) for 1–4 h.
2. Pass incubated material through a 50 μm filter.
3. Wash protoplasts with 0.6 M sorbitol or by flotation on Ficoll (see *Protocol 1*, step 4).

2.5 Evaluating freshly isolated protoplasts

Figures 1–4 provide examples of typical protoplasts from different tissues.

2.5.1 Viability

Healthy protoplasts appear spherical and display cytoplasmic streaming. A number of relatively simple tests are available to assess the viability of freshly isolated protoplasts. The most common test uses fluorescein diacetate (FDA), which produces fluorescence inside the plasma membrane of viable protoplasts (13). Protoplasts are suspended in a 0.01% (w/v) solution of FDA (prepared from stock solution of 5 mg/ml FDA in acetone) in the appropriate osmoticum and examined under UV light in a fluorescence microscope.

2.5.2 Presence of cell wall remnants

Cell wall remnants are difficult to detect with conventional bright field microscopy. Calcofluor White M2R binds to cell walls and fluoresces brightly under UV illumination. Protoplasts are stained with 0.1% (w/v) Calcofluor in the appropriate osmoticum for 10 min, washed, and examined in a fluorescence microscope.

2.6 Protoplast culture

This chapter describes methods for isolating viable protoplasts and for using them without subsequent culture; a thorough discussion of culture media and methods is beyond the scope of this chapter. One commonly used protoplast culture medium is described in Section 3.2 and the reader is referred to the references 1, 2 and 8 for other media and a complete description of culture methods.

2.7 Optimizing protoplast isolation

One of the protocols described above should yield some protoplasts from a wide variety of plant tissues. A number of modifications of these basic methods will improve protoplast yields for individual plants. The quality of the source material is most important. When isolating protoplasts from callus and cell suspensions it is generally best to use frequently subcultured material in the early log phase of the growth cycle. The highest yields of algal protoplasts are also achieved with rapidly growing and dividing material. When using plant tissues and organs it is important to grow the donor plants under carefully controlled conditions and to select plants of a certain stage of development. The physiological condition of the source material markedly influences the yield and viability of the isolated protoplasts. The best results are often obtained when plants are grown in controlled environment chambers. Optimal protoplast yields from leaves may also require pretreatments of excised leaves or intact shoots just prior to the enzyme treatment

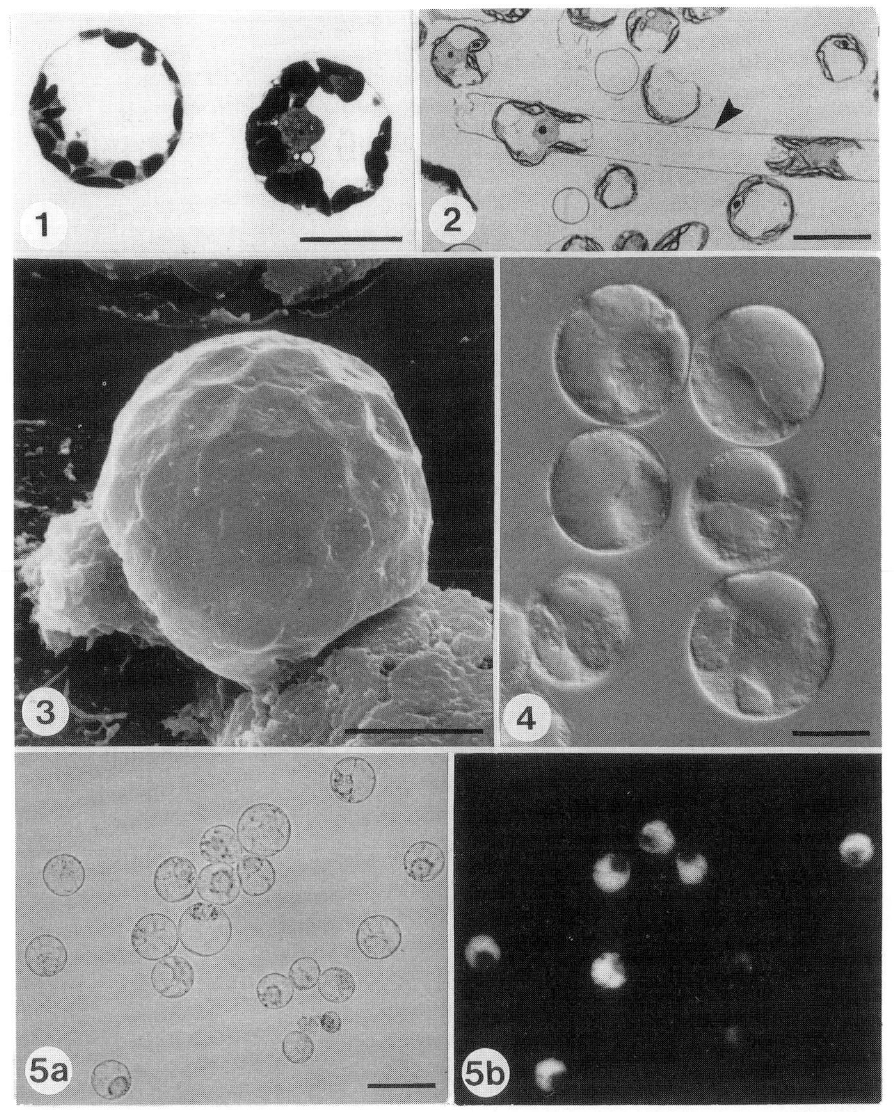

Figure 1. Light micrograph showing pea leaf protoplasts prepared for microscopy according to reference 8. Bar = 20 μm.

Figure 2. Light micrograph showing release of protoplasts from a filament (arrow) of *Ulothrix*. Bar = 20 μm. Specimens prepared for microscopy according to methods in reference 8.

Figure 3. Scanning electron micrograph of pea leaf protoplast, prepared for microscopy according to reference 8. Bar = 10 μm.

Figure 4. Differential interference contrast micrograph of white spruce protoplasts isolated from somatic embryos. Bar = 20 μm. (Reproduced from reference 31, with permission).

Figure 5. Light micrograph of soybean protoplasts permeabilized by electroporation in the presence of calcein (a). Bright field, bar = 50 μm; (b) fluorescence, same magnification as (a).

(e.g. pea, *Protocol 2*). The basic conditions for incubation such as light, temperature, pH, and time will also have to be optimized. The type of osmoticum (e.g. sugars, salts, sorbitol, or mannitol) and its concentration can also affect protoplast yield. The most suitable osmotic potentials are generated by 0.2–0.7 M solutions of these compounds. Some tissues or cells require a preplasmolysis period for optimal yields (see *Protocol 3*). In some cases the addition of $CaCl_2 \cdot 2H_2O$ to the isolation mixture stabilizes the protoplasts. However, this treatment may cause adhesion between the plasma membrane and cell wall and result in protoplast damage. Protoplast yields may also be improved by isolating in a complete protoplast culture medium (e.g. Kao medium (*Table 2*)) rather than the simple osmotica described in *Table 1*. Perhaps the most important variables to consider relate to the enzymes used. The types of enzymes and their combinations and concentrations must be optimized for each tissue. In addition, the purity of individual enzymes may prove important. Some laboratories routinely desalt all enzymes used for protoplast isolation. One method for desalting enzymes is presented in *Protocol 5*.

Table 2. Composition of Kao medium[a]

Ingredient	**Amount per litre of medium**[b]
(a) Mineral salts	
NH_4NO_3	600
KNO_3	1900
$CaCl_2 \cdot 2H_2O$	600
$MgSO_4 \cdot 7H_2O$	300
KH_2PO_4	170
KCl	300
Sequestrene 330 Fe	28
(b) Micronutrients	
KI	0.75
H_3BO_3	3.00
$MnSO_4 \cdot H_2O$	10.00
$ZnSO_4 \cdot 7H_2O$	2.00
$Na_2MoO_4 \cdot 2H_2O$	0.25
$CuSO_4 \cdot 5H_2O$	0.025
$CoCl_2 \cdot 6H_2O$	0.025
(c) Sugars	
glucose	68 400
sucrose, fructose, ribose, xylose, mannose, rhamnose, cellobiose, sorbitol, mannitol	125 of each
(d) Organic acids[c]	
pyruvic acid, disodium salt	5
citric acid	10
malic acid	10
fumaric acid	10

Table 2. (*Continued*)

Ingredient	Amount per litre of medium[b]
(e) Vitamins[d]	
inositol	100
nicotinamide	1
pyridoxine–HCl	1
thiamine–HCl	10
D-calcium pantothenate	0.5
folic acid	0.2
p-aminobenzoic acid	0.01
biotin	0.005
choline chloride	0.5
riboflavin	0.1
ascorbic acid	1
vitamin A	0.005
vitamin D_3	0.005
vitamin B_{12}	0.01
(f) Hormones	
2,4-D	0.2
zeatin	0.5
napthaleneacetic acid	1
(g) Other components	
vitamin-free casamino acids	125
coconut water[e]	10 ml
double-distilled water	1000 ml
pH adjusted to 5.7 (NaOH)	

[a] Various groups of components (e.g. organic acids, micronutrients, and water-soluble vitamins) can be prepared as concentrated stock solutions.
[b] Amounts are in milligrams unless otherwise stated.
[c] Dissolved in water and pH adjusted to 5.5 before addition to medium.
[d] Vitamins A, D3, and B12 must be added from a concentrated stock solution dissolved in 70% ethanol.
[e] From mature fruits. Heat liquid to 60 °C for 30 min and filter.

Protocol 5. Purification (desalting) of enzymes

Enzymes can be desalted on a 4.8 cm × 60 cm column of Bio-gel P6 (Bio-Rad) or Sephadex G25 (200–400 mesh) at 4 °C.

1. Wash column thoroughly with distilled water.
2. Dissolve 10 g of crude enzyme in 100 ml water and centrifuge (12 000 *g*, 10 min).
3. Transfer supernatant to column, add 5 ml water followed by 3–4 ml of 3 M NaCl, and then elute enzyme with water.
4. Check the fractions by using 0.1 M $AgNO_3$ for detection of Cl^-. Combine the protein fractions which are Cl^--free and lyophilize. Store the dried enzymes at 4 °C.

3. Protoplast permeabilization

3.1 Introduction

Permeabilization has been most often used for the uptake and expression of cloned genes. Internalized DNA is degraded and so expression is usually transient. In a small proportion of cells one or more copies of the gene may be integrated into the host cell chromosomes and result in stable transformation and expression.

Apart from DNA, permeabilization is also capable of facilitating uptake of proteins, carbohydrates, and small molecules. There have been few efforts to probe cellular processes by introduction of materials other than DNA. Examples of various chemical permeabilization techniques and applications are given by Felix (14).

Electroporation has been widely used to permeabilize plant protoplasts and produces quantitatively greater uptake of external solutes than common alternatives such as PEG or hypotonic shock treatment (6). However, application of electric shocks to protoplasts is an inherently stressful process and there is a trade-off between lysis and effective permeabilization of the membrane. Fluorescent dyes such as calcein may be used to provide a rapid, visual indication of permeabilization under various conditions (see reference 6 and *Figure 5*, *a* and *b*). If the objective is to obtain maximal expression of cloned genes, preliminary experiments to identify the optimal voltage should be carried out in the presence of DNA. Optimal conditions for uptake of other macromolecules and low molecular weight substances may differ from those determined for DNA.

3.2 Chemicals

(a) calcein (Sigma, C-0875)

(b) DNA: the following protocol has been used with plasmid pUC8CaMVCAT (this 4.96 kb plasmid was obtained from Dr Virginia Walbot of Stanford University and is analogous to pBR322CaMVCAT described in reference 15) and pBI221 (Clontech Inc.) and can be used with any plant expression vector up to at least 10 kb in size. The key feature of both vectors is the expression of a gene coding for an easily detectable enzyme activity such as *cat* (coding for chloramphenicol acetyl transferase in pUC8CaMVCAT) or *uidA* (coding for β-glucuronidase in pBI221). In both cases, expression is driven by the cauliflower mosaic virus 35S promoter.

(c) culture medium: with the commonly used protoplast culture medium developed by Kao (16 and *Table 2*) SB-1 protoplasts achieve a plating efficiency of up to 65%. Kao medium provides a nutritionally rich environment that minimizes stress during protoplast isolation and in subsequent treatments. It should be filter-sterilized (not heat-sterilized!)

immediately after preparation and can then be stored at 4 °C or frozen until use.

3.3 Apparatus

A variety of commercial equipment is available for electroporation. The method described here uses the Bio-Rad system but other equivalent equipment can be easily substituted. The following is required.

- Bio-Rad Gene Pulser and Capacitance Extender
- Bio-Rad Gene Pulser cuvettes (0.4 cm gap between electrodes)
- sterile polystyrene test tubes (Corning, 15 ml total volume)
- sterile glass Pasteur pipettes (wide-bore or with the elongated tip removed)
- adjustable pipettors with disposable sterile tips. The narrow end of the tips should be removed
- bench-top centrifuge
- microscope equipped for observation of fluorescent samples

3.4 Preparation of plasmid DNA

Large quantities of DNA are usually required for transient expression studies and therefore plasmids are purified by caesium chloride gradient centrifugation (17).

3.5 Isolation of protoplasts

A method for obtaining protoplasts from soybean cell suspension culture (SB-1) is given above (*Protocol 1*). Protoplasts should be electroporated as soon as possible after isolation as cell wall deposition (which occurs rapidly once the digestion enzymes are removed) reduces macromolecule uptake.

3.6 Basic electroporation method

Protocol 6. Electroporation of soybean (SB-1) protoplasts

1. Wash freshly isolated protoplasts twice in Kao culture medium. Resuspend in medium and immediately divide into aliquots containing 200–300 μl packed volume of protoplasts.
2. Centrifuge protoplasts (80 *g*, 3 min) and remove supernatant. Add 800 μl culture medium, then 50 μg plasmid DNA dissolved in 10 μl TE buffer (10 mM Tris–HCl, 1 mM EDTA, pH 7.6), mix and transfer to electroporation sample chambers (Gene Pulser cuvettes). If permeabilization is to be visually observed, calcein (5 mM) should be included in the medium in which the protoplasts are resuspended. Cool samples on ice for 5 min.

Protocol 6. *Continued*

3. Resuspend protoplasts in medium and subject samples to two 400 V/cm pulses separated by 1 min. The Capacitance Extender is set at 960 μF giving a pulse length (a measure of the time that the protoplasts are subject to the applied voltage) of approximately 45 msec.
4. Immediately after the second pulse, cool the samples on ice for 5 min. Transfer samples to fresh tubes and dilute with Kao medium to a final volume of 3–4 ml. Transfer protoplasts and medium to Petri dishes and incubate for 24 h.

The proportion of surviving protoplasts permeabilized can be monitored immediately by fluorescence microscopy if calcein is included in the electroporation medium (*Figure 5a* and *b*). The protoplasts must be thoroughly washed by centrifugation (three times with Kao medium) to eliminate background fluorescence before microscopic observation (6).

5. Assay marker gene activity. Expression can usually be measured from 6 to 72 h after electroporation, but 24 h is optimal for this system.
   (a) Preparation of protoplast extracts and chloramphenicol acetyl transferase assays are performed by standard methods (18; see also references 15, 19, 20).
   (b) A description of common GUS vectors and assay procedures is given by Jefferson (21).
   (c) Stable transformation and permanent expression occur in a small fraction of protoplasts. Cells in which these rare events have taken place can be rescued by use of a gene whose expression confers resistance to an otherwise toxic selective agent (e.g. 20, 22, 23).

3.7 Optimizing electroporation conditions

3.7.1 Voltage

The most important variable is the applied voltage. The effective range is usually 250–450 V/cm. For example, we have found optimal voltages to be 400 V/cm for soybean, 250 V/cm for corn suspension culture protoplasts, and 300–350 V/cm for tobacco suspension culture protoplasts (unpublished results). Note that the electric field to which the protoplasts are subjected should be expressed as volts per centimetre (the Bio-Rad Gene Pulser displays the total voltage irrespective of the distance between electrodes) in order to compare treatments in different laboratories using different equipment. If the voltage is too high all the cells will be killed, but a degree of cellular mortality is necessary for optimal uptake and expression. Typically, a voltage sufficient to kill about 50% of the protoplasts gives optimum expression. In all cases, the Capacitance Extender is set to 960 μF to maximize the pulse length.

3.7.2 Amount and structure of plasmid DNA

The level of expression is proportional to amount of plasmid DNA in the sample. At least 10 μg is normally required and amounts up to 100 μg are sometimes added. Increases in transient expression of up to 10-fold have been observed by linearizing plasmids prior to transfection (19). However, this has not always been found to be beneficial (e.g. see reference 20). Addition of genomic carrier DNA is usually of marginal utility.

3.7.3 Health and number of protoplasts

Gene expression increases with the number of protoplasts in the sample until there is a ratio of roughly 1:1 of protoplasts to medium. Protoplasts of high viability are essential for high expression and reproducible results. Protoplasts should preferably be isolated from cells 1–2 days after serial transfer to fresh medium using the mildest digestion procedure possible (lowest concentration of enzymes and shortest digestion time). Reproducible results can only be obtained if the source cells and protoplast isolation procedure are identical from experiment to experiment. Higher levels of gene expression may be obtained by use of synchronized protoplasts in electroporation (20).

3.8 Alternative permeabilization protocols

The most common alternative method for introduction of DNA into plant protoplasts involves PEG treatment (e.g. 23). Many other cell permeabilization protocols are available that are effective for smaller molecules (14).

4. Protoplast fusion

4.1 Introduction

Electrical or PEG-mediated membrane fusion in adjacent protoplasts can lead to novel hybrid cells. In many cases, membrane fusion is followed by nuclear fusion. Although culture and regeneration of these cells to produce fertile plants is often difficult, this technique represents an approach to producing novel hybrids that cannot be obtained by conventional sexual crosses (4, 5).

The method described in *Protocol 7* can be adapted to a variety of systems and is similar to that described in reference 16. With these specific parental protoplasts efficiency of heterokaryocyte formation varies widely in the range 0.1–10%. All manipulations are performed under sterile conditions in a laminar flow hood.

4.2 Chemicals

- PEG 1540, gas chromatography grade (BDH, B 15100–30)
- 2-(*N*-morpholino)ethanesulphonic acid (Mes buffer) (Sigma, M-8250)
- silicone oil
- sorbitol
- glucose
- glycine
- $CaCl_2 \cdot 2H_2O$

4.3 Apparatus

- laminar flow cabinet
- plastic Petri dishes, 100 mm × 15 mm
- glass Pasteur pipettes
- adjustable pipettors with sterile, disposable plastic tips
- inverted microscope
- glass coverslips (22 mm × 22 mm)
- polystyrene test tubes, 15 ml volume (Corning)

4.4 Basic fusion method

Protocol 7. Fusion of soybean (SB-1) and pea leaf protoplasts

- sorbitol solution (0.55 M sorbitol, 5 mM Mes buffer, and 15 mM $CaCl_2$), pH adjusted to 5.8
- fusion solution (PEG 1540, 50 g in a total volume of 100 ml, and 15 mM $CaCl_2$)
- dilution solution A (0.3 M glucose, 0.1 M glycine), pH adjusted to 10.5
- dilution solution B (0.3 M glucose, 0.1 M $CaCl_2$)

1. Place three or four dry-sterilized coverslips on to small drops (2–3 μl) of silicone oil in culture dishes (60 mm × 15 mm).
2. Prepare pea leaf protoplasts as described in *Protocol 2* and soybean SB-1 protoplasts as described in *Protocol 1*, then wash each once with sorbitol solution. Mix approximately equal volumes of protoplasts (typically about 100–200 μl packed protoplast volume of each type).

Remove a small aliquot and observe relative amounts of protoplasts under an inverted microscope. If necessary adjust amounts to give a ratio of approximately 1:1. After centrifugation (80 *g*, 3 min), resuspend the protoplasts in approximately 2 ml of sorbitol solution. There should be a dense suspension of protoplasts to facilitate cell–cell contact. Formation of somatic hybrids either by electric pulses or PEG treatment is favoured by tight membrane contact between adjacent protoplasts. A 1:1 ratio of the parental protoplasts produces the optimal proportion of heterologous versus homologous fusions.

3. Using a wide-bore Pasteur pipette or an adjustable pipettor, resuspend protoplasts then deposit 150–200 μl of suspension in five separate drops on to each coverslip and leave for 5–10 min for the protoplasts to settle. Add approximately 10–12 drops of the PEG solution (400–500 μl). The PEG solution should be inserted between the protoplast droplets so that all of the individual drops become linked and after the final addition there is a continuous layer. This procedure allows the PEG to diffuse slowly over the protoplasts whilst mixing with the sorbitol solution. The ratio of PEG solution volume to original protoplast suspension should be approximately 3:1.

4. The protoplasts are now left to adhere for 20 min (optimum time must be determined by trial and error and could range between 5 and 40 min). The protoplasts should be monitored under an inverted microscope. During the adhesion stage, the protoplasts will appear highly distorted and aggregated into large clumps (*Figure 6*).

5. Mix equal volumes of dilution solutions A and B (giving final concentrations of 0.3 M glucose, 0.05 M glycine, 0.05 M $CaCl_2$, pH 10.5). Gently add this solution (10–12 drops; 500 μl) around the edge of the coverslip. Incubate the protoplasts for 10 min then add 5–10 drops more. The protoplasts should begin to re-circularize at this stage as disaggregation occurs. Hybrids (both homokaryocytes and heterokaryocytes) first become apparent during the dilution stage and so this must be performed slowly. The occurrence of rapid disaggregation and re-circularization is a sign that adhesion is not sufficient for optimal hybrid formation.

6. After 5 min add culture medium slowly from the edge of the coverslip until it flows from the coverslip. After a further 5 min, remove excess medium and add fresh medium. Protoplasts should adhere tightly to the glass coverslip. Repeat washing five times.

7. Examine protoplasts for hybrids containing the prominent soybean nucleus and the pea chloroplasts.

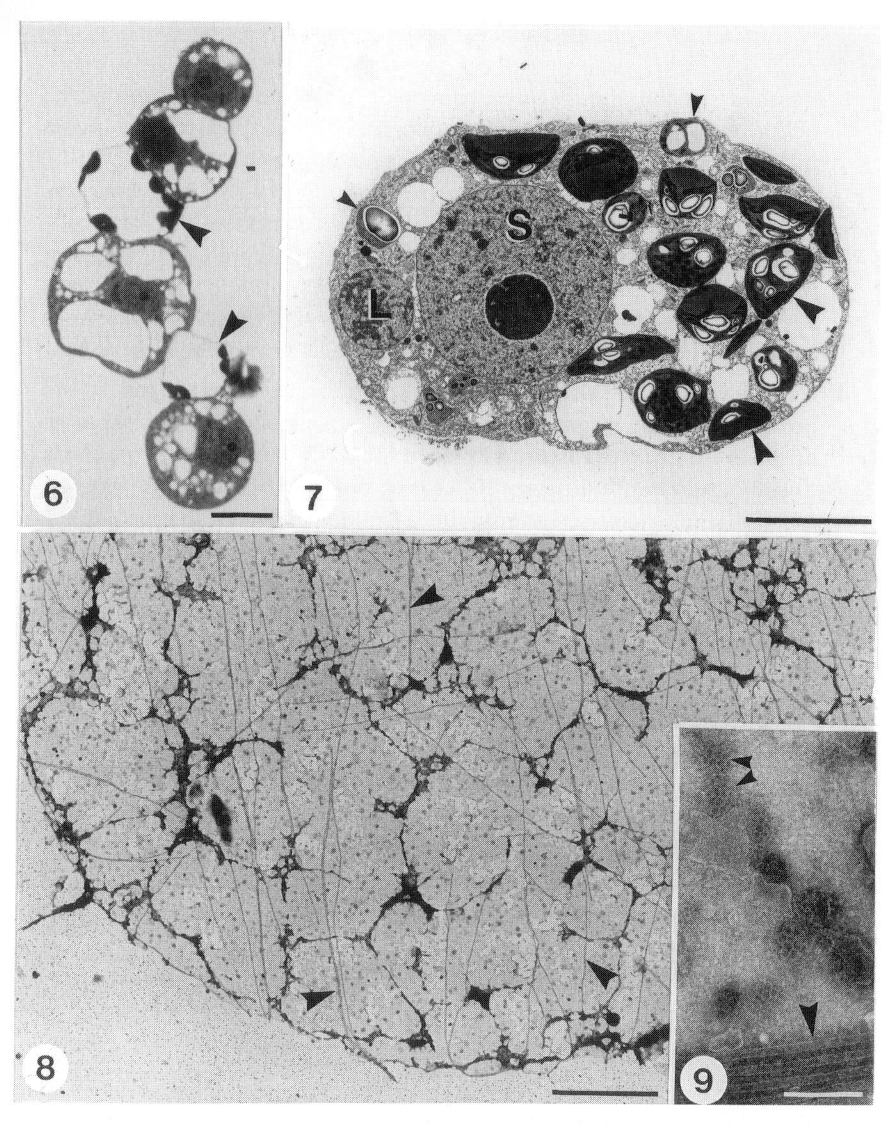

Figure 6. Light micrograph showing PEG-induced aggregation of protoplasts from pea leaf (arrows) and *Vicia* suspension cultures prepared for microscopy according to reference 8. Bar = 20 μm.

Figure 7. Electron micrograph of *Vicia* fusion product showing leucoplasts (small arrows) and nucleus (S) from the suspension protoplast and chloroplasts (large arrows) and nucleus (L) from the leaf protoplast. Bar = 10 μm. Specimens were prepared for microscopy according to methods in reference 8. (Reproduced from reference 32 with permission).

Figure 8. Electron micrograph of a tobacco protoplast plasma membrane fragment showing many microtubules (arrows). Bar = 2 μm.

Figure 9. Electron micrograph of negatively stained tobacco protoplast plasma membrane fragment showing microtubules (large arrow) and coated pits (small arrows). Bar = 0.2 μm.

4.5 Optimizing the basic method

4.5.1 Purification and molecular weight of PEG

The ability of PEG to promote adhesion and fusion increases with its molecular weight, therefore PEG 4000 can be used instead of PEG 1540 to provide a higher fusion efficiency although it is also more toxic. Purification of the PEG by ion-exchange has been used to minimize toxicity (24), but should not be required if PEG of the highest available purity is employed (BDH).

4.5.2 Time of adhesion and speed of dilution

Some improvement in fusion efficiency may be gained by prolonging the adhesion stage and performing a slower and more gradual dilution. On the other hand, if there is unacceptable loss of viability but a high propensity to fuse, dilution may be performed with culture medium. This will result in a reduced fusion efficiency with a higher overall viability.

4.6 Alternative fusion protocols

(a) Adhesion of protoplasts at the interface of a glucose solution and a high density sucrose–PEG mixture (25) produces a high fusion efficiency and the protoplasts do not adhere to the glass coverslip.

(b) Protoplasts can also be fused by high voltage pulses (26). In this method, the protoplasts are brought into contact by application of a preliminary aligning voltage before the fusion pulse is applied. Electrical fusion utilizes a specialized (and often expensive) apparatus.

(c) If the parental protoplasts are highly fusogenic, it may be convenient to perform fusion in a test-tube (27). As before, high protoplast concentration and slow addition of the PEG and dilution solutions are required. Subsequent manipulations of the protoplasts are easier since they do not adhere to a glass coverslip but the proportion of hybrids formed is generally lower than with *Protocol* 7.

4.7 Selecting hybrids

Hybrids are difficult to identify unless there is a clear visual difference between the parents (*Figure* 7). Even so, parental characteristics will become less pronounced during culture, making heterokaryocytes impossible to follow over time. There are several approaches to dealing with this problem. Some practical suggestions are made below and more detailed information can be found in reference 5.

4.7.1 Staining parental protoplasts

As a short-term aid to visual identification of fusion products, one of the parental protoplasts may be labelled with a fluorescent dye such as fluorescein isothiocyanate (4).

4.7.2 Selection systems for culturing hybrids

Parental protoplasts may be eliminated if they are susceptible to a herbicide (or other toxin) to which the hybrids are resistant. An alternative involves pretreating each parent with a lethal physical or chemical agent. After fusion, only hybrids can survive since the metabolic deficiencies of each parent are complemented in the hybrid cytoplasm by metabolites from the other parent. This technique has been used to produce novel *Brassica oleraceae* × *Brassica napus* hybrids (28). In this case, *B. oleraceae* protoplasts were treated with iodoacetate and *B. napus* with γ radiation. The treatments were calibrated so that unfused parental protoplasts were killed.

4.7.3 Microculture of individual hybrids

Techniques have been developed for the culture of individual protoplasts (29). In this case, individual hybrids can be removed with a micro-pipette and cultured.

5. Preparation of plasma membrane fragments for electron microscopy

5.1 Introduction

The plant plasma membrane is difficult to examine in intact cells. The availability of protoplasts permits studies of both the external and internal surfaces of the plasma membrane (30). By attaching freshly isolated protoplasts to either glass slides or electron microscope grids and osmotically bursting them it is possible to expose the inner surface of large fragments of the plasma membrane. After careful washing, components of the plant cortical cytoskeleton (microtubules, actin microfilaments) and clathrin-coated pits and vesicles are revealed (*Figures 8* and *9*). This type of preparation has been useful for examining the distribution of microtubules with both fluorescence light microscopy and by negative staining for transmission electron microscopy. Ultrastructural studies of membrane fragments have also contributed important information regarding the morphology of clathrin-coated pits in plants. The protocol presented in this section describes the method for preparing plasma membrane fragments for study by electron microscopy. Details of the light microscope method are described by Van der Valk *et al.* (7).

5.2 Chemicals

- sorbitol
- Mes buffer
- $CaCl_2 \cdot 2H_2O$
- NaH_2PO_4
- Pipes buffer
- $MgSO_4$
- EGTA
- glutaraldehyde, 50%
- polylysine (Sigma, P-1274)
- uranyl acetate
- potassium phosphotungstic acid (PTA)

5.3 Apparatus

- 200 mesh electron microscope grids coated with Formvar with or without carbon (SPI Supplies, Toronto)
- glass welled staining dish
- fine forceps
- Parafilm 'M' (Canlab, P1150–4)

Protocol 8. Preparation of plasma membrane fragments for electron microscopy (see reference 7)

Prepare the following protoplast wash buffer:

- sorbitol at same molarity as used in isolation
- 1.5 mM Mes buffer
- 3.0 mM $CaCl_2 \cdot 2H_2O$
- 0.35 mM NaH_2PO_4

Prepare the following microtubule stabilizing buffer (MtSB) and adjust pH to 6.9:

- 100 mM Pipes buffer
- 1.0 mM $MgSO_4$
- 2.0 mM EGTA

1. Float coated grids face down on drops of 0.1% polylysine on a sheet of Parafilm.

Protocol 8. *Continued*

2. Rinse grid in drop of wash buffer and then submerge grid face up in drop of freshly isolated protoplasts suspended in wash buffer. Leave for 10 min to allow protoplasts to settle on the grid.
3. Gently transfer grid to large volume of wash buffer in staining dish to dislodge unbound protoplasts. During transfer retain a drop of wash buffer on the grid surface to prevent collapse of protoplasts.
4. Carefully remove grid from wash buffer and float face down on about 2 ml of MtSB in staining dish. Leave for about 20 min to allow bursting of protoplasts.
5. Carefully wash membrane fragments in a stream of MtSB to remove cytoplasmic remnants.
6. Transfer to 2% glutaraldehyde for 10 min. Rinse well with MtSB and air dry.
7. Stain with either 2% aqueous uranyl acetate (pH 4.0) or 1% aqueous PTA (pH 7.0) for 1 min, blot with filter paper, and air dry. Examine in an electron microscope.

5.4 Optimizing the preparation of plasma membrane fragments

It is important to use freshly isolated protoplasts for isolating plasma membrane fragments. The type of protoplast will also influence the quality of membrane fragments produced. Highly vacuolated protoplasts generally work best because the thin layer of cortical cytoplasm can usually be removed by washing. However, within a single protoplast preparation there may be considerable variation in quality of membranes. Protoplast bursting can also be achieved by using low concentrations of Triton X-100 but this treatment may remove most of the plasma membrane and disturb the distribution of microtubules (7). If staining with PTA is uneven, it can usually be improved by inclusion of bovine serum albumin (0.01–0.1%) in the stain solution.

References

1. Eriksson, T. R. (1985). In *Plant protoplasts* (ed. L. C. Fowke and F. Constabel), pp. 1–20. CRC Press, Boca Raton, FL.
2. Giles, K. L. (ed.) (1983). *Plant protoplasts* (Supplement 16 of *International Review of Cytology*). Academic Press, New York.
3. Fowke, L. C., Griffing, L. R., Mersey, B. G. and Tanchak, M. (1985). In *Plant protoplasts* (ed. L. C. Fowke and F. Constabel), pp. 39–52. CRC Press, Boca Raton, FL.

4. Constabel, F. and Cutler, A. J. (1985). In *Plant protoplasts* (ed. L. C. Fowke and F. Constabel), pp. 53–65. CRC Press, Boca Raton, FL.
5. Harms, C. T. (1985). In *Plant protoplasts* (ed. L. C. Fowke and F. Constabel), pp. 169–203. CRC Press, Boca Raton, FL.
6. Cutler, A. J. and Saleem, M. (1987). *Plant Physiol.*, **83**, 24.
7. Van der Valk, P., Rennie, P. J., Connolly, J. A., and Fowke, L. C. (1980). *Protoplasma*, **105**, 27.
8. Wetter, L. R. and Constabel, F. (ed.) (1982). *Plant tissue culture methods*. National Research Council of Canada, Saskatoon.
9. Wang, H., Cutler, A. J., and Fowke, L. C. (1989). *J. Cell Sci.*, **92**, 575.
10. Attree, S. M. and Sheffield, E. (1986). *Plant Cell Rep.*, **5**, 288.
11. Attree, S. M., Dunstan, D. I., and Fowke, L. C. (1989). *Can. J. Bot.*, **67**, 1790.
12. Marchant, H. J. and Fowke, L. C. (1977). *Can. J. Bot.*, **55**, 3080.
13. Widholm, J. M. (1972). *Stain Technol.*, **6**, 169.
14. Felix, H. (1982). *Anal. Biochem.*, **120**, 211.
15. Fromm, M., Taylor, L. P., and Walbot, V. (1985). *Proc. Natl. Acad. Sci. USA*, **82**, 5824.
16. Kao, K. N. (1982). In *Plant tissue culture methods* (ed. L. R. Wetter and F. Constabel), pp. 49–56. National Research Council of Canada, Saskatoon.
17. Sambrook, J., Fritsch, E. F., and Maniatis, T. (ed.) (1989). *Molecular cloning, a laboratory manual.* Cold Spring Harbor Laboratory Press, Cold Spring Harbor, NY.
18. Crabb, D. W. and Dixon, J. E. (1987). *Anal. Biochem.*, **163**, 88.
19. Ballas N., Zakai, N., Friedberg, D., and Loyter, A. (1988). *Plant Mol. Biol.*, **11**, 517.
20. Okada, K., Takebe, I., and Nagata, T. (1986). *Mol. Gen. Genet.*, **205**, 398.
21. Jefferson, R. A. (1987). *Plant Mol. Biol. Rep.*, **5**, 387.
22. Rhodes, C. A., Pierce, D. A., Mettler, J. J., Mascarenhas, D., and Detmer, J. J. (1988). *Science*, **240**, 204.
23. Krüger-Lebus, S. and Potrykus, I. (1987). *Plant Mol. Biol. Rep.*, **5**, 289.
24. Kao, K. N. and Saleem, M. (1986). *J. Plant Physiol.*, **122**, 217.
25. Kao, K. N. (1986). *J. Plant Physiol.*, **126**, 55.
26. Zachrisson, A. and Bornman, C. H. (1986), *Physiol. Plant.*, **67**, 507.
27. Fowke, L. C. (1982). In *Plant tissue culture methods* (ed. L. R. Wetter and F. Constabel), pp. 63–6. National Research Council of Canada, Saskatoon.
28. Kao, H. M., Keller, W. A., Gleddie, S., and Brown, G. C. (1992). *Theor. Appl. Genet.*, **83**, 313.
29. Kao, K. N. (1977). *Mol. Gen. Genet.*, **150**, 225.
30. Fowke, L. C. (1986). In *Hormones, receptors and cellular interactions in plants* (ed. C. M. Chadwick and D. R. Garrod), pp. 217–39. Cambridge University Press.
31. Attree, S. M., Bekkaoui, F., Dunstan, D. I., and Fowke, L. C. (1987). *Plant Cell Reports*, **6**, 480.
32. Rennie, P. J., Weber, G., Constabel, F. and Fowke, L. C. (1980). *Protoplasma*, **103**, 253.

9

Chemical analysis of components of the primary cell wall

S. C. FRY

1. Introduction

The primary† cell wall is an important and intriguingly complex component of the plant cell. It is tough but flexible, and has a defined porosity and plastic extensibility. It is often very thin (typically 100 nm thick), although collenchyma cell walls are conspicuous exceptions (up to 10 μm thick). The primary wall is an extraprotoplasmic fabric composed of several major homo- and hetero-polysaccharides and smaller amounts of glycoproteins and sometimes of lignin. The constituent polymers, except cellulose and lignin, are mainly water-soluble after extraction, but within the wall are cross-linked by a range of covalent and non-covalent bonds. The quantity and quality both of the individual polymers and of the cross-links define the physical and biological properties of the cell wall. The primary cell wall contains numerous enzymes which are capable of modifying it post-synthetically. Concerning its biological significance, the primary cell wall

- prevents the osmotic bursting of the protoplast
- limits the expansion of the cell (and thus the growth of the plant)
- provides a significant barrier to penetration by potential pathogens
- mediates the adhesion of adjacent cells within a tissue
- contains the latent form of endogenous signalling molecules (oligosaccharins)
- provides one of most definitive means of distinguishing cell types

For all these reasons, the primary cell wall is of considerable significance to plant cell biologists. In the present chapter I have assembled a range of relatively straightforward, effective chemical methods for the analysis of cell wall components. More detailed coverage is found in references 1–3.

† A primary cell wall is a wall, or wall layer, whose microfibrillar skeleton was laid down while the cell was still growing (i.e. irreversibly increasing in volume). Such walls, or wall layers, are still classed as primary even if lignin is deposited within them after the cessation of growth. Secondary walls are those whose microfibrils were laid down after the cessation of growth.

Table 1. Assays for major classes of sugar residues

	Total carbohydrates	Hexoses	Pentoses	Uronic acids	6-Deoxyhexoses
To	400 μl	500 μl	500 μl	200 μl	200 μl
of an aqueous solution or suspension containing					
	2–15 μg	5–50 μg	1–10 μg	1–20 μg	1–100 μg
of	carbohydrate	hexose	pentose	uronic acid	6-deoxyhexose
residues, add	10 μl A	–	67 μl B	–	–
then 1 ml of	conc. H_2SO_4[a]	0.2% anthrone in conc. H_2SO_4[a]	0.1% $FeCl_3{\cdot}6H_2O$ in conc. HCl[a]	0.5% borax in conc. H_2SO_4[a]	86% H_2SO_4[a]
The acid should be dispensed quite forcibly into the centre of the aqueous sample so as to cause maximum mixing; this is conveniently done from a dispenser bottle. The components should then be vortexed thoroughly to form a single phase. Incubate the					
tubes	on the bench	at 100 °C[b]	at 100 °C[b]	at 100 °C[b]	at 100 °C[b]
for	10 min	5 min	20 min	5 min	10 min
then cool in a bowl of cold tap water, add					
	–	–	–	20 μl C	20 μl D
mix thoroughly, incubate for a further					
	–	–	–	5 min	120 min
and then read the absorption at					
	485 nm	620 nm	665 nm	520 nm[c]	369 and 427 nm[d]
in a spectrophotometer or colorimeter. A standard curve should be constructed using a relevant representative of the group, e.g.					
	D-glucose	D-glucose	D-xylose	D-galacturonic acid	L-fucose

[a] **Caution**—corrosive acid. Solution becomes hot: wear gloves and face mask.
[b] In a boling water bath and with a glass marble on top of the test tube.
[c] For improved reliability, determine the difference in A_{520} before and after the addition of solution C.
[d] Determine A_{396} minus A_{427}.

A: 80% (w/w) phenol in water (available from BDH).
B: 6% (w/v) orcinol in ethanol.
C: 0.15% (w/v) *m*-hydroxybiphenyl in 1 M NaOH.
D: 3% (w/v) L-cysteine-HCl.

2. Semi-quantitative assays of classes of residues within wall polymers

Cell walls are usually isolated after homogenization of the tissue in an aqueous buffer, de-proteinization in phenol/acetic acid/water, and de-starching in 90% dimethylsulphoxide (1, 2). If the simpler method of preparing an alcohol-insoluble residue (1) is used, the presence of starch and non-wall proteins should be taken into account.

For many analytical purposes it is sufficient to determine the proportions of the major classes of building blocks (hexoses, pentoses, uronic acids, etc.) in the sample. The following methods permit such analyses and are all suitable both for polymers solubilized from cell walls and for the entire walls themselves. The analyses are simple and can provide precise quantitative data for comparing *qualitatively similar* samples. However, each spectrophotometric assay gives a different colour-yield for the different compounds within a class: an extreme example is that the 'total carbohydrate' method is three times more sensitive for xylose than for fucose.

2.1 Assay of classes of sugar residues

All the assays in *Table 1* include a treatment with hot concentrated acid which will hydrolyse both soluble and insoluble polysaccharides and glycoproteins to yield free monosaccharides, which then participate in a colour reaction. They can all be described in a generalized routine (*Table 1*) (4, 5).

2.2 Assay of amino acid and hydroxyproline residues

Since some (glyco)proteins are covalently bound to the cell wall and thus not amenable to assay by standard methods such as the Bradford, Lowry, or biuret tests, the only satisfactory means of measuring total wall protein is by assay of the amino and imino acids released by hydrolysis (*Protocol 1*) (6, 7).

Protocol 1. Assay of amino acid and hydroxyproline residues after hydrolysis of peptide bonds.

You will need the following reagents:

- 3 M NaOH
- solution E, i.e. 21% acetic acid, pH adjusted to 5.0 with 10 M NaOH; immediately before use, add 0.02 vol. of 0.049% NaCN
- 3% ninhydrin in 2-methoxyethanol
- 50% (v/v) propan-2-ol
- solution F, i.e. 180 μl neat bromine dissolved in 50 ml ice-cold 1.25 M NaOH and stored at 4 °C for 3–90 days (pre-cooled to 0 °C)

Protocol 1. *Continued*

- 16% (w/v) Na_2SO_3
- 5% *p*-dimethylaminobenzaldehyde in propan-1-ol
- 6 M HCl

You will need the following equipment:

- screw-capped Pyrex tubes[a]
- oven or heating block at 110 °C
- boiling water bath
- bench centrifuge
- colorimeter or spectrophotometer set at 560 and/or 570 nm

1. Pipette 250 μl of aqueous solution or cell wall suspension containing 0.1–2.0 μg of hydroxyproline residues or 0.4–8.0 μg of total protein, into a screw-capped Pyrex tube[a] and add 50 μl of 3 M NaOH.
2. Cap the tube securely (with a well-fitting Teflon liner in place) and heat at 110 °C for 4 h to hydrolyse (glyco)proteins. Cool. Proceed to section A or B.

A. *α-Amino acids*

3. Add 150 μl of solution E followed by 150 μl of 3% ninhydrin in 2-methoxyethanol.
4. Cap the tube and incubate in a boiling water bath for 15 min.
5. Remove from the water bath; while the samples are still warm, add 1.5 ml of 50% propan-2-ol and shake vigorously.
6. Cool, spin down any turbidity, and read A_{570} of the (blue) supernatant. If absorbance is off-scale, add more 50% propan-2-ol.

B. *Hydroxyproline*

3. Cool the sample from step 2 in an ice-bucket, add 300 μl of ice-cold solution F, shake, and incubate on ice for 3–10 min.
4. With the sample still on ice, add 15 μl of 16% (w/v) Na_2SO_3, shake, add 300 μl of 5% *p*-dimethylaminobenzaldehyde in propan-1-ol, shake, add 150 μl of 6 M HCl, and shake again.
5. Incubate in a 95 °C water bath for 2.5 min.
6. Cool the tube in tap water, spin down any turbidity, and read the A_{560} of the (pink) supernatant.

[a] Suitable tubes are the 14 × 100 mm Pyrex® culture tubes which are supplied with Teflon-lined screw-caps (Bibby). Check that each tube has a smooth, flat-surfaced mouth.

2.3 Total phenolics and lignin

Phenolic side-chains occur in small amounts on glycoproteins (tyrosine residues) and polysaccharides (feruloyl, *p*-coumaroyl, *p*-hydroxybenzoyl groups, etc.), where they act as potential sites of covalent cross-linking to form isodityrosine, diferulate, 4,4′-dihydroxytruxillate, etc. Phenolics are also major components of cutin and suberin, and lignin is a wholly phenolic polymer—these three substances occur in large amounts in and/or on some specialized cell walls, where they provide strength, waterproofing, resistance to microbial digestion, etc.

Owing to the diversity of compounds within this class, there is no fully satisfactory method for the determination of total phenolics. No assay has been found in which equal concentrations of different phenols give equal readings. However, the available methods are useful for survey work and especially for quantitative comparison of qualitatively similar samples. Some phenolics can be solubilized from cell walls by treatment with cold NaOH (*Protocol 9*), others require hot NaOH (e.g. at 170 °C, in a sealed metal bomb), and others, especially lignin, are not solubilized even by these drastic conditions. Material solubilized in aqueous solutions can be assayed by Folin and Ciocalteu's phenol reagent (*Protocol 2*) (8); total phenolics, including lignin, can also be assays after solubilization in acetyl bromide (*Protocol 3*) (9).

Protocol 2. Folin and Ciocalteu's phenol reagent

You will need the following reagents:

- Folin and Ciocalteu's phenol reagent (available from BDH)
- saturated aqueous Na_2CO_3

You will need the following equipment:

- colorimeter or spectrophotometer set at 750 nm (700 nm is also acceptable, although the assay then has lower sensitivity)

1. Mix 250 μl of clear aqueous solution (containing 0.5–25 μg of ferulic acid equivalents) with 250 μl of half-stength Folin and Ciocalteu's phenol reagent.
2. After 3 min at 25 °C, add 500 μl of saturated aqueous Na_2CO_3 and mix.
3. Incubate at 25 °C for 30 min and read A_{750}.

Protocol 3. Acetyl bromide method for assaying total phenolics

You will need the following reagents:

- acetyl bromide:glacial acetic acid (1:3, v/v) (**caution**, hazardous reagents)
- solution G, prepared by adding 500 ml of glacial acetic acid slowly and with cooling to 90 ml of 2 M NaOH
- 7.5 M hydroxylamine hydrochloride
- glacial acetic acid

You will need the following equipment:

- 70 °C water bath
- UV spectrophotometer set at 280 nm

1. Mix the dry sample (containing 10–500 μg phenolic material) with 1 ml of acetyl bromide:glacial acetic acid (1:3, v/v) (**caution**) in a glass tube.
2. Cap loosely and incubate at 70 °C with occasional shaking for 30 min.
3. Cool to about 15 °C, add 6 ml of solution G, mix, add 100 μl 7.5 M hydroxylamine hydrochloride (to destroy bromine and polybromide, which absorb at 280 nm), mix again, briefly spin down any turbidity, and read A_{280}[a] (the reading should be taken within 3 min of cooling the sample).
4. If the reading is off-scale, dilute the sample with glacial acetic acid.

[a] Acetylated lignin at 10 μg/ml in the final solution gives an A_{280} of about 0.24; other phenolics (including protein-bound tryosine) probably give values in a similar range.

2.4 Methyl and acetyl esters

The GalA residues of pectins are partially methylesterified, and the degree of esterification will determine the cation-exchange properties, gel-forming ability, and influence on wall porosity of pectic polysaccharides. It has been suggested that pectins are secreted from the protoplast in a highly esterified form and then partially de-esterified by the action of wall-bound pectin methylesterase. Analytically, the methyl groups are readily released (by cold alkali) as methanol, which can be assayed enzymically (*Protocol 4*).

Protocol 4. Methyl esters[a]

You will need the following reagents:

- 200 mM KOH
- 1 M KH_2PO_4
- an aqueous solution containing 20 mM 4-aminoantipyrine and 120 mM phenol
- 200 U/ml horse-radish peroxidase (RZ ≈ 3; Sigma Chemical Co.)
- 400 U/ml alcohol oxidase (e.g. from *Pichia pastoris*; Sigma Chemical Co.)

You will need the following equipment:

- bench centrifuge
- colorimeter or spectrophotometer set at 546 nm

1. Set up two replicate 0.75 ml samples (a, and b) of the polysaccharide solution of cell wall suspension containing 50–500 nmol of methyl ester.
2. To (a) add 200 μl of 200 mM KOH, incubate in a capped tube at 25 °C for 1 h to saponify the esters, then add 66 μl of 1 M KH_2PO_4; to (b) add 266 μl of a prepared mixture of 200 mM KOH/1 M KH_2PO_4 (3:1 (v/v)).
3. Bench-centrifuge (a) and (b) to pellet any insoluble cell wall material.
4. To 500 μl of each supernatant, add 50 μl of a solution containing 20 mM 4-aminoantipyrine and 120 mM phenol; mix; add 50 μl of horse-radish peroxidase solution (200 U/ml; Sigma Chemical Co.); mix again.
5. Add 40 μl of a solution of alcohol oxidase (400 U/ml) mix immediately, incubate at 25 °C for exactly 10 min; read A_{546} of (a) minus A_{546} of (b).

[a] I am very grateful to Dr Bernd Brümmer (Diversa GmbH, Hamburg) for providing information for this assay.

For uniformly ^{14}C-labelled cell walls or polysaccharides, a straightforward method is available for the assay of both [^{14}C]methyl and *O*-[^{14}C]acetyl ester groups (10). Suspend a sample of the ^{14}C-labelled material in 1 ml of 100 mM NaOH at 25 °C for 1 h, then add 3 ml of absolute ethanol, incubate at 4 °C for 1 h, and bench-centrifuge. Take three 1.0 ml samples (a–c) of the supernatant.

(a) To (a) add 250 μl glacial acetic acid, cap the vial tightly and store.

(b) To (b) add 250 μl glacial acetic acid, dry the contents of the vial *in vacuo* or under a stream of N_2 (**caution**: [^{14}C]acetic acid and [^{14}C]methanol will be volatilized), then re-dissolve the residue in 1.25 ml of freshly prepared ethanol/acetic acid/water (3:1:1 by volume).

(c) Dry the contents of vial (c) (**caution**: [^{14}C]methanol will be volatilized), then re-dissolve the residue in 1.25 ml of fresh ethanol/acetic acid/water (3:1:1).

(d) Add 12.5 ml of water-miscible scintillant to each vial and assay for ^{14}C:

- c.p.m. (c) minus c.p.m. (b) gives [^{14}C]acetate (which is volatile only at low pH)
- c.p.m. (a) minus c.p.m. (c) gives [^{14}C]methanol (which is volatile at high pH)
- c.p.m. (b) (i.e. labelled material released in low molecular weight form by NaOH, but not volatile at low pH) will be mainly [^{14}C]ferulate and related ester-linked phenolics.

3. Release of individual building blocks

For more detailed characterization and more precise quantification of polysaccharides and glycoproteins, it is necessary to conduct hydrolysis under conditions that permit recovery of the products. The hot concentrated acids used in *Table 1* are unsuitable because they degrade monosaccharides.

3.1 Acid hydrolysis

Polysaccharides and (glyco)proteins can be hydrolysed by hot acid to yield the constituent monosaccharides and/or amino acids—a fundamental procedure for the characterization of these polymers. Glycosidic bonds are generally more susceptible to hot acid than are peptide bonds. This is fortunate because the released monosaccharides are more acid-labile than free amino acids. For example, 2 M trifluoroacetic acid (TFA) for 1 h is sufficient to hydrolyse most polysaccharides (*Protocol 5*) whereas 6 M HCl for ~20 h is usually used for proteins—conditions that severely degrade sugars. For hydrolysis of proteins, the cell walls are heated in a solution containing 10 mM phenol and 6 M HCl at 110 °C in a sealed glass tube for 20 h (1). The HCl is removed *in vacuo* and the amino acids (often contaminated with brown degradation products of wall carbohydrates) are re-dissolved in dilute aqueous ammonia.

Protocol 5. Acid hydrolysis of non-cellulosic polysaccharides

You will need the following reagent:

- 2 M trifluoroacetic acid (TFA)

You will need the following equipment:

- oven, block-heater, or autoclave set at 120 °C
- vacuum concentrator (e.g. SpeedVac, Savant Inc.) or stream of nitrogen gas
- screw-capped Pyrex tubes[a]

1. Suspend or dissolve the sample in 2 M TFA (about 10 mg dry weight of sample per ml) in a screw-capped tube.
2. Heat at 120 °C for 1 h.
3. Cool the tube. Centrifuge down the insoluble material (cellulose, proteins, lignin, cutin, etc.). Recover the supernatant.
4. If necessary (but see comments on paper chromatography—Section 4.1.1), dry the sample to a syrup in a vacuum concentrator or under a stream of N_2.

[a] For suitable tubes, see *Protocol 1*, footnote *a*.

Microfibrillar cellulose is not efficiently hydrolysed by 2 M TFA, which cannot penetrate the crystalline structure; therefore special conditions are required for cellulose (and chitin etc.). For example, in the Saeman method (11), the walls are first dissolved in 72% (w/w) H_2SO_4 at room temperature for 1–3 h; the acid is then diluted 21-fold in water and the mixture heated at 120 °C for 1 h. The cellulose precipitates during the dilution step but in a form susceptible to acid hydrolysis. An alternative to Saeman hydrolysis is the use of Driselase (Section 3.2), which can also be used on the insoluble residue obtained after hydrolysis in 2 M TFA.

3.2 Enzymic hydrolysis

Driselase is a mixture of enzymes secreted by the fungus, *Irpex lacteus* (1). Its usefulness stems from the fact that it contains the enzymes required to hydrolyse, under very mild conditions, all the major polysaccharide of the primary cell wall to yield mono- and small oligosaccharides (*Table 2*). Driselase also contains pectin methylesterase (PME) so that homogalacturonans yield galacturonic acid whether or not they were methyl-esterified. Feruloyl and *p*-coumaroyl ester bonds remain intact during

Table 2. The major Driselase digestion products of some plant polymers

Polymer	**Major Driselase hydrolysis products**
Cellulose	Glc > Glc-β(1→4)-Glc
Arabinoxylans	Ara > Xyl-β(1→4)-Xyl > Xyl > 4-*O*-Me-GlcA-Xyl-Xyl[a], GlcA-Xyl-Xyl[a], Fer-Ara-Xyl[b], Fer-Ara-Xyl_2[b]
(1→3),(1→4)-β-D-Glucan (mixed-linkage glucan)	Glc > Glc-β(1→4)-Glc, Glc-β(1→3)-Glc, traces of higher oligosaccharides
(1→3)-β-D-Glucan (callose)	Glc > Glc-β(1→3)-Glc, traces of higher oligosaccharides
Xyloglucan	Xyl-α(1→6)-Glc > Glc, Gal > Fuc
Homogalacturonan	GalA, (trace of GalA-α(1→4)-GalA)
Rhamnogalacturonan-I	GalA, Ara, Gal, Rha (trace of GalA-α(1→4)-GalA)
Rhamnogalacturonan-II	GalA, (trace of GalA-α(1→4)-GalA), larger uncharacterized oligosaccharides
Chenopodiaceae pectins	GalA > Ara, Gal > Rha (trace of GalA-α(1→4)-GalA), Fer-Ara-Ara[b], Fer-Gal-Gal[b]
Extensin	Ara (only ~20–25% of the total Ara content, therefore possibly only the α-linked residues)
Arabinogalactan-proteins	Ara, Gal
Starch	Glc, (trace of Glc-α(1→4)-Glc), larger oligosaccharides
Inulin	Not digested

[a] Tentative identification.
[b] Fer = feruloyl ester; the corresponding *p*-coumaroyl esters may also be obtained. For other abbreviations, see *Figure 1*.

Driselase digestion, and thus diagnostic hydroxycinnamoyl-oligosaccharides can be isolated (12). Driselase also lacks galacturonoylesterase activities other than PME and thus provides an interesting tool with which to look for galacturonoyl–sugar ester linkages (13), potential novel cross-links in the primary cell wall.

Many growing plant cell walls are essentially completely solubilized by Driselase. However, some walls, especially lignified ones, are relatively resistant. Therefore, when analysis of hydrolysis products has to be quantitative, it should be established what proportion of the wall carbohydrates is solubilized by Driselase: this can be done by assay of any remaining insoluble material (*Table 1*).

Driselase as bought (e.g. from Sigma Chemical Co.) contains insoluble diatomaceous earth (~70–80% w/w) and soluble low molecular weight compounds, including phenolics and sugars. The diatomaceous earth is readily removed by bench-centrifugation of the suspension obtained by stirring Driselase in water; if necessary, all the sugar and most of the phenolics can be removed by a simple desalting procedure (*Protocol 6*) prior to use on cell walls (*Protocol 7*). However, for many applications, especially for the analysis of radiolabelled cell walls, the presence of contaminating non-radioactive solutes is not a problem and crude Driselase can be used.

Protocol 6. Partial purification of Driselase

You will need the following reagents:

- solid Driselase (Sigma Chemical Co.)
- 50 mM acetate buffer (Na^+, pH 5.0)
- solid ammonium sulphate
- 52% (w/v) ammonium sulphate

You will need the following equipment:

- magnetic stirrer
- refrigerated high-speed centrifuge
- 200 ml bed-volume column of Sephadex G-25 (in cold room if possible (a fraction collector is useful but, since the Driselase elutes fairly rapidly, manual collection of the appropriately coloured fraction is adequate)
- freeze-drier

1. Stir the Driselase (10 g) for 15 min at 0 °C in 100 ml of 50 mM acetate (Na^+) buffer, pH 5.0.
2. Bench-centrifuge the suspension at approximately 2000 *g* for 10 min to pellet the diatomaceous earth. Collect the clear brown supernatant and re-cool to 0 °C[a].
3. Add 52 g solid $(NH_4)_2SO_4$ per 100 ml supernatant, slowly, with constant stirring until the crystals have dissolved. Stand the solution at 0 °C for 1 h.
4. Centrifuge the turbid suspension at ≥ 20 000 *g* at 4 °C for 30 min to pellet the proteins. Thoroughly resuspend the pellet in 50 ml 52% (w/v) $(NH_4)_2SO_4$ and repeat the centrifugation.
5. Dissolve the pellet in 10 ml of water and pass the solution through a column of Sephadex G-25 (bed volume 200 ml; equilibrated and eluted (about 1–2 ml/min) with water). Reject the first colourless solution, then collect ~25 ml of the fast-eluting brown solution. Freeze-dry. Store the powder desiccated at <0 °C. It is stable for several years.

[a] For analysis of radioactive samples, or non-radioactive samples in which the low molecular weight contaminants of Driselase are inconsequential, the solution can be used at this stage (store frozen at −20 °C); the buffer in step 1 may be replaced by acetic acid/pyridine/H_2O (1:1:98), which has the advantage of being volatile.

Protocol 7. Driselase digestion of cell walls[a] or isolated polysaccharide

You will need the following reagents:

- buffer (pyridine/acetic acid/water (1:1:98, v/v/v) containing 0.5% (w/v) chlorbutol)[b]
- 0.5% Driselase (purified) dissolved in the above buffer
- 90% (w/v) formic acid

You will need the following equipment:

- incubator (with shaker or tube rotator if possible) at 25 or 37 °C

1. Mix the dry sample (10 mg) with 0.5 ml of purified Driselase in buffer (or 0.5 ml of the crude solution from step (2) of *Protocol 6*) and incubate with gentle agitation in a screw-capped tube at 25–37 °C for 16–48 h.
2. Simultaneously set up an enzyme-only control (without cell walls) to check for Driselase-autolysis products, and a walls-only control (with buffer lacking Driselase) to check for possible wall-autolysis and the presence of low molecular weight contaminants in the wall preparation.
3. Stop the reaction by addition of 50 μl of 90% formic acid[c]. Store the hydrolysis products frozen if they are not to be analysed immediately.

[a] For walls that prove to be relatively resistant to Driselase, a pretreatment in alkali may aid digestion. To the dry cell walls, add 0.25 ml 0.1 M NaOH, incubate at 25 °C for 1 h, add 3 μl HOAc (which brings the pH to ~4.7), and finally add the Driselase as described.

[b] To minimize microbial contamination; note that chlorbutol is detected by Dionex's pulsed amperometric detector; the samples should therefore be dried prior to analysis on a Dionex HPLC.

[c] The use of formic acid has the great advantage (e.g. over simply freezing or drying the reaction mixture) that the solution can be applied directly to chromatography paper without the Driselase attacking the paper. In the absence of formic acid (and even in the presence of acetic acid), Driselase is able to attack the paper during the loading of a chromatogram, generating spots of xylose, glucose, xylobiose, and cellobiose. In the presence of formic acid, a trace of mannose is the only sugar visible in driselase-only controls; this arises from autolysis of the Driselase.

3.3 Alkaline hydrolysis

Most glycosidic bonds are very resistant to alkali. However, alkali is useful for the hydrolysis of peptide bonds and for the release of ester-linked substituents such as ferulic and *p*-coumaric acid.

Alkaline hydrolysis (*Protocol 8*) splits peptide bonds, releasing mainly free amino acids; it has the advantage over acid hydrolysis that tryptophan is not degraded. Carbohydrate moieties attached to hydroxyproline remain attached, and thus extensins yield characteristic hydroxyproline-oligoarabinosides (Hyp-Ara$_n$s where $n = 1$–4); arabinogalactan-proteins yield hydroxyproline-polysaccharides in which the polysaccharide undergoes relatively little degradation. Carbohydrate moieties *O*-linked to serine or threonine are removed by β-elimination, and those *N*-linked to aspargine are removed by hydrolysis. Asparagine and glutamine are (as in acid hydrolysis) converted to aspartic acid and glutamic acid, respectively.

Protocol 8. Alkaline hydrolysis of peptide bonds

You will need the following reagents:

- saturated aqueous barium hydroxide (~0.18 M)[a]
- a cylinder of nitrogen gas
- 1% bromothymol blue
- dry ice

You will need the following equipment:

- oven or block heater at 110 °C
- one 50 ml glass beaker per sample
- microfuge

1. Suspend 5 mg of cell walls or glycoprotein in 0.5 ml of $Ba(OH)_2$[a] solution, flush the tube with N_2, cap securely, and heat at 110 °C for 5–10 h.
2. Cool, add 12 µl 1% bromothymol blue, and (especially important if the volumes being used are larger than 0.5 ml) transfer the solution into a flat-bottomed vessel such that the depth of the solution is <5 mm.
3. Cover the vessel with Parafilm, push a 1 ml plastic pipette tip through this and fill the pipette tip with crushed dry ice (solid CO_2). Shake or stir until the colour change (blue → yellow) indicates that the solution has been neutralized: $Ba(OH)_2 + CO_2 \rightarrow H_2O + BaCO_3\downarrow$.
4. Freeze and thaw the solution to enhance the precipitation of $BaCO_3$.
5. Microfuge to pellet the insoluble matter and collect the supernatant; resuspend the pellet in a further 0.5 ml H_2O and microfuge again. Pool the two supernatants; store frozen.

[a] **Poison**. Barium is used because of the ease with which it is removed from solution after hydrolysis. Store the stock $Ba(OH)_2$ solution, with an excess of undissolved solid $Ba(OH)_2$, in a well-stoppered bottle to exclude CO_2.

Alkaline hydrolysis will release ferulic, *p*-coumaric, caffeic, sinapic and *p*-hydroxybenzoic acids from their polysaccharide- or lignin-esterified form in the cell wall (*Protocol 9*). It will also release simple oxidative coupling products e.g. diferulic acid and light-generated dimers (truxillic and truxinic acids). It will not, under the conditions described, release phenolics that are ether-linked to polysaccharides or lignin.

Protocol 9. Alkaline hydrolysis of hydroxycinnamoyl esters

You will need the following reagents:

- 0.1 M NaOH
- a cylinder nitrogen gas
- 2 M and 10 mM trifluoroacetic acid (TFA)
- butan-1-ol

You will need the following equipment:

- bench centrifuge
- vacuum concentrator (e.g. SpeedVac) or stream of N_2

1. To 5 mg cell walls, add 0.5 ml of 0.1 M NaOH[a]. Quickly purge the tube with N_2, cap tightly, and incubate at 25 °C in the dark for 1 h.
2. Add 50 μl of 2 M TFA and mix well.
3. Add 0.5 ml of butan-1-ol.
4. Shake vigorously, then bench-centrifuge to separate the two phases. Put aside the upper (organic) phase.
5. To the aqueous phase, add a further 0.5 ml of butanol and repeat step 4.
6. Pool the two butanol solutions and add 0.5 ml 10 mM TFA; repeat step 4.
7. Dry the butanol solution *in vacuo* or under a stream of N_2. Re-dissolve in a small volume of acetone; store in a tightly-sealed vial in the dark at −20 °C.

[a] For heavily suberized or cuticularized cell walls, de-wax the walls by washing in chloroform/methanol (2:1) and use 0.1 M NaOH in 90% methanol.

4. Separation and assay of individual building blocks

4.1 Mono- and disaccharides

4.1.1 Paper chromatography

All the common monosaccharides produced by acid hydrolysis of cell wall polysaccharides, as well as the major disaccharides produced by Driselase digestion, can be separated using the simple but effective technique of paper chromatography (1). Its sensitivity is good; for example, as little as 0.4 μg of arabinose can be detected by the aniline hydrogen-phthalate stain. No derivatization and little or no work-up of samples are required. Up to 80 samples can conveniently be analysed in a standard chromatography tank overnight. Paper chromatography has the great advantage over column methods that *all* non-volatile compounds in the sample will be present on the chromatogram somewhere between R_f values of 0 and 1; in contrast, a compound may bind irreversibly to a chromatography column. For radioactive samples, planar (i.e. paper and thin-layer) chromatography also has the advantage of presenting the analytes in a form (a flat sheet) ideal for detection of radioisotopes (by autoradiography for near-perfect spatial resolution, or by scintillation counting for low activity samples): the whole available sample can be assayed for radioactivity and still recovered for further analysis if necessary (1).

i. Sample preparation

For non-radioactive samples, an aqueous solution of each hydrolysate is applied to Whatman no. 1 paper as a 1 cm diameter spot. Allow 2.5 cm (centre-to-centre) between samples. Ideally, aim for the sample to contain 5–100 μg of each sugar of interest. For low specific activity radioactive samples, it may be helpful to apply the sample as a streak rather than a spot. Common contaminants of biological samples, e.g. protein and salts, are relatively well tolerated (especially if butan-1-ol/acetic acid/water is the first or only solvent used), minimizing sample work-up. Driselase should be denatured with formic acid (*Protocol* 7) but can then be applied directly to the paper. TFA hydrolysates (*Protocol* 5) can be applied directly to chromatography paper (but not TLC plates): the acid evaporates without compromising the quality of the chromatogram. However, H_2SO_4 hydrolysates should be neutralized (e.g. with $BaCO_3$) prior to application because this non-volatile acid burns holes in paper when dried.

ii. Development

Suitable running solvents (compositions by volume) include:

- BAW (butan-1-ol/acetic acid/water, 12:3:5) for 16 h by the descending or ascending method

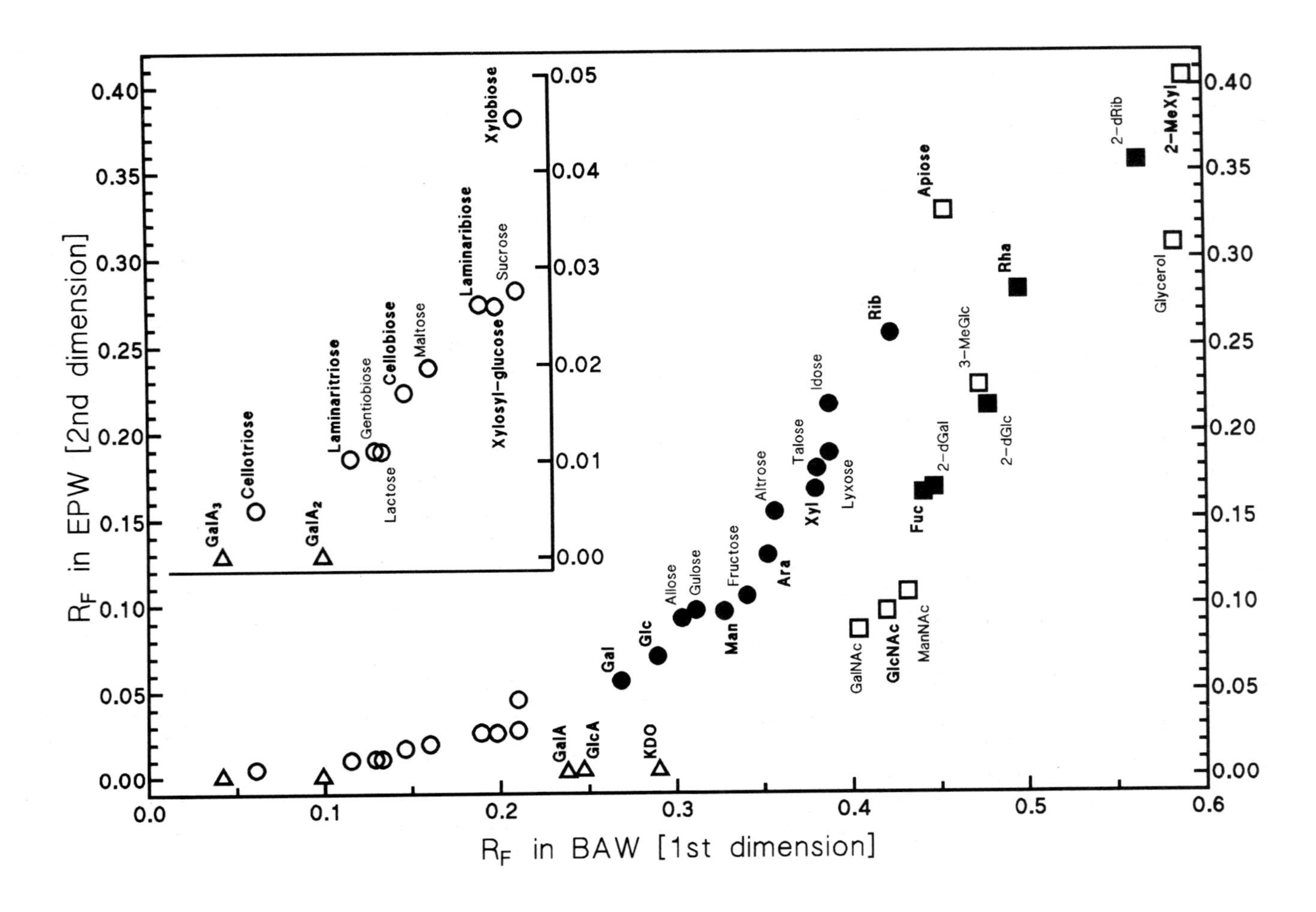

R_F in EPW [2nd dimension]
R_F in BAW [1st dimension]
0.00
0.05
0.10
0.15
0.20
0.25
0.30
0.35
0.40
0.0
0.1
0.2
0.3
0.4
0.5
0.6
0.01
0.02
0.03
0.04
0.05
GalA₃
Cellotriose
GalA₂
Laminaritriose
Gentiobiose
Lactose
Cellobiose
Maltose
Laminaribiose
Xylosyl-glucose
Sucrose
Xylobiose
GalA
GlcA
KDO
Gal
Glc
Allose
Gulose
Man
Fructose
Ara
Altrose
Xyl
Talose
Idose
Lyxose
GalNAc
GlcNAc
ManNAc
Rib
Fuc
2-dGal
Apiose
3-MeGlc
2-dGlc
Rha
2-dRib
Glycerol
2-MeXyl

- EPW (ethyl acetate/pyridine/water, 8:2:1), descending, for 16 h, the end of the paper being serrated to allow the solvent to drip off evenly
- BAW followed by EPW in the same dimension, with drying of the paper between the runs
- excellent resolution is achieved by two-dimensional chromatography, first in BAW and then, with the running direction rotated by 90°, in EPW, although only one sample can then be analysed per sheet.

The R_f values of major cell wall components are given in *Figure 1*.

iii. Staining

Reducing sugars are conveniently stained with aniline hydrogen-phthalate. Dissolve 16 g phthalic acid in 490 ml acetone plus 20 ml water and then add 490 ml diethyl ether (this mixture is stable for years); immediately before use, add 100 ml of the mixture to 0.5 ml aniline, dip the paper, dry for at least 3 min in a well-ventilated fume hood, and then heat the paper in an oven at 105 °C for 5 min. The reagent gives different colours with different classes of sugar, assisting identification:

- uronic acids stain orange
- hexoses stain greyish brown (faint spots show up more clearly as bright fluorescence under long wavelength UV)

Figure 1. R_f values of some major sugars, in descending chromatography on Whatman no. 1 paper (from reference 1, with corrections and additions). BAW, butan-1-ol/acetic acid/water (12:3:5, freshly prepared) for 16 h; EPW, ethyl acetate/pyridine/water (8:2:1). In EPW, the solvent is allowed to drip off the end of the paper so that low R_f compounds are better resolved. The time allowed for this will depend on the compounds of particular interest, e.g. 24 h for most monosaccharides or 72 h for disaccharides; as a guide, rhamnose migrates about 30–40 cm in 24 h. In BAW, in contrast, the development time should be kept constant because the solvent composition gradually changes during the run (butyl acetate is formed) and thus R_f values vary slightly with time. ○, Neutral oligosaccharide; ● neutral unmodified monosaccharide; △, acidic; ■ deoxymonosaccharide; □, other. Names in bold type are the major sugars encountered during the analysis of cell walls; those in light type are some related compounds. Abbreviations: Ara, arabinose; 2-dGal, 2-deoxygalactose); 2-dGlc, 2-deoxyglucose; 2-dRib, 2-deoxyribose; Fuc, fucose (= 6-deoxygalactose); Gal, galactose; GalA, galacturonic acid; $GalA_2$, galacturonosyl-α-(1→4)-galacturonic acid; $GalA_3$, galacturonosyl-α-(1→4)-galacturonosyl-α-(1→4)-galacturonic acid; GalNAc, *N*-acetylgalactosamine; Glc, glucose; GlcA, glucuronic acid; GlcNAc, *N*-acetyl-glucosamine; KDO, 2-keto-3-deoxy-*manno*-octulosonic acid; Man, mannose; ManNAc, *N*-acetylmannosamine; 3-MeGlc, 3-*O*-methylglucose; 2 MeXyl, 2-*O*-methylxylose; Rha, rhamnose (= 6-deoxymannose); Rib, ribose; Xyl, xylose; Xylobiose, xylosyl-ß-(1→4)-xylose; Xylosyl-glucose, xylosyl-α-(1→6)-glucose. Optical isomerism (D- versus L-) of the sugars is not indicated since this has no discernible effect on R_f. Inset shows the R_2f values of the di- and oligosaccharides on an expanded EPW axis.

- pentoses stain red
- 6-deoxyhexoses stain chocolate brown (faint spots show up more clearly as bright fluorescence under long wavelegnth UV)
- ketoses and apiose stain yellow
- disaccharides generally stain the colour expected of the reducing terminal monosaccharide moiety.

If the staining is carried out under carefully controlled conditions, the intensity of staining can usefully be quantified with a densitometer; alternatively, the spots can be quantified on a logarithmic scale by visual comparison with known loadings of authentic sugars.

iv. Confirmation of identity

Where the combination of R_f and colour of staining leaves any doubt about identity, the use of internal markers is extremely useful. When analysing non-radioactive material, spike the sample with a mixture of pure, high specific activity [^{14}C]sugars (1 kBq each, i.e. an unstainable amount) prior to loading, and locate these by autoradiography prior to staining. When analysing radioactive material, spike the sample with authentic non-radioactive sugars and stain these either after autoradiography or (after scintillation-counting in a water-immiscible scintillant) on strips of the paper washed in toluene and re-dried. Inclusion of *internal* markers in this way provides a very searching means of testing the identity of the unknown with the marker, since a mismatch of the centres of the spots by only 2–3 mm (out of a total paper length of ~450 mm) is sufficient to disprove the suspected identity.

4.1.2 Other chromatographic methods

Other methods of separating monosaccharides include:

(a) High-pressure liquid chromatography (HPLC) on a Dionex CarboPac PA1 column connected to a pulsed amperometric detector (14). Sensitivity is greater than for paper chromatography and quantification of non-radioactive (though not of radioactive) samples is simpler; resolution of monosaccharides is usually slightly better than in paper chromatography. Sample work-up is somewhat more demanding since it is recommended that Driselase and TFA are removed prior to sample injection.

(b) Gas chromatography (GC) after derivatization (with $NaBH_4$ and acetic anhydride) to yield alditol acetates (15). Note that fructose, which may occur in vacuolar polysaccharides that contaminate cell walls, is derivatized to a mixture of glucitol- and mannitol-hexaacetates, and is thus not distinguished from a mixture of glucose and mannose (unless a more involved method is used, e.g. derivatization with NaB^2H_4 and acetic anhydride followed by GC-MS to identify the position of the ^2H atom).

Glucose is also not distinguished from gulose because D-glucitol and L-gulitol are identical; similarly, arabinose is not distinguished from lyxose since L-arabinitol and L-lyxitol are identical. In contrast, fructose, glucose, arabinose, gulose, and lyxose are all separated from each other by paper chromatography and HPLC.

4.2 Amino acids

Methods are widely available for the analysis of the common protein amino acids (16). Hyp-Ara$_n$s are not available as markers but are readily identified by paper electrophoresis in pH 2.0 buffer, when they are usually the most slowly migrating cations (17, 18) and stain a characteristic orange colour with isatin/ninhydrin. Isodityrosine, one of the very few cationic phenols likely to be encountered, is also readily visualized by paper electrophoresis at pH 2.0 as a compound moving just faster than tyrosine (19); like tyrosine, but unlike dityrosine, isodityrosine is not fluorescent under long wavelenth UV in the presence of NH_3 vapour but does give a dark blue spot after spraying with Folin and Ciocalteu's phenol reagent followed by exposure of the paper to NH_3 vapour (1).

4.3 Phenolics

Many of the common phenolic acids of the plant cell wall can be resolved by TLC on silica-gel plates in benzene/acetic acid (9:1). The plate should be exposed to long wavelength UV light *during* development so that the hydroxycinnamic acids are maintained as an equilibrium mixture of the *cis*- and *trans*-isomers, which are thus not resolved (20). Detection of all phenolics is possible by their ability to form a dark spot against the fluorescent background of certain TLC plates (sold 'with fluorescent indicator'); alternatively spray the plate with Folin and Ciocalteu's phenol reagent: *o*- and *p*-dihydroxy phenols (e.g. caffeic acid) stain immediately; monohydroxy (e.g. ferulic and *p*-coumaric acids) and *m*-dihydroxy phenols only show full staining after exposure of the sprayed plate to NH_3 vapour.

Excellent resolution of the phenolic acids is also achieved by reversed phase HPLC on a C_8 or C_{18} column with a 30 min linear gradient of water/methanol/acetic acid (94:5:1 to 49:50:1) and detection by UV absorbance at 270 nm (21).

5. Recommended methods for assay of selected polymers

With the foregoing as general materials and methods, it is now possible to outline some favoured approaches to the identification and quantification of the major polysaccharides and glycoproteins of the primary cell wall.

5.1 Cellulose

The method of Updegraff (22) is widely used for the quantification of cellulose. The dry sample, containing 0.2–20 mg cellulose, is incubated for 30 min at 100 °C in 3 ml of acetic acid/water/nitric acid (8:2:1) to hydrolyse non-cellulosic carbohydrates. The remaining insoluble cellulose is washed thoroughly in water and assayed by the anthrone method (*Table 1*).

Alternatively, the cell walls are hydrolysed in TFA (*Protocol 5*) and the residue is then digested with Driselase (*Protocol 7*); the glucose produced is assayed chromatographically (*Figure 1*).

5.2 Hemicelluloses

The two major hemicelluloses of the primary cell wall are xyloglucan and arabinoxylan. TFA hydrolysis would not distinguish the source of the xylose and is thus not recommended. However, xyloglucan and arabinoxylan contain α- and β-linked xylose, respectively, and this leads to different products upon Driselase digestion (*Table 2*). The recommended method is therefore Driselase digestion (*Protocol 7*) followed by chromatographic analysis of the Xyl-α(1→6)-Glc (from xylogucan) and xylose plus Xyl-β(1→4)-Xyl (from arabinoxylan) (*Figure 1*). Fucose is another sugar largely derived from xyloglucan.

5.3 Pectins

It is usual to distinguish homogalacturonan, rhamnogalacturonan I (RG-I), RG-II (3). However, it appears likely that these three entities are domains within a larger pectin macromolecule. An assessment of total pectic material is readily made by measurement of total galacturonic acid (GalA) released after TFA hydrolysis (*Protocol 5*); rhamnose is also highly typical of pectins. If it is important to distinguish the three major domains, a pure pectinase (endopolygalacturonase) is required. Pectin should be saponified (with 0.5 M Na_2CO_3 at 0 °C for 18 h) and then neutralized and preferably desalted prior to digestion with pure pectinase. This enzyme cleaves homogalacturonan into $GalA_{1-3}$ fragments and thereby releases RG-I and RG-II, which can be separated from each other and from $GalA_{1-3}$ by gel-permeation chromatography on Sephadex G-75 or Bio-Gel P-60 in a volatile buffer such as acetic acid/pyridine water (1:1:18). the distribution of GalA between the fractions is assayed by the *m*-hydroxybiphenyl test (*Table 1*). After drying *in vacuo*, the fractions are hydrolysed with TFA (*Protocol 5*) and the characteristic products identified.

A simple indication of RG-II content, which does not require the use of pure pectinase, is given by the yield of 2-*O*-methylxylose and 2-*O*-methylfucose after TFA hydrolysis.

5.4 Glycoproteins

Extensins and arabinogalactan proteins (AGPs) are considered here, both of which are classes of hydroxyproline-rich glycoproteins though with very different properties and functions. Extensins are highly basic, possess only short (usually mono- to tetra-saccharide) sugar side-chains and may become covalently bound within the cell wall (probably via isodityrosine bridges (19, 23)); AGPs are acidic, have polysaccharide side-chains, and do not appear to become firmly bound within the cell wall (they are plasmalemma-bound and/ or soluble in the apoplast) (24).

Extracted glycoproteins can be separated into these two classes (basic and acidic) by ion-exchange chromatography (23) or by precipitation with β-glucosyl Yariv antigen (available commercially from Biosupplies Australia) (24), and the distribution of hydroxyproline residues assayed (*Protocol 1*). Separation is also possible by gel electrophoresis: the Yariv antigen stains AGPs; extensins give an unusual magenta with Coomassie Blue owing to their high pI.

Hydroxyproline residues (*Protocol 1*) of glycoproteins which remain associated with the cell wall after extensive washing (e.g. with salts, detergents, phenol, cold acids, and cold alkalis) are likely to belong to extensin. As confirmation, the products of $Ba(OH)_2$ hydrolysis (*Protocol 8*) should include Hyp-Ara$_n$s, isodityrosine, tyrosine, serine, and lysine.

Acknowledgements

The excellent technical assistance of Mrs Janice Miller is acknowledged. The work was supported by a European Community 'BRIDGE' contract.

References

1. Fry, S. C. (1988). *The growing plant cell wall: chemical and metabolic analysis*. Longman, Harlow.
2. Selvendran, R. R. and O'Neill, M. A. (1987). In *Methods of biochemical analysis*, Vol. 32 (ed. D. Glick), pp. 25–153. Wiley, NY.
3. O'Neill, M., Albersheim, P., and Darvill, A. G. (1990). In *Methods in plant Biochemistry*, Vol. 2 *Carbohydrates* (ed. P. M. Dey), pp. 415–441. Academic Press, London.
4. Dische, Z. (1962). In *Methods in carbohydrate chemistry*, Vol. 1. *Analysis and preparation of sugars*. (ed. R. L. Whistler and M. L. Wolfrom), pp. 475–514. Academic Press, New York.
5. Blumenkrantz, N. and Asboe-Hansen, G. (1973). *Anal. Biochem.*, **54**, 484.
6. Rosen, H. (1957). *Arch. Biochem. Biophys.*, **67**, 10.
7. Kivirikko, K. I. and Liesmaa, M. (1959). *Scand. J. Clin. Lab. Invest.*, **11**, 128.
8. Forrest, G. L. and Bendall, D. S. (1969). *Biochem. J.*, **113**, 741.
9. Johnson, D. B., Moore, W. F. and Zank, L. C. (1961). *TAPPI*, **44**, 793.

10. Gray, D. F., Fry, S. C., and Eastwood, M. A. (1993). *Brit. J. Nutr.*, **69**, 177.
11. Adams, G. A. (1965). In *Methods in carbohydrate chemistry*, Vol. 5 *General polysaccharides*. (ed. R. L. Whistler), pp. 269–75. Academic Press, New York.
12. Fry, S. C. (1982). *Biochem J.*, **203**, 493.
13. Brown, J. A. and Fry, S. C. (1993). *Carbohydr. Res.*, **240**, 95.
14. Fry, S. C. (1991). In *Methods in plant biochemistry*, Vol. 5 *Amino acids, proteins and nucleic acids*. (ed. L. J. Rogers), pp. 307–31. Academic Press, London.
15. Blakeney, A. B., Harris, P. J., Henry, R. I., and Stone, B. A. (1983). *Carbohydr. Res.*, **113**, 291.
16. Joseph, M. H. and Mardsen, C. A. (1986). In *HPLC of small molecules: a practical approach* (ed. C. K. Lim), pp. 13–28. IRl Press, Oxford.
17. Heath, M. F. and Northcote, D. H. (1971). *Biochem. J.*, **125**, 953.
18. Murray, R. H. A. and Northcote, D. H. (1978). *Phytochemistry*, **17**, 623.
19. Fry, S. C. (1982). *Biochem. J.*, **204**, 449.
20. Fry, S. C. (1983). *Planta*, **157**, 111.
21. Hall, R. D., Holden, M. A. and Yeoman, M. M. (1987). *Plant Cell Tissue & Organ Culture*, **8**, 163.
22. Updegraff, D. M. (1969). *Anal. Biochem*, **32**, 420.
23. Biggs, K. I., and Fry, S. C. (1990). *Plant Physiol.*, **92**, 197.
24. Komalavilas, P., Zhu, J.-K, and Nothnagel, E. A. (1991). *J. Biol. Chem.*, **266**, 15956.

10

Immunofluorescence techniques for analysis of the cytoskeleton

K. C. GOODBODY and C. W. LLOYD

1. Introduction

The cytoskeleton extends throughout the cytoplasm of plant cells forming a complex network of actin filaments, microtubules, and intermediate filament antigens; all three types of element have been shown to co-distribute to varying extents. During division the cytoskeleton is responsible for segregating the chromosomes and for positioning the new cell plate. During interphase the cytoskeleton is organized differently and serves other functions such as orienting the deposition of cell wall microfibrils, organizing the contents of the cytoplasm, and cytoplasmic streaming. The succession of structures formed by cytoskeletal filaments is important to cellular morphogenesis and the application of immunological techniques has greatly increased our understanding of this process by allowing the distribution of the various elements to be determined in a wide range of plant cells. Indirect immunofluorescence microscopy is now a reasonably straightforward technique that allows the three-dimensional complexity of cytoskeletal networks to be examined and appreciated. By studying populations of dividing cells, the spatial changes in cytoskeletal organization have been observed throughout the cell cycle and this has allowed something of cytoskeletal dynamics to be inferred. However, the more recent microinjection of cytoskeletal probes into living plant cells has provided direct images of the changing cytoskeleton.

Immunofluorescence techniques have proved very informative in studies of the structure and dynamics of the cytoskeleton. There are, however, some particular problems specific to plant cells which must be addressed if immunofluorescence techniques are to be employed.

Some plant cell walls fluoresce upon illumination under the epifluorescent microscope; this autofluorescence can be a major problem and can interfere with immunofluorescence patterns. Chlorophyll-containing chloroplasts, together with other unknown constituents of the cytoplasm, can also autofluoresce. Likewise, an area next to a wound, where phenolics have been released, can be difficult to study.

In larger cells, most of the volume is dominated by vacuoles which make fixation and sectioning difficult. There is a greater chance of encountering division figures in small, densely cytoplasmic, meristematic cells, but if the disadvantages of vacuolated cells can be overcome, they can be used with advantage to study the spatial arrangement of mitotic structures since the nucleus will migrate to the cell centre, suspended by conspicuous trans-vacuolar strands, in order to divide.

Unless the cell wall is punctured or removed completely, antibodies are unable to penetrate the cytoplasm. In many cell types, an additional obstacle to overcome is the waxy cuticle. This is particularly prominent on some types of leaves. Cutinases can be effective in the removal of the cuticle, but their general lack of availability is a barrier to study. The cell wall, however, is easily digested using a range of commercially available enzymes. This can be performed before or after fixation, and need only be a partial digestion to permeabilize the cell to antibodies. The cell membrane also needs to be permeabilized to allow antibody penetration. This can be achieved by physically sectioning through the cell, or, if whole cells are required, then the use of detergents will disrupt the membrane sufficiently, and this also helps reduce cytoplasmic autofluorescence. Air drying or the use of solvents can also help permeabilize membranes, but these are both rather destructive techniques and can cause distortion artefacts.

Plant cells have a high osmotic potential, i.e. water tends to try to move into them. However, the pressure of the wall prevents the net movement of additional water into the cell. Consequently, equilibrium of salt concentration is not reached and water continues to 'try' to move into the cell, maintaining a constant pressure (i.e. turgor) on the cell from the inside. When the cell wall is removed, unless the surrounding medium has the same osmotic potential as the protoplast, water will move through the cell membrane. This will result in the protoplast either bursting or collapsing, depending on the direction of the water flow. Therefore, protoplasts should be maintained in isosmotic medium until fixed. This is usually achieved by addition of mannitol or sorbitol. Although 0.3 M is commonly used, the precise concentration should be determined for each cell type by protoplasting in a range of osmotica from perhaps 0.1 to 0.8 M, to see which allows the protoplasts to retain their integrity without excessive shrinkage or swelling.

Since plant cells do not naturally adhere to surfaces, they need to be aided in this matter to enable their processing for the microscope. Large sections of tissue are reasonably easy to encourage to stay on a slide, but single cells or protoplasts from suspension cultures or squashes are more difficult. Air drying encourages cells to flatten and, hence, to adhere better. Indeed, air drying is a standard method (1) of also permeabilizing cells without further extraction. This is satisfactory with highly cytoplasmic cells but larger, vacuolated cells are distorted in the process. It is best to let such cells attach to poly-L-lysine-treated slides without drying. If an acetone-washed slide is coated with a 1 mg/ml solution of poly-L-lysine for at least 10 min and then rinsed with water, cells subsequently allowed to settle on these slides become firmly attached. Sometimes, however, it can be beneficial to air dry cells partially to aid penetration of antibodies, although there is invariably a trade-off with the quality of structural preservation.

Lastly, fixation of plant cells is important for good labelling. There are only a few possibilities concerning the choice of fixative. Solvents are best avoided because of the severe dehydration, and therefore distortion, that they cause. In general, aldehydes are the most reliable fixatives for plant studies. Formaldehyde is usually considered the best fixative for plant cells. A 4% (w/v) solution is recommended, but 8% can be used for some tissues, and 0.1% (v/v) glutaraldehyde is added to formaldehyde by some workers. However, glutaraldehyde can cause severe background fluorescence problems, even after reducing free aldehyde with borohydride. The length of time needed for adequate fixation varies with the size of the sample in question. For individual cells or protoplasts, around 15 min is reasonable, but for tissues fixed whole before sectioning, hours may be necessary.

2. Immunofluorescence microscopy

Essentially, immunofluorescent labelling is a detection method making use of the ability of antibodies to recognize specific antigens with the primary antibodies then located in the specimen by secondary antibodies to which fluorochromes have been conjugated. Fluorochromes excited by light of a certain wavelength emit light of a longer wavelength which is detected; epifluorescence illumination is preferable (see Chapter 1). There are two types of immunofluorescent labelling: direct and indirect. The first refers to the fluorochrome being directly linked to the antibody which will recognize the antigen and is rarely used now. The latter refers to the use of a fluorochrome-conjugated secondary antibody to detect the unlabelled first antibody bound to its antigen. Indirect immunofluorescence provides a stronger, amplified signal since more molecules are labelled. Also, secondary antibodies linked to a range of fluorochromes are available commercially, avoiding the necessity to purify and label all the primary antibodies. The main drawback of this method is that non-specific background is sometimes amplified. The two most commonly used fluorochromes are fluorescein isothiocyanate (FITC) (which gives a green emission) and rhodamine (red). There are several other fluorochromes but none is quite as widely available as these two. Most fluorescent microscopes are equipped with FITC/rhodamine/UV filter sets. The emission spectra of FITC and rhodamine do not overlap, making it possible to perform double labelling experiments using two separate antibodies, each labelled with a separate fluorochrome, on the same specimen.

2.1 Outline of method

There are five stages:

(a) Cell or tissue preparation: the cells or tissue sections must be attached to a support (most commonly by the use of poly-L-lysine) to permit labelling.

(b) Fixation: the antigen must be properly preserved for successful labelling.

(c) Permeabilization: digestion of the plant cell wall and/or permeabilization of the plasma membrane must occur to allow sufficient antibody penetration.

(d) Antibody binding: incubations with both primary and secondary antibodies allow localization of the antigen, and then unbound antibody is removed by washing.

(e) Detection using epifluorescence microscopy.

2.2 Adapting methods to a new cell type

When first confronted with a specimen to process for immunofluorescence several parameters may have to be varied. If the sample is a suspension culture, then *Protocol 3* is recommended, in which protoplasts are made first. The most likely variables that could need altering are the cocktail of enzymes (try the suggested ones first, then experiment with different concentrations of each, or try different enzymes) and the osmoticum. If the suggested amount of mannitol or sorbitol does not yield stable, spherical protoplasts, try a range of molarities. If a sample of cells is required, rather than protoplasts, then very briefly treat the cells with enzyme before proceeding with the method for protoplasts. Alternatively, the same method could be used, but fixing the cells before digesting their walls.

The size and shape of cells in a tissue will influence the method employed. In meristematic tissue, where the cells will be small and very cytoplasmic, a cryostat can be used to produce thin sections, or the tissue can be squashed and air dried as in *Protocol 2*. For the larger cells of leaves or stems a different approach is necessary. These tissues typically consist of large vacuolated cells with thick cell walls and sometimes a cuticle. Hand sectioning has been used successfully for such tissue but only cut cells will be stained and the large cells must not be allowed to dry (*Protocol 6*), otherwise they will collapse. As suggested in the protocol, if antibody labelling fails then brief enzyme treatments, partial air drying, and the use of detergents such as Triton X-100 should be tried. Different plant cells often require very different techniques.

2.3 FITC conjugates for labelling

FITC is excited by light of wavelength 450–500 nm (blue), with a maximum absorption at 490 nm. It emits at 500–550 nm (green). Rhodamine absorbs green light (wavelengths between 520 and 560 nm) and emits red light (550–600 nm). The overlapping of these excitation and emission spectra may cause a problem when double labelling. The green light emitted by FITC could excite the rhodamine molecules, possibly allowing both elements to be viewed simultaneously, and cause fading of the rhodamine fluorescence. Such 'bleed-through' can be overcome by ensuring that the molecule with the longest excitation wavelength is viewed first when using multiple fluorescent tags. Then, after taking photographs, one can switch to the shorter wavelengths.

2.4 Antibodies

2.4.1 Source, storage, and dilutions

Commercial antibodies to cytoskeletal components are available from companies such as Sera Labs, Amersham, Sigma, and Polysciences. All will have been raised against animal cytoskeletal proteins but some are known to cross-react with plant proteins. A recommended monoclonal anti-tubulin antibody is YOL 1/34 (sold as culture supernatant by Sera Labs) which will reliably label microtubules preserved by any of the techniques in this chapter if used at 1/100 dilution. Secondary antibodies must be chosen to recognize the species in which the primary antibody was raised. For example, if using YOL 1/34, which was raised in rat, then the secondary antibody which is conjugated to the fluorochrome must be anti-rat, e.g. FITC-linked, rabbit anti-rat IgG. Recommended suppliers of secondary antibodies are Dako and Sera Labs but others might work equally well. The titre of each antibody will vary and therefore a useful concentration must be empirically determined. Test a dilution series of 1/10, 1/100, 1/1000, then refine the scale to obtain the greatest dilution giving clear specific staining of the primary antibody. Controls, in which the secondary antibody is used at that dilution without the primary antibody, should not produce non-specific background staining. Commercial secondary antibodies are likely to work admirably at quite high dilutions; to begin with, try 1/100. Antibodies should be diluted in 3% (w/v) bovine serum albumin (BSA) in microtubule-stabilizing buffer (MTSB) with 0.01% (w/v) sodium azide as a preservative. They can be stored without loss of activity at 4 °C for months, but undiluted stocks should preferably be stored in small aliquots at −20 °C or −80 °C. After dilution, fluorochrome-conjugated antibodies should always be centrifuged for 5 min at 10 000 *g*, 4 °C, to allow any unlinked, coagulated fluorochrome to be pelleted.

2.4.2 Double labelling

When performing double labelling experiments, it is important to ensure that the two primary antibodies were raised in different species so that different secondary antibodies can be successfully employed, otherwise there will be no separation of the two labelled antigens. It should also be borne in mind that anti-rat antibodies may recognize a mouse antibody. Some researchers choose to incubate the specimen with both primary antibodies at once, then wash and apply both secondary antibodies together, others prefer to perform all incubations separately but these operations must incorporate appropriate controls.

2.4.3 Controls

To check that the primary antibody reaction is specific the antibody should be pre-adsorbed with purified antigen; this should result in loss of any specific staining pattern. Further, a slide with only the secondary antibody should be prepared to test that there is no contribution from the secondary antibody. To ensure that preparative techniques were successful, it is useful to include a reliable antibody such as a commercial anti-tubulin antibody (e.g. YOL 1/34 supernatant from Sera Labs at 1/100, followed by FITC-linked anti-rat IgG from Dako at 1/100). If there is no specific staining, it is likely that the steps leading up to antibody labelling were not effective and effort should be directed toward better preservation of the antigen. When using several different antibodies, great care should be taken to avoid cross-contamination. Keep different antibodies on separate slides and use fresh pipette tips for each when adding solutions or washing. Controls are most important for double labelling experiments. As well as testing individual secondary antibodies, check combinations of different secondary antibodies without primaries. Also, test that no cross-reactivity is observed when various primary and secondary antibodies are cross-paired.

2.4.4 Sample preparation

It is perhaps noteworthy that certain antibodies may only work if the antigen is preserved in a specific way. If an antibody was raised against a formaldehyde-fixed structure, it is likely that it will only recognize that form of antigen. Some antibodies have been found to require the dehydrating effect of a methanol treatment in addition to aldehyde fixation. Microtubules are generally fixed adequately with 4% (w/v) formaldehyde. Actin is more difficult to fix in that the larger actin cables are preserved by formaldehyde, but the finer filaments may not remain. F-actin also sometimes requires a methanol treatment to help reveal the epitope for antibody labelling. Methods that have been devised to help overcome these problems are discussed in Section 2.6. Intermediate filament antigens provide another problem. The larger bundles are stabilized by formaldehyde, and so are the finer networks if the specimen is not allowed to dehydrate. In general, it is a good idea to start with 4% (w/v) formaldehyde, then try alternatives only if this fails.

2.5 Chromatin staining

Chromatin is generally stained with DNA-specific dyes which are excited by UV light. One such dye is DAPI (4,6-diamidine-2-phenylindole dihydrochloride), which is prepared as a 0.5 mg/ml stock in water and stored at 4 °C. A 1 μg/ml final concentration of DAPI (Sigma Chemical Co.) should be used on slides after antibody labelling, for 1–10 min, followed by a final wash in MTSB. This will give good staining of nuclei and individual chromosomes of dividing cells.

2.6 Actin microfilament labelling

Not all animal anti-actin antibodies cross-react with actin from plants but the commercial anti-actin antibody from Amersham works well using standard fixation conditions employed for anti-tubulin labelling. Fine actin filaments are notoriously difficult to preserve, especially with aldehyde fixatives, and alternative methods have been developed to combat this. One technique is to treat unfixed, detergent-permeabilized cells with rhodamine-phalloidin (2). Briefly, equal volumes of cell suspension and extraction buffer (MTSB, 5% (v/v) dimethylsulphoxide (DMSO), 0.025% (v/v) Nonidet P40, 100 mM mannitol) are mixed then made 10^{-6} M with rhodaminyl lysine phallotoxin. An alternative method uses *m*-maleido-benzoyl *N*-hydroxysuccinimide ester (MBS), to cross-link and help stabilize actin filaments before they are subjected to brief aldehyde fixation (3). In this method, suspension cells are treated with 100 μM MBS in phosphate-buffered saline (PBS: 20 mM sodium phosphate buffer, pH 7.0, 150 mM NaCl) containing 0.05% Triton X-100, for 30 min, and then fixed with 3.7% formaldehyde. Finally, an approach which combines elements of these two methods is to add MBS to the detergent extraction step, then immediately add rhodamine-phalloidin (4). Whole leaves are cut and mounted on a slide with extraction buffer containing 300 μM MBS. The buffer is MTSB with 0.05% (v/v) Nonidet P40. Rhodamine-phalloidin is immediately added to the slide at 10^{-7} M and this is left in the dark for 10 min before viewing. This method stabilizes the finest actin filaments of unfixed samples for long enough to enable photographs to be taken. If double labelling with an antibody is required, the MBS should be added before conventional fixation as in the Sonobe and Shibaoka method (3), and the rhodamine-phalloidin should be added after antibody labelling.

2.7 Anti-fade mountants

After a short time of excitation, fluorochromes become quenched and their emission fades (for a general discussion of dye photobleaching see Chapter 2).

There are ways to postpone the onset of this problem. The exposure of the specimen slides to light should be minimized by storing them in the dark at 4 °C and by including an anti-fade reagent in the mounting medium. All anti-fade reagents are anti-oxidants and include 10% 1,4-diazobicyclo-(2,2,2)-octane (DABCO) and freshly made 5% *n*-propyl gallate in 90% glycerol. CITIFLUOR contains DABCO and is recommended for use with FITC molecules. It is available from City University, London, UK. It is made in several forms: AF1 contains PBS and glycerol and is for general use when mounting slides. Most of the liquid on the slide should be removed, then a drop of anti-fade should be placed on the slide, followed by a coverslip. Slides can be made semi-permanent by sealing around the edge of the coverslip with nail varnish, and storing them in the dark at 4 °C, or even −20 °C.

3. Materials

3.1 Formaldehyde

A solution prepared from powdered paraformaldehyde (Sigma or BDH) acts as a better fixative than formalin. Generally, the solution can be prepared in large quantities and frozen. For routine work, frozen formaldehyde is an adequate fixative, but for the best results, especially when fixing larger blocks of tissue, it is advisable to use freshly made formaldehyde.

Protocol 1. Preparation of formaldehyde

1. For an 8% (w/v) final solution, make a 16% (w/v) solution in water but do not yet add the total amount of water. Add drops of 8 M KOH until the pH is above 10.0 while heating to about 60 °C inside a fume hood. Heating at alkaline pH depolymerizes the paraformaldehyde.
2. Once the solution is completely clear, allow it to cool before using H_2SO_4 to bring the pH to 6.9 (do not use HCl, as this makes a carcinogenic product)[a]
3. Make the solution to its final volume in water, then filter it through a 0.45 μm filter to remove any particulate matter. This solution can then be used as fresh formaldehyde by diluting with 2 × MTSB, or it can be stored in aliquots at −20 °C for a few months[b].

[a] This method is more sensible than making the solution directly in buffer, as this requires far more ions to raise the pH to the level necessary to dissolve the paraformaldehyde.

[b] Do not thaw and re-freeze unused aliquots. After thawing, it may be necessary to warm the aliquot slightly to clear the solution.

3.2 Cell wall digesting enzymes

Enzymes used routinely in this chapter are listed below:

- cellulase 'Onozuka' R-10 and R-S (Yakult Honsha Co., Tokyo)
- hemicellulase (Sigma)
- pectolyase Y23 (Yakult Honsha Co., Tokyo)
- Driselase (Sigma)

Note that enzymes should be weighed out in a fume hood as the dusts are extremely hazardous to health. Even the most expensive enzymes are still fairly crude so it is a good idea to centrifuge enzyme solutions at about 3200 *g* for 5 min, using only the supernatant for the experiment. If making protoplasts overnight, keep the enzyme mixture at 55 °C for 5 min to inactivate proteases (5). Alternatively, add protease inhibitors or BSA to act as a protein substrate for any proteases. All enzymes must be kept dry and can be stored at 4 °C for many months. Stocks should be frozen at −20 °C for long-term storage. To avoid condensation, stocks should be allowed to warm to room temperature before opening.

3.3 Buffers and solutions

(a) **Microtubule stabilizing buffer** (MTSB): microtubule stabilizing buffer consists of 50 mM Pipes (pH 6.9), 5 mM $MgSO_4$, and 5 mM EGTA. It is a good idea to keep stocks of 1 M Pipes, 1 M $MgSO_4$, and 200 mM EGTA ready for use. These, together with MTSB, can be stored at 4 °C for several months. All three stock solutions should be brought to pH 6.9 using KOH. The Pipes should be the free acid form and not the disodium salt (Sigma). Always store stock solutions of EGTA in plastic, not glass, as a concentrated solution could erode the glass with time. For every 200 ml of 1 M Pipes, approximately 20 g of solid KOH will be needed to enable the Pipes to dissolve.

(b) **Poly-L-lysine**: molecular weight > 250 000 (Sigma); used at 1 mg/ml in water. Store stock solution at 5 mg/ml in aliquots at −20 °C. Can be reused about three times.

(c) **DAPI** (4,6-diamidine-2-phenylindole dihydrochloride) (Sigma): 0.5 mg/ml stock in water. Store at room temperature, in the dark. **Beware**: this is a carcinogen.

(d) **MBS** (*m*-maleido-benzoyl *N*-hydroxysuccinimide ester) (Pierce): 100 mM stock in DMSO. Store at 4 °C. Use at 100–300 μM so that the sample is never exposed to a DMSO concentration greater than 0.3%.

(e) **Rhodamine-phalloidin** (Molecular Probes, Cambridge Biosciences): a stock solution of 3.3 μM should be prepared in methanol, then diluted

1/1000 in 0.5 M Pipes before use. Use about 3 μl per well to give about 10^{-6} to 10^{-7} M. **Beware**: very toxic.

(f) **Mes wash**: 0.5% (w/v) Mes (pH 5.8), 80 mM $CaCl_2$, 0.3 M mannitol. This can be made in large quantities and autoclaved for storage at room temperature in small bottles, using a fresh bottle for each experiment.

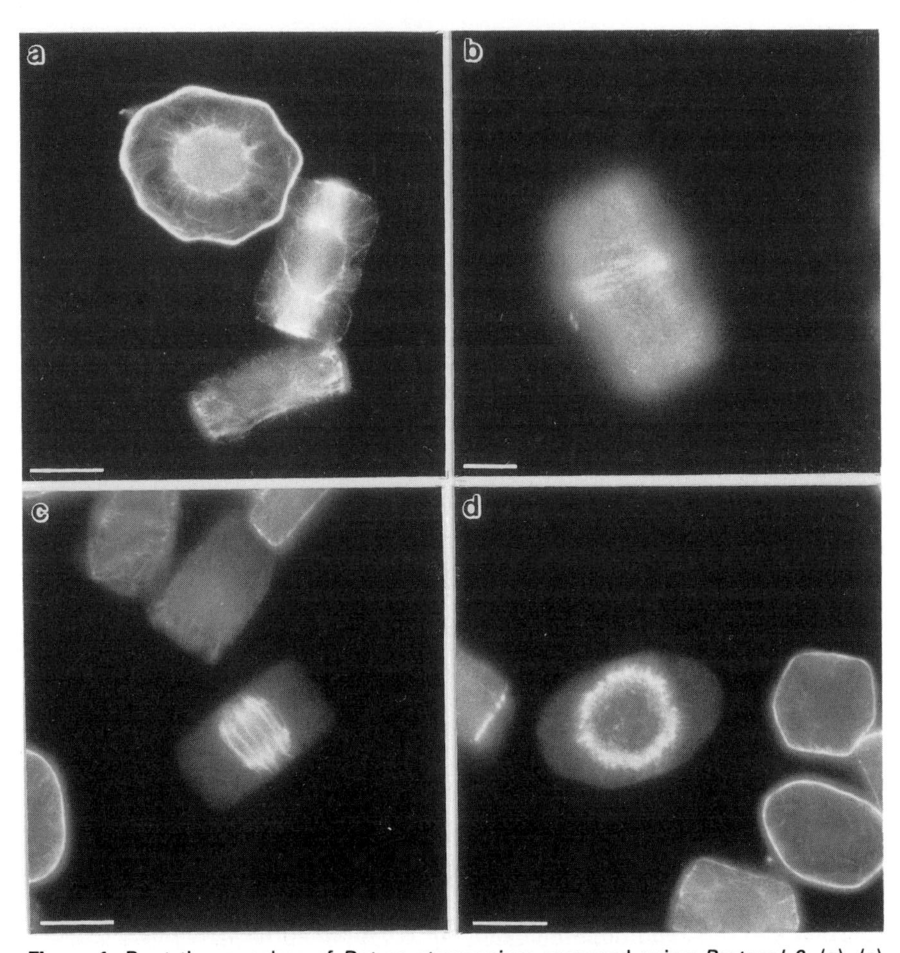

Figure 1. Root tip squashes of *Datura stramonium* prepared using *Protocol 2*. (a), (c), and (d) were all labelled with YOL 1/34, an antibody against tubulin. In (a), a group of three cells is seen. One shows the interphase array of microtubules surrounding the cell. The other two cells in (a) both have pre-prophase bands labelled with the anti-tubulin antibody. One is seen to be longitudinal, while in the polygonal cell, microtubules are seen radiating from the central nucleus. (b) A pre-prophase band labelled with an antibody against plant intermediate filament antigens which also recognizes cytokeratin 8 in appropriate animal cells (12). In (c), a clear metaphase spindle has been labelled, while in (d), a late phragmoplast is shown *en face*. Scale bar = 15 μm.

4. Methods

4.1 Dividing tissues

Protocol 2. The cytoskeleton in root tip cells

Root tip cells are possibly the easiest of systems in which to gain good microtubule staining. The following is a reliable technique based upon that first described by Wick *et al.* (1). Although phosphate buffer, as in the original method, is satisfactory for root tip cells, it can fragment microtubules in cells such as root hairs, and for this reason the phosphate has been replaced by Pipes buffer (6).

- MTSB
- 4% (w/v) formaldehyde in MTSB
- 2% (w/v) cellulase ('Onozuka' R-10) in MTSB
- primary antibody in 3% BSA, MTSB
- secondary antibody in 3% BSA, MTSB
- DAPI
- anti-fade mountant
- multiwell slides (Flow Labs) and coverslips

1. Germinate approximately 30 seeds of onion, wheat, or *Datura stramonium* on moistened tissue paper in a sealed plastic box placed in the dark at 25 °C for about 4 days.
2. Fix the tips (about 2–3 mm excised with a razor blade) of the germinated roots in 1 ml of 4% formaldehyde in MTSB for 45 min at room temperature in an Eppendorf tube. Larger root tips such as pea should be cut longitudinally to aid penetration of fixatives and enzymes. Wash the fixative out of the tissue with at least three buffer changes over 45 min.
3. Digest the cell wall by replacing the MTSB with 1 ml of 2% (w/v) cellulase 'Onozuka' R-10 in MTSB for 6 min[a].
4. Wash out the enzyme by exchanging for MTSB at least three times over 45 min.
5. Squash the prepared root tips gently, preferably on a multitest slide (Flow Labs), using a small rod (such as the blunted end of a Perspex cuvette stirrer) and only the minimum amount of buffer necessary to prevent the root tip from immediately sticking to the slide, or moving around. Do not over-macerate. A single press can be sufficient to split the tip, allowing cells to float out when the tip is moved around by

Protocol 2. *Continued*

forceps. Observe this process under a light microscope, selecting wells which contain many good rectangular cells and rejecting those that do not. Use fine forceps to remove any large debris and allow the isolated cells to air dry (at least 15 min).

6. Apply the primary antibody at the required dilution in 3% BSA/MTSB, sufficient to cover each well (approximately 20 μl), and place the slide in a closed plastic container with a moist tissue in the bottom (to maintain humidity) for 1 h at room temperature.

7. Wash the slide carefully by removing the antibody by aspiration and flushing the wells at least three times with MTSB for 5 min each. Note that some poorly attached cells will be lost during the rigours of washing. Hussey *et al.* (5) used a microsyringe for washing to prevent detergent-extracted cytoskeletons from being dislodged.

8. Incubate the slide with the secondary antibody diluted in 3% BSA/MTSB, in the same manner as the primary for 1 h, this time excluding the light (by covering the container in foil) to prevent fading of the fluorochrome.

9. Wash three or four times in MTSB, and finally in DAPI stain (if required) by adding 15 μl of 1 μg/ml solution to each well for 1 min, then rinsing wells once more with MTSB. Remove as much buffer as possible before adding a few drops of anti-fade mountant between the wells and placing a coverslip on top. Press down lightly on the slide with a folded tissue to remove excess anti-fade, then seal the edges of the coverslip with nail varnish.

[a] This timing is fairly important. It may be less for softer tissue (about 4 min minimum for soft onion roots) or more for firmer and larger roots e.g. maize (up to around 15 min).

4.2 Suspension-cultured cells

Access to a suspension culture of plant cells offers the advantage of a regular supply of easily maintained cells, many of which will be undergoing division, and are reliably stainable. Several different cultures are easy to process for cytoskeletal staining, including *Daucus carota* (carrot), BMS (*Zea mays* v. Black Mexican sweetcorn), and BY-2 (*Nicotiana tabacum* v. Bright Yellow-2).

Protocol 3. The cytoskeleton in protoplasts

Take care not to let the protoplasts dry out at any stage during this protocol and change solutions gently to avoid washing protoplasts away.

- MTSB
- enzyme mixture: 1.5% cellulase 'Onozuka' RS, 0.7% hemicellulase, 0.1% pectolyase Y23 in Mes
- 4% (w/v) formaldehyde in MTSB, 0.3 M mannitol
- immunolabelling reagents, slides etc. as *Protocol 2*
- orbital shaker
- bench centrifuge and tubes

1. Prepare 10 ml of the following enzyme mixture in Mes wash: 1.5% (w/v) cellulase 'Onozuka' RS, 0.7% (w/v) hemicellulase, 0.1% (w/v) pectolyase Y23.
2. Using 3–5 day old cells, allow about 25 ml of cell culture to settle, or collect by centrifuging at 150 *g* for 5 min. Replace medium with enzyme mix and incubate for 1–3 h at 25 °C on orbital shaker.
3. Check for the presence of spherical protoplasts under the microscope before proceeding.
4. Centrifuge at 150 *g* for 5 min and remove the enzyme solution by gentle aspiration. Add 20 ml Mes wash, invert, centrifuge again, and remove the supernatant.
5. Resuspend the washed protoplasts in 5 ml Mes wash and place drops of the suspension on to poly-L-lysine-treated slides. Allow to settle for 10 min, then remove excess protoplasts with a pipette.
6. Fix protoplasts for 15 min in 4% (w/v) formaldehyde/MTSB/0.3 M mannitol.
7. Extract protoplasts by replacing fixative with 1% Triton X-100/MTSB/ 0.3 M mannitol for 10 min[a].
8. Wash slides five times with MTSB (or until liquid stays in wells), taking care that protoplasts are not washed away.
9. Immunolabelling: apply the primary and secondary antibodies directly to the wells (see *Protocol 2*); mount the slide with anti-fade and seal the coverslip with nail varnish.

[a] Steps 6 and 7 can be reversed i.e. extraction can be done before fixing, but this may lead to many protoplasts being washed away. However, it does allow better antibody penetration and often bursts protoplasts leaving residual plasma

Protocol 3. *Continued*

membrane discs with cytoskeletal structures remaining attached. These 'footprints', freed of cytoplasmic contents, allow very clear labelling of cortical microtubules. Protoplasts can be burst deliberately not only by using detergents but also by placing them in hypotonic MTSB (without added mannitol) to obtain footprints, exposing cortical cytoskeletal elements.

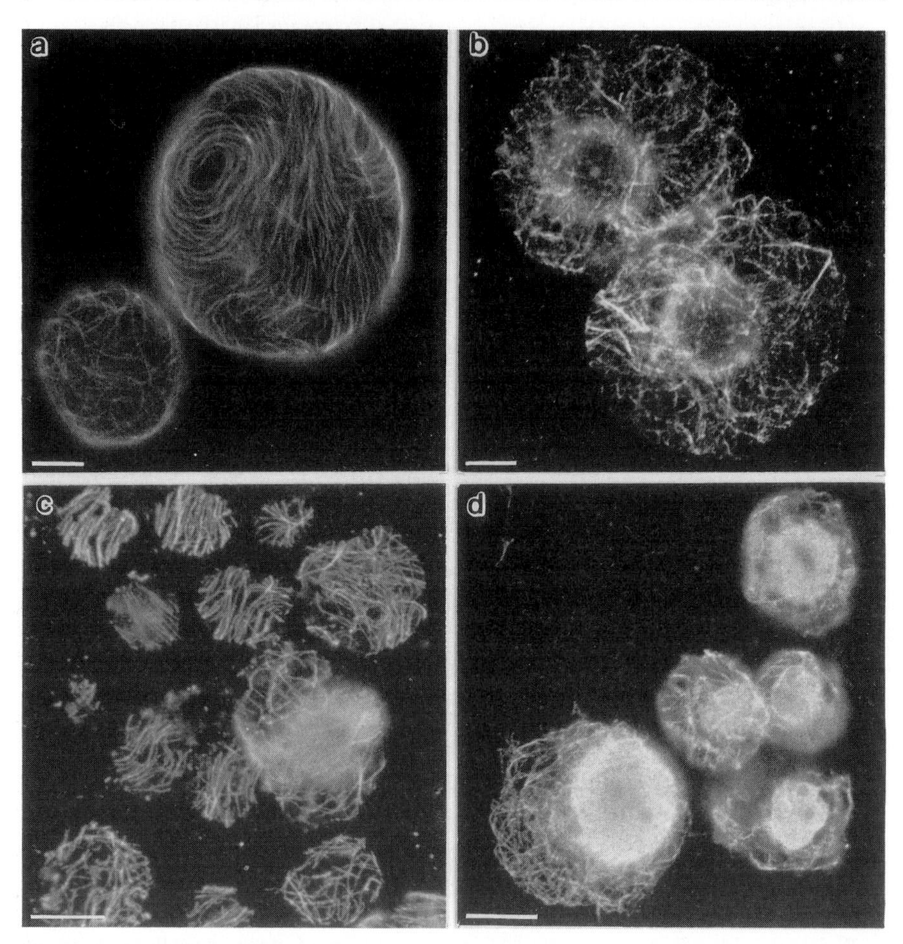

Figure 2. Protoplasts obtained from suspension cultured cells, prepared using methods based on *Protocol 3*. (a) BY-2 protoplasts labelled with YOL 1/34 show a swirling interphase pattern of microtubules at the cortex of each protoplast. (b) A well-extracted pair of BMS protoplasts, exhibiting clear networks of filaments labelled with AFB, an antibody raised against carrot fibrillar bundles (13) which cross-reacts with animal intermediate filaments such as vimentin. (c) Footprints of carrot protoplasts show microtubule-like patterns of labelling when incubated with anti-IFA, a universal antibody against intermediate filament antigens. (d) BY-2 protoplasts, extracted before fixing, and labelled with ME101. This is an antibody raised against the human intermediate filament protein, peripherin. Scale bar = 15 μm.

When studying cell shape, for instance, cells must be fixed to preserve the positions of cytoskeletal elements in intact cells, as these may alter during the course of conversion to spherical protoplasts. However, some cells are difficult to immunolabel satisfactorily if fixed first, as the antibodies simply cannot penetrate the fixed cell walls. This is the case for BMS cells which must be enzyme-treated, albeit briefly, before fixation. In other cases, addition of detergent to the fixative can be helpful. However, this is not always successful, and so an alternative is to digest the cell walls briefly (perhaps for 10 min) before fixing, so that a proportion of cells will permit penetration of antibodies but will not have been fully protoplasted. These alternatives are described in *Protocols 4*A and B below.

Protocol 4. Pretreatments for cultured cells

- 8% (w/v) formaldehyde in Mes
- enzyme mixture as in *Protocol 3*
- wash buffers, immunolabelling reagents, and equipment as in *Protocol 3*

A. *Fixation before enzyme digestion*

1. Allow about 25 ml of cell culture to settle, or collect by centrifugation at 150 *g* for 5 min. Replace medium with 10 ml of 8% (w/v) formaldehyde/Mes wash. Mix by inverting the tube, then leave to settle for 45 min.
2. Remove fixative by aspiration, replace with 25 ml Mes wash, and invert tube. Gently centrifuge the cells (50 *g*) for 5 min. Repeat this wash step twice more.
3. Prepare 10 ml of the following enzyme mixture in Mes wash: 1.5% (w/v) cellulase 'Onozuka' RS, 0.7% (w/v) hemicellulase, 0.1% (w/v) pectolyase Y23. Replace the wash buffer from the fixed cells with this enzyme mixture and incubate for 1–3 h with gentle agitation.
4. Centrifuge at 150 *g* for 5 min and remove the enzyme solution by gentle aspiration. Add 20 ml Mes wash, invert, centrifuge again, and remove the supernatant.
5. Resuspend the washed cells in 5 ml Mes wash and place drops of the suspension onto poly-L-lysine-treated slides. Allow to settle for 10 min, then remove excess cells with a pipette.
6. Extract cells with 1% Triton-X-100/MTSB for 10 min.
7. Wash slides five times with MTSB, taking care that the cells do not float away.
8. Immunolabel as in *Protocols 2* and *3*, applying primary and secondary antibodies directly to the wells.

Protocol 4. *Continued*

B. *Brief enzyme digestion before fixation*

1. Prepare 10 ml of the following enzyme mixture in Mes wash: 1.5% (w/v) cellulase 'Onozuka' RS, 0.7% (w/v) hemicellulase, 0.1% (w/v) pectolyase Y23.
2. Allow about 25 ml of cell culture to settle, or collect by centrifuging at 150 *g* for 5 min. Replace medium with enzyme mix and incubate for about 10–15 min at 25 °C on an orbital shaker. The duration of this incubation may have to be varied if immunolabelling is found to be unsuccessful, but should not be lengthened too much otherwise too many cells will be fully protoplasted.
3. Check cells under the microscope, to ensure that the great majority of cells are indeed still intact.
4. Proceed as in *Protocol 3* for protoplasts, beginning at step 4.

The following method is mentioned only in passing, as being a useful technique for preparing a large number of slides for antibody screening. The amounts given in this protocol should provide at least enough cells for a dozen multitest slides, and can easily be scaled up. This method relies on fixing a large volume of cells then partially digesting the cell walls. The cells are attached to slides by air drying, which also aids antibody penetration, and the slides are finally dipped in methanol before washing and staining. This method does not allow very good preservation of microtubules but is a quick, useful method for antibody screening since the prepared slides can be air dried then frozen in a sealed container with silica gel after stage 6 for future use.

Protocol 5. Quick preparation of cultured cells (for antibody screening)

- 8% (w/v) formaldehyde in MTSB
- 2% (w/v) Driselase, 5 mM EGTA in MTSB
- wash buffers, immunolabelling reagents, and equipment as in *Protocol 3*

1. Working with a volume of approximately 30 ml of 5-day-old culture cells, either allow them to settle for 10 min or collect by centrifugation as in *Protocol 3*. Remove the medium and resuspend the cells in 10 ml of 8% (w/v) formaldehyde/MTSB. Mix by inversion of the tube, then leave for 40 min to settle.

2. Remove fixative by aspiration, replace with 25 ml MTSB, and invert tube. Gently centrifuge the cells (50 *g*) for 5 min. Repeat this wash step twice more.
3. Replace the MTSB with 10 ml 2% (w/v) Driselase/5 mM EGTA[a]. Invert tube and leave to settle for 30 min. This step allows some of the cellulose cell wall to be removed without losing the shape of the fixed cells.
4. Repeat step 2 to wash out the enzyme.
5. Resuspend the cells in a few ml of MTSB to give a fairly thick suspension. Place a drop of cells in each well of a multitest slide and leave them to settle for 10 min. Remove excess liquid leaving a monolayer of cells at the bottom of each well. Allow the slides to air dry completely[b].
6. Permeabilize cells by dipping each slide quickly into a beaker of methanol, then straight into a beaker of MTSB. Finally wash the wells three times with MTSB until the liquid remains only inside the wells.
7. Apply the primary and secondary antibodies directly to the wells (see *Protocol 2*). Mount the slide with anti-fade and seal the coverslip with nail varnish.

[a] Different enzymes may be required for some cell lines, but Driselase works for most.
[b] Air drying will take about an hour, but is quicker if done inside a laminar flow cabinet or fume hood. This makes the cells adhere to the wells, and also begins to permeabilize them to antibodies. It does, however, make the cells collapse.

4.3 Tissues

Protocol 6. Tissue sections

The following protocol was developed for staining stem sections of *Datura stramonium*, but suggestions are made as to how it can be adapted for other tissues.

- 4% (w/v) formaldehyde, 1% DMSO, 0.1% NP40 in MTSB
- wash buffers, immunolabelling reagents, and equipment as in *Protocol 3*

1. Fix tissue blocks in 4% (w/v) formaldehyde/1% (v/v) DMSO/0.1% (v/v) NP40/MTSB (cut large tissues such as stem into <1 cm pieces). For large, dense tissues, vacuum infiltrate for 5 min, slowly release the vacuum, and repeat. Leave in fixative for 1–5 h.
2. Wash four times over an hour by replacing fixative with MTSB.

Protocol 6. *Continued*

3. Section by hand using a flexible, double-sided razor blade or a vibratome, keeping the tissue wet all the time. If the sample contains very small cells which are not susceptible to collapsing, such as in root meristems, then use a cryostat for sectioning, followed by air drying. If the cells are very large and the tissue has been hand-sectioned, then an additional way to aid antibody penetration is to cross-hatch the specimen with a sharp razor blade, i.e. score the sections with the blade from the inside of the tissue, to the epidermis, without cutting right through (7). Often only these cut cells or cells at the edge of the segment will stain, but this method avoids the distortions caused by air drying and separation by enzymes. This method was exploited by Flanders *et al.* (7), who, using computer-aided image reconstruction of confocal sections, required to retain the three-dimensional shape of large, vacuolated cells.
4. An optional, short (10 min) cellulase treatment could now be used, and/or air drying of the sections. Both increase antibody penetration but are best avoided as large cells often collapse. Sometimes a very brief drying of the sections (just so that the surface begins to dry) can be useful and should be considered if labelling has failed without this procedure. This technique was used for tissue culture explants of *Nautilocalyx lynchii* by Goodbody *et al.* (8).
5. For immunolabelling, thin sections are placed in multitest slides, but thick hand sections require special slides with deeper wells, made by gluing four pieces of coverslip onto a slide, forming an open square into which the tissue can be placed, to prevent the sections moving from the centre of the slide. All antibody incubations should be carried out in these slides. Proceed as in *Protocol 2*, but for thick hand sections longer incubations are necessary, maybe even overnight at 4 °C for the primary antibody, then 3 h at room temperature with the secondary antibody, with several 20 min washes.
6. Mount the slides with anti-fade and seal the coverslip with nail varnish.

4.4 Electron microscopy

Protocol 7, adapted from that of Traas *et al.* (9), allows the cortical cytoskeleton to be examined at the ultrastructural level. It has been used to locate intermediate filament antigens, which were found, by immunofluorescence, to co-distribute with microtubules in a punctate manner (10).

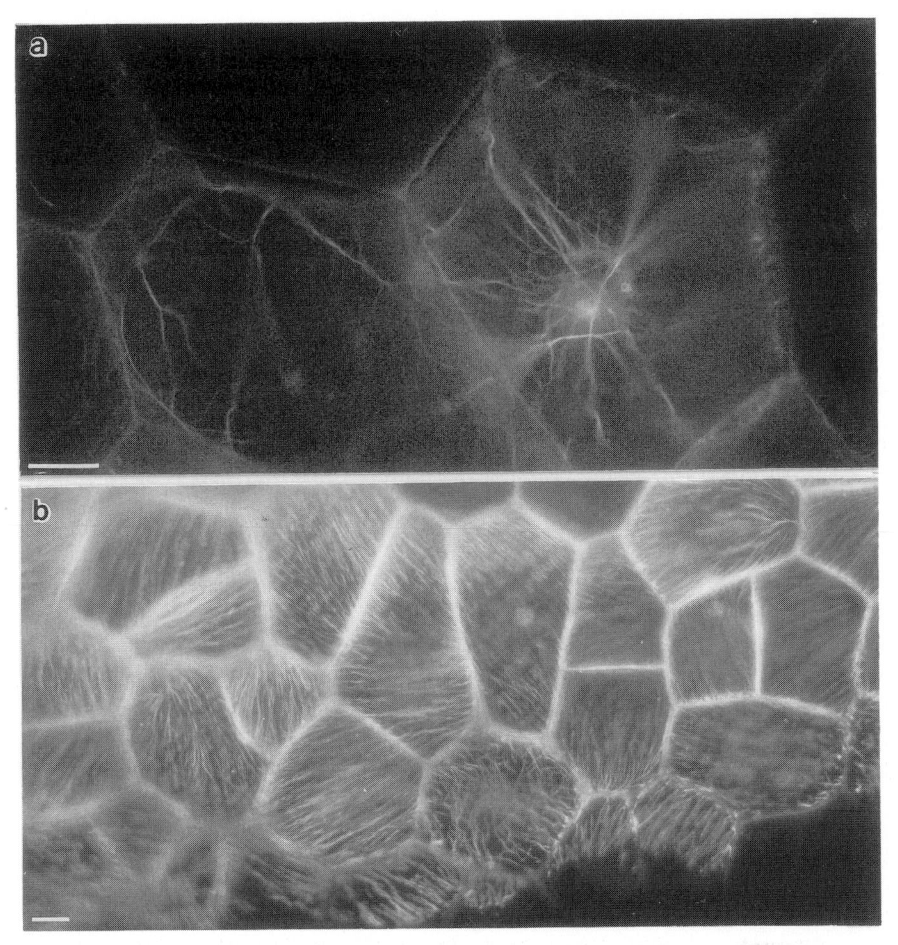

Figure 3. Tissues prepared from whole plants. (a) Wounded leaves of *Tradescantia albovittata*, unfixed but cut and incubated with extraction buffer containing MBS. Rhodamine-phalloidin was used to label the actin microfilaments within these large, regularly shaped cells (4). Actin can clearly be seen surrounding and radiating from the nucleus in one cell. (b) A hand section from the base of a petiole of *Nautilocalyx lynchii*, prepared according to *Protocol 6*, and stained with YOL 1/34 to show the microtubules. Scale bar = 5 μm.

Protocol 7. Cleaved protoplasts

- glutaraldehyde or formaldehyde stock
- wash buffers, immunolabelling reagents (replace fluorochrome-conjugated with colloidal gold-conjugated secondary antibody) and equipment as in *Protocol 3*

Protocol 7. *Continued*

- mannitol
- DMSO
- 2% aqueous uranyl acetate
- poly-L-lysine-treated carbon/pyroxylin-coated EM grids
- plastic Petri dishes

1. Prepare protoplasts as in *Protocol 3* and allow them to settle on poly-L-lysine-coated carbon/pyroxylin-coated gold grids for 20 min by immersing the grids, coated side up, in a Petri dish containing the washed protoplasts.
2. Fix for 30 min in either 4% formaldehyde or 0.2% glutaraldehyde (from 25% stock solution (TAAB)) in MTSB/0.3 M mannitol/5% DMSO.
3. Wash grids three times by immersing them in successive Petri dishes full of MTSB.
4. Briefly dip the grids in water, then invert them on to poly-L-lysine-coated slides under water. Withdraw the grids carefully, cleaving open the protoplasts and leaving areas of plasma membrane and their associated cytoskeletal elements still attached to the grids. Some discs of membrane will also remain attached to the slides and these can now be immunolabelled for fluorescence microscopy as in *Protocol 3*, in parallel with the ultrastructural study.
5. To reduce non-specific binding, incubate grids with 5% BSA in MTSB for 10 min.
6. For immunolabelling, invert grids on to a drop of antibody diluted in 5% BSA in MTSB in a moist chamber for at least 1 h. Secondary antibodies for the EM are gold-conjugated. Treat them as for fluorochrome-conjugated antibodies, but useful dilutions will not be as high: usually a 1/10 dilution is necessary.
7. Wash grids thoroughly by several immersions in MTSB after each incubation, then finally wash in distilled water before negative staining with 2% uranyl acetate for 30 sec. Air dry the grids then view in the transmission electron microscope.

5. Troubleshooting

Assuming that the microscope is set up correctly, the following are suggestions for overcoming lack of staining.

(a) **Cell attachment**: if no cells remain on the slides, then, when repeating the method, check that sufficient are initially attached by the poly-L-lysine.

With care, cells should be attached by poly-L-lysine but alternatives may be necessary. One ingenious method of attaching difficult cells was exploited by Doonan *et al.* (11) on moss filaments. To prevent the separation by cellulase of cells in files, colonies grown on Cellophane were blotted on to coverslips derivatized with amino-silane. Subsequent glutaraldehyde treatment covalently coupled the cells to the glass and retained developmental continuity. Such elaborate methods are, however, rarely necessary since the most likely reason that cells are removed is by over-vigorous washing. Be extremely gentle when aspirating solutions from the slide.

(b) **Antibody concentration**: test monoclonal and polyclonal antibodies of unknown titre at the following dilutions: undiluted, 1/10, 1/25, 1/50, and 1/100. If staining is too strong, try 1/200, 1/500, and 1/1000. For commercial antibodies, try 1/10 to 1/1000. Always dilute the antibody in buffer containing a high concentration of non-specific proteins, e.g. BSA, non-fat dry milk, or serum from the same species as the labelled antibody. If there is no staining at any dilution, check that the sample was covered completely by the antibody solution and that the incubation was long enough. Forty-five minutes for each antibody is usually sufficient.

(c) **Fixation**: if a reliable antibody such as YOL 1/34 anti-tubulin fails to work, then check the fixation conditions. Formaldehyde at 4% is recommended for most plant studies but 8% may be necessary for some tissues. Glutaraldehyde is a better fixative than formaldehyde but is generally avoided for plant tissue since it causes strong autofluorescence. However, a low concentration of glutaraldehyde (e.g. 0.2%) can be added to the formaldehyde. The free aldehydes are then reduced by the use of three successive washes of the fixed specimen in freshly made 0.5 mg/ml sodium borohydride, although this can reduce antigenicity. The use of solvents such as methanol to fix plant tissues is best avoided. Their rapid dehydrating effects often cause damage to the tissues.

(d) **Autofluorescence** (see also Chapter 2): if an antigen is obscured by autofluorescence, it may be possible to reduce this by degrading the wall with cellulases. It is important that there is no wound tissue near the subject cells and that the sample does not come from a highly coloured tissue or one rich in chloroplasts. Doonan *et al.* (11) were able to extract autofluorescent pigments from moss filaments using 3% (v/v) Nonidet P40 or 1% Triton X-100 for 1–4 h. However, it was essential that DMSO was included in the fixative (8% (w/v) formaldehyde) for subsequent extraction to be effective.

(e) **Non-specific staining**: it is important that antibody concentrations are optimal. The secondary antibody should be used at a dilution at which it does not contribute to the staining pattern. If other structures in addition to the target structures continue to stain, consider adsorbing the antibody at 4 °C with an acetone powder of plant material to remove the non-specific component. If this fails, the antibody may need to be affinity purified. Always test the antibody by western blotting to check that it specifically recognizes the antigen. However, some antibodies stain well but do not blot, and *vice versa*, since the antigen is presented differently in these two procedures.

(f) **Cross-reactions between antibodies**: where two different antibodies label the same cytoskeletal network, it is clearly important to ensure that each antibody is independently capable of producing the same staining pattern before confirming the coincidence of staining by double immunofluorescence. Controls include separate immunofluorescence assays of antibody A and antibody B. Before embarking on double immunofluorescence, it is important to check that the secondary antibody for A does not recognize B and *vice versa*. If, in such cross-over controls, a conjugated secondary antibody is found to recognize the other 'wrong' antibody, it may be possible to block the reaction with non-immunized serum from the same animal species as the primary antibody. If ineffective, change one of the primary antibodies for another from a different species e.g. use a polyclonal rabbit anti-tubulin antibody instead of a rat anti-tubulin antibody. Ensure that different antibodies are kept separate and use clean pipette tips to wash each slide.

(g) **Fading**: mount slides with an anti-fade mountant such as Citifluor AF1. Use a fast film such as TMAX 400 (Kodak), shot at 800 ASA, and photograph interesting cells immediately. Keep slides in the dark so that the fluorochromes are not excited unnecessarily. Store slides when not in use at 4 °C in the dark.

(h) **High background**: increase the length and frequency of the washing steps if high background is a problem. It is often helpful to include a mild detergent in the washes such as 0.05% (v/v) Tween 20. Optimize the antibody concentrations, especially that of the primary, and centrifuge the secondary antibody before use. If necessary, pre-incubate the slides with a blocking solution such as 3% BSA in MTSB, or a non-immunized serum.

(i) **Bleed-through**: special care must be taken, when viewing doubly labelled slides, that the emission spectrum of one fluorochrome is not exciting the other, giving rise to bleed-through of another signal. As a rule of thumb, work from the longest to the shortest wavelength e.g. rhodamine then FITC.

Finally, be prepared to try a variety of preparative procedures to suit a specimen, or to suit a particular antibody. Because of the variables that affect plant tissue, there is unlikely to be one standard method that will work without modification.

References

1. Wick, S. M., Seagull, R. W., Osborn, M., Weber, K., and Gunning, B. E. S. (1981). *J. Cell Biol.*, **89**, 685.
2. Traas, J. A., Doonan, J. H., Rawlins, D. J., Shaw, P. J., Watts, J., and Lloyd, C. W. (1987). *J. Cell Biol.*, **105**, 387.
3. Sonobe, S. and Shibaoka, H. (1989). *Protoplasma*, **148**, 80.
4. Goodbody, K. C. and Lloyd, C. W. (1990). *Protoplasma*, **157**, 92.
5. Hussey, P. J., Traas, J. A., Gull, K., and Lloyd, C. W. (1987). *J. Cell Sci.*, **88**, 225.
6. Lloyd, C. W. and Wells, B. (1985). *J. Cell Sci.*, **75**, 225.
7. Flanders, D. J., Rawlins, D. J., Shaw, P. J., and Lloyd, C. W. (1990). *J. Cell Biol.*, **110**, 1111.
8. Goodbody, K. C., Venverloo, C. J., and Lloyd, C. W. (1991). *Development*, **113**, 931.
9. Traas, J. A. (1984). *Protoplasma*, **119**, 212.
10. Goodbody, K. C., Hargreaves, A. J., and Lloyd, C. W. (1989). *J. Cell Sci.*, **93**, 427.
11. Doonan, J. H., Cove, D. J., and Lloyd, C. W. (1985), *J. Cell Sci.*, **75**, 131.
12. Ross, J. H. E., Hutchings, A., Butcher, G. W., Birgitte Lane, E., and Lloyd, C. W. (1991). *J. Cell Sci.*, **99**, 91.
13. Hargreaves, A. J., Dawson, P. J., Butcher, G. W., Larkins, A., Goodbody, K. C., and Lloyd, C. W. (1989). *J. Cell Sci.*, **92**, 371.

11

Isolation of endo- and plasma membranes

D. G. ROBINSON, G. HINZ, and K. OBERBECK

1. Introduction

Subcellular fractionation has, as its general principle, the application of three consecutive procedures. The first, homogenization, always results in the abolition of cellular integrity. Depending on the forces involved, individual compartments or organelles may also no longer remain intact. The second step entails imposing a new system of order among the randomly dispersed organelles in the homogenate. This can usually be achieved through separations based on size (gel filtration), mass (differential or rate zonal centrifugations), density (isopycnic centrifugation), or surface charge (electrophoresis, phase partitioning). Finally, the subcellular units or fractions obtained have to be checked for their homogeneity. For this, optical (electron microscopy of negatively stained or thin-sectioned samples), chemical (enzyme tests), and immunological (western blotting) methods can be employed.

The ultimate aim of subcellular fractionation is to provide pure, homogeneous samples of a particular organelle without contamination from other cellular constituents. This goal is seldom achieved and often leads to poor recoveries of the organelle in question. Where the structural integrity of the organelle is important, particular care has to be taken that shear forces in homogenization and pellet resuspension steps be kept to a minimum. This again will adversely affect the yields. Quite often, however, subcellular fractionation merely serves the purpose of determining the intracellular distribution of an enzyme or substance. For this pure fractions are not necessary; a similar behaviour in terms of centrifugation characteristics, e.g. co-sedimentation or similar density to known organelle markers, is usually sufficient to allow positive identification. *Table 1* lists the plasma membrane and various endomembrane components of the plant cell with respect to their equilibrium density in sucrose (the most frequently used medium for their separation) and to the marker enzymes or substances used for their identification.

Table 1. Endo- and plasma membranes of the plant cell: equilibrium densities and marker enzymes/substances

Membrane	**Equilibrium density in sucrose**		**Marker enzyme/ substance**
	g/cm^3	**% w/w**	
Clathrin-coated vesicles	1.18–1.26	40–55	as-ni-ATPase polypeptides; 190 kDa clathrin polypeptide
Golgi apparatus	1.12–1.15	27–34	IDPase; GS I
Intact vacuoles	various depending on content		AP; α-M
Microbodies	1.21–1.25	46–53	CAT
Nucleus (nuclear envelope)	1.30–1.32	60–65	RNA polymerase; CCR
Plasma membrane	1.13–1.18	30–40	cs-vi-ATPase; GS II
Rough ER	1.13–1.18	30–40	CCR; RNA
Secretory vesicles	1.11–1.13	26–30	IDPase
Smooth ER	1.08–1.12	21–27	CCR
Tonoplast (from mesophyll and root storage tissue)	1.07–1.09	18–22	as-ni-ATPase; cs-PPase
Tonoplast (from seed storage tissue and maize root and coleoptile)	1.17–1.19	37–42	as-ni-ATPase; cs-PPase TP-25 (seeds only)

Abbreviations: α-M, α-mannosidase; AP, acid phosphatase; as-ni-ATPase, anion-stimulated, nitrate-inhibitable ATPase; CCR, NAD(P)H-dependent cytochrome *c*-reductase; cs-vi-ATPase, cation-stimulated, vanadate-inhibitable ATPase; CAT, catalase; GS I/II, glucan synthase I or II; IDPase, inosine diphosphatase; cs-PPase, cation-stimulated pyrophosphatase; RNA polymerase, DNA-dependent RNA polymerase; TP-25, 25 kDa tonoplast integral protein

2. General methodology

2.1 Homogenizing media

In general homogenizing media must be prepared so that differences from the *in vivo* situation will be minimal upon cell breakage. Because, with plants, the latter event nearly always results in vacuolar rupture, special precautions have to be made to counteract the negative consequences of vacuolar dilution and the release of a number of deleterious substances. Thus, it is usual to include larger amounts of osmoticum, e.g. sucrose or mannitol (0.2–0.4 M), than is the case for homogenizing media for animal material. Buffer strengths are also often higher (20–50 mM; sulphonic acid, 'Good' buffers, are most frequently used) and higher concentrations of chelators (EDTA, or EGTA; 1–5 mM) may also be necessary.

In the past, insufficient efforts have been undertaken to combat proteolysis

Table 2. Antiproteolysis agents which can be included in homogenizing media

Agent	Mode of activity	Recommended final concentrations	Source
Aprotinin[a]	serine protease inhibitor	100 μg/ml	Sigma A-1153
Bestatin[b]	exopeptidase inhibitor	0.1–0.2 mM	Sigma B-8385
E-64[a,b]	thiol protease inhibitor	1–3 μM	Sigma E-3132
EDTA[a]	metalloprotease inhibitor	1–3 mM	Sigma E-9884
Leupeptin[a]	serine and thiol protease inhibitor	1–5 μM	Sigma L-2884
Phenylmethylsulfonyl fluoride[b]	serine and thiol protease inhibitor	0.2–1 mM	Sigma P–7626
o-Phenanthroline[a]	metalloprotease inhibitor, aminopeptidase inhibitor	1–3 mM	Sigma P-9375

[a] H_2O soluble; [b] soluble in methanol (stock solutions 10^3–10^4 times final concentrations)

in plant homogenates. We recommend that a 'cocktail' of anti-proteases always be included in homogenizing media. The right combination of these substances and the required amounts can only be determined by trial and error, using, for example, azocasein as a standard substrate in a proteolysis assay (1). *Table 2* gives details of some agents which can be employed for this purpose (2). In addition, the inclusion of up to 0.2% (w/w) fatty acid-free bovine serum albumin (BSA) as a competitive substrate for proteases can also be recommended, in particular since it also serves as an excellent scavenger for free fatty acids (3). Phenolic oxidases can also be a problem in plant cell fractionation, but their action (causing 'browning' due to the production of quinones) is relatively easy to overcome or prevent. One can include in the homogenizing medium either polyvinylpyrrolidone (1–2% (w/v)), which adsorbs the end products, or ascorbate (1 mM), which reduces the quinones, or, probably most effective of all, thiourea or metabisulphite (2 mM), which inhibit phenol oxidase activity (4). If deemed important, lipolytic activity (5) can be counteracted by the addition of 10 mM glycerol-1-phosphate (inhibitor of phosphatidic acid phosphatase) and/or 4% (w/v) each of choline and ethanolamine (prevention of phospholipid degradation). However, one should be aware that such additions to homogenizing media are not universally beneficial. Indeed, from the point of view of maintaining electrochemical gradients across membranes they can be most detrimental (6).

2.2 Methods of homogenization

Achieving high recovery and maximizing organelle integrity are not necessarily mutually exclusive features of plant cell fractionation, but become more and more irreconcilable with increasing organelle size and cell wall strength. The spectrum of homogenizing methods for plant tissues is broad, ranging from low-shear procedures such as chopping with a hand-held razor blade, through mortar grinding (with or without acid-washed sand as an abrasive), to more brutal ones e.g. using motorized mincers (Polytron, Waring blender). In addition, unicells (e.g. algae, or suspension-cultured higher plants) may have to be subjected to rapid (1000 strokes/min) shaking with glass beads (diameter 0.3–0.5 mm) or high pressure-induced adiabatic expansion (using French or Jeda presses) in order to produce homogenates.

Providing they can be made quickly and cheaply, protoplasts (see Chapter 8) are an ideal starting point for plant cell fractionation. They can be induced to rupture by gently pressing them through a hypodermic syringe (the diameter of the protoplasts determining the size of the needle), or by subjecting them to an osmotic shock (by resuspending them in a hypotonic medium). The residual shear forces still operative in these methods can be overcome by briefly treating protoplast suspensions with low concentrations of detergent (*Protocol 1*). Whereas this procedure allows the structural preservation of even the largest of organelles (7), it inevitably involves the loss of the plasma membrane.

Protocol 1. Detergent-assisted homogenization of protoplasts

Reagents:

- antiproteases (see *Table 2*)
- bovine serum albumin (BSA, Sigma B-2518)
- digitonin (Serva 19550)
- dithiothreitol (DTT, Sigma D-0632)
- ethylene diaminotetra-acetic acid (EDTA, Serva 11280)
- N-[2-hydroxyethyl]piperazine-N′-[2-ethane sulphonic acid] (Hepes, Biomol 05288)
- polyethylene glycol 600 (PEG-600, Sigma P–3390)
- sucrose (Sigma S-5016)
- inorganic chemicals (KCl, KH_2PO_4, K_2HPO_4, $MgCl_2$, and NaOH)

Solutions:

- detergent solution (0.004% (w/v) digitonin in 5 mM KH_2PO_4/K_2HPO_4 buffer pH 6.5 containing 6% (w/w) PEG-600 and 4 mg/ml BSA)
- homogenizing medium (0.4 M sucrose, 1 mM DTT, 0.1 mM EDTA, 3 mM $MgCl_2$, 10 mM KCl, in 40 mM Hepes–NaOH pH 7.5, plus antiproteases as needed)

Procedure:

1. Resuspend protoplasts in 25 ml prechilled detergent solution.
2. Bring quickly (in less than 40 sec) to 30 °C in a shaking water bath.
3. Incubate for 30 sec to 2 min (time depends on protoplast type and must be determined before).
4. Return to ice bath. Centrifuge at 500 g^a for 2 min.
5. Add 1–2 ml prechilled homogenizing medium to pellet.
6. Homogenize by resuspending pellets with the help of a vortex mixer (two short bursts are usually sufficient).

[a] Except where stated, g values in isolation protocols are always given as g_{max}.

2.3 Methods of organelle separation

By far the most frequently employed method is that of (ultra)centrifugation. Other methods, e.g. free-flow electrophoresis (8), are gaining in importance especially now that the necessary apparatus is no longer so expensive. However, since they are not yet universally available, we will not describe their operation here. Nor do we feel it necessary to give a discourse on ultracentrifugation, there being numerous excellent texts (e.g. references 9 and 10) on this technique already available. For the same reasons, and since apparatuses for preparing and fractionating density gradients are relatively cheap and/or easy to make, we will not dwell on the construction or operation of such basic laboratory equipment.

After homogenization the brei is usually filtered through 2–4 layers of cheesecloth or Miracloth (Calbiochem 475855), and then precentrifuged at around 3000 g for 5 min to sediment large particles (pieces of cell wall, starch granules, plastids, and unbroken cells). Mitochondria and nuclei are often completely sedimented by a subsequent centrifugation of 5000–10 000 g for 15–30 min. The organelles which interest us here are all present in the supernatant of this latter centrifugation. In order to concentrate these membranes a total membrane pellet can be prepared by spinning this supernatant at 100 000 g for 1 h on to a cushion of dense liquid, e.g. 60% (w/w) sucrose. The interface is then removed with a pipette. In this way the dangers of resuspending pellets are avoided.

The procedures which follow depend on the organelle to be isolated. These can encompass variations both in gradient techniques (rate zonal versus isopycnic, continuous 'linear' versus discontinuous 'step' gradients), as well as in gradient media (low molecular weight sugars, e.g. sucrose or glycerol; high molecular weight polysaccharides, e.g. dextran, or Ficoll®; suspensions of colloidal silica, e.g. Percoll®; iodinated compounds, e.g. metrizamide, renografin, or Nycodenz®). Despite its drawbacks (high osmotic strength, and high viscosity) sucrose still enjoys great usage as a gradient medium, but is gradually being replaced by the, albeit more expensive, iodinated media. Irrespective of the medium used, gradient fractions should be diluted at least 3-fold with homogenizing medium (minus sucrose if present) in order to pellet out (at 100 000 *g* for 1 h) the organelles in question.

3. Isolation of plasma membrane

Plasma membrane fractions of up to 95% purity can be obtained by free-flow electrophoresis as well as aqueous two-phase partitioning. As judged by latency tests (measurements of cation-stimulated vanadate-inhibitable H^+-ATPase (cs-vi-ATPase) in the presence and absence of detergent), the vesicles of plasma membrane isolated by the former method can separate into fractions with either cytoplasmic side-out or cytoplasmic side-in orientation (11). Phase-partitioning, by contrast, gives rise to a more or less uniform population of cytoplasmic side-in vesicles (12). A variation of the basic two-phase procedure, which involves freezing and thawing of the partially purified plasma membrane vesicle fraction, can be used to produce a population of cytoplasmic side-out vesicles (13). Many of the vesicles in these preparations are tightly sealed and can be separated from leaky ones by centrifuging on to cushions of high molecular weight dextran (14). Sealed vesicles will not penetrate into the dextran layer but collect at the interface whilst vesicles leaky to dextran will sediment. Sandelius and Morré (8) have discussed the relative merits of these two methods for plasma membrane isolation and have compared them with fractions obtained by sucrose density centrifugations. Certainly the former are superior in terms of purity, with the phase-partitioning procedure probably giving rise to the least contamination. The principles and practice of phase-partitioning have been described on numerous occasions (e.g. references 8 and 15) and require little embellishment here. *Protocol 2* presents this standard method, which, depending on the species of plant used, may have to be altered with respect to the ratio of PEG to dextran, and to the concentration of potassium ions. When successfully used, plasma membrane derived vesicles partition preferentially to the PEG-rich upper phase, while other membranes concentrate in the lower (dextran) phase. Monitoring the behaviour of marker enzymes will confirm this.

Protocol 2. Isolation of plasma membranes by aqueous phase-partitioning

Reagents:

- EDTA, sucrose, and inorganic chemicals (see *Protocol 1*)
- dextran T 500 (DEX-500, Sigma D-5251)

- PEG-3350 (Union Carbide)
- 2-amino-2-(hydroxymethyl)-1, 3-propanediol (Tris, Sigma T 1563)

Solutions:

- stock solution (in H_2O) of 20% (w/w) DEX-500
- stock solution (in H_2O) of 40% (w/w) PEG-3350
- psk buffer (20 mM KH_2PO_4/K_2HPO_4 buffer pH 7.8 containing 1 M sucrose and 8 mM KCl).
- wash buffer (50 mM Tris–HCl pH 7.5 containing 250 mM sucrose and 1 mM EDTA).

Procedure:

All manipulations are to be carried out at 1–4 °C.

1. Prepare a large volume of two-phase mixture by weight (75.6 g DEX-500 stock, 37.8 PEG-3350 stock, 60 g psk buffer, 66.6 g distilled H_2O). Pour into separating funnel, shake vigorously, and leave to stand overnight. Separate upper and lower phases. Discard interface.
2. Prepare a small volume of two-phase mixture by weight (7.5 g DEX-500 stock, 3.78 g PEG-3350 stock, 4.75 g psk buffer, 2.91 g distilled H_2O).
3. Prepare total membrane preparation of plant concerned (see section 2.3). Resuspend in 5–7 ml psk buffer. (Protein concentration 2–3 mg/ml).
4. Add 5 g membranes to the small volume two-phase mixture in 30 ml Corex tubes (Du Pont). Shake vigorously (by inverting about 20 times).
5. Centrifuge for 5 min at 1500 *g* in a swing-out rotor. Remove upper phase. Add lower phase from the large volume two-phase mixture and make up to 24 g. Shake vigorously.
6. Repeat step 5 twice. Collect separately upper and lower phases. The final upper phase (U_3) is enriched with plasma membranes.
7. Dilute upper phases to 75 ml with wash buffer. Dilute lower phases 8- to 10-fold with wash buffer. Centrifuge for 30 min at 80 000 *g*.

4. Isolation of tonoplast

As judged by the distribution of marker enzymes (anion-stimulated nitrate inhibitable H^+-ATPase (as-ni-ATPase) and cation-stimulated pyrophosphatase (cs-PPase)) in isopycnic sucrose density gradients of total membrane preparations, tonoplast vesicles band most frequently at low densities (around 21% (w/w) sucrose; see reference 16). However, this is not always so, there being several cases (17, 18) where the peak activities of these marker enzymes lie much deeper in the gradient. Unfortunately, fractions removed from sucrose gradients corresponding to these activities are rarely pure in tonoplast vesicles. Substantial contamination by endoplasmic reticulum (ER) and Golgi apparatus membranes (see *Figure 1* for distribution of appropriate marker enzyme activities) is to be expected for gradients prepared under low Mg^{2+} conditions. Only under high Mg^{2+} conditions, when the ER tends to lie primarily at densities higher than 30% (w/w) sucrose and the tonoplast marker enzyme activities remain at around 21% sucrose, are there good prospects of minimal contamination by other endomembranes. Tonoplast vesicles have also been isolated from total membrane populations by free-flow electrophoresis (19, 20). They are physiologically competent (sealed, cytoplasmic-side out, and capable of pumping protons) and appear to be relatively pure.

When absolutely clean tonoplast preparations are required there appears to be no other recourse than to isolate intact vacuoles. Two methods are available for this. One (see *Protocol 3*) entails the mechanical release of vacuoles and can only be applied to compact storage tissue (21). For this a special, motor-driven, tissue slicing apparatus (depicted in 21) with rotating, single-edged razor blades is necessary. This is a simple piece of apparatus, which can easily be made in any workshop. The other method involves the production of protoplasts which are then ruptured either by controlled osmotic swelling and lysis (22) or by forcing them through a hypodermic syringe (23). Vacuoles are then separated in step gradients of Ficoll by either sedimentation or flotation. Our protocol (*Protocol 4*) is based on the method of Martinoia *et al.* (23) and leads to vacuole preparations within 10 min of protoplast rupture. The preparation of tonoplast membranes from both types of vacuole preparation is achieved by rupturing the vacuoles and isolating the tonoplast by flotation in a sucrose density gradient (22). This procedure is included in *Protocol 4*.

Protocol 3. Large scale preparation of vacuoles from storage tissue

Reagents:

- EDTA, Tris (see *Protocol 1*)
- mercaptobenzothiazole (Sigma M-2274)
- metrizamide (Sigma M-4512)
- sorbitol (Serva 25230)

Solutions:

- 'collection medium' (50 mM Tris–HCl pH 7.6 containing 5 mM EDTA, 0.1 mg/ml mercaptobenzothiazole, and 1 M sorbitol)
- 'isolation buffer' (10 mM Tris–HCl pH 7.6 containing 1 mM EDTA and 1.5 M sorbitol).
- flotation gradient media (15%, 10%, and 2.5% (w/w) metrizamide each dissolved in isolation buffer).

Procedure:

1. Load tissue slicer with 5 mm thick slabs of storage tissue (e.g. red beet or sugar beet). About 400–500 g can be processed per homogenizing cycle. Pour 1 litre of collection medium into catching pan. Operate slicer at 150 r.p.m. at room temperature.
2. Separate tissue pieces from collecting medium by filtering through two layers of Miracloth. Return pieces to tissue slicer for a second passage into 1 litre of fresh collection medium. Filter again. Combine first and second filtrates (roughly 1750 ml).
3. Chill on ice. Carry out all subsequent operations at 1–4 °C.
4. Centrifuge at 2000 *g* for 10 min. Discard supernatant.
5. Gently resuspend pellets in 15 ml isolation buffer containing 15% (w/w) metrizamide.
6. Construct 13 ml step gradients for flotation in 15 ml centrifuge tubes (from bottom to top: 3 ml crude vacuole suspension, 5 ml 10% (w/w) metrizamide, 2.5 ml 2.5% (w/w) metrizamide, 2.5 ml isolation buffer).
7. Centrifuge at 650 *g* for 10 min. Vacuoles collect at interface of 10% and 2.5% metrizamide solutions.

Protocol 4. Preparation of tonoplast membranes from isolated vacuoles

Reagents:

- DTT, EDTA, sorbitol, sucrose, Tris, and inorganic chemicals (see *Protocols 1–3*)
- Ficoll 400 (Pharmacia F-4375)

Solutions:

- protoplast resuspension medium (15 mM KH_2PO_4/K_2HPO_4 buffer pH 7.6 containing 2 mM EDTA, 0.4 M sucrose, and 2.5% (w/w) Ficoll 400)

Protocol 4. *Continued*

- flotation solutions (for vacuole separation): solution A, 15 mM KH_2PO_4/K_2HPO_4 buffer pH 7.6 containing 2 mM EDTA, 0.2 M sucrose, and 0.2 M sorbitol; solution B, 15 mM KH_2PO_4/K_2HPO_4 buffer pH 7.6 containing 0.4 M sorbitol
- flotation solutions (for tonoplast isolation): solution C, room temperature-saturated sorbitol in gradient buffer (1 mM EDTA–Tris pH 7.2 containing 1 mM DTT); solution D, 30% (w/w) sucrose in gradient buffer; solution E, 40% (w/w) sucrose in gradient buffer; solution F, 15% (w/w) sucrose in gradient buffer

Procedure:

All manipulations are to be carried out at 1–4 °C

1. Prepare and purify protoplasts (e.g. from mesophyll).
2. Take up protoplasts in resuspension medium (density 2–5 $\times 10^6$ protoplasts/ml).
3. Pass protoplast suspension through a large hypodermic needle (10 cm × 0.2 mm) using a syringe. Depending on the degree of protoplast rupture this step can be repeated twice if necessary.
4. Prepare 12 ml step gradient in 15 ml tubes (from bottom to top: 5 ml protoplast homogenate, 5 ml flotation solution A, 2 ml flotation solution B).
5. Centrifuge in swing-out rotor for 2 min at 200 *g* then for 3 min at 1000 *g*. Vacuoles collect in solution B.
6. Sonicate vacuole suspension (20 sec; 1 W). Repeat if necessary in order to achieve 100% rupture.
7. Add flotation solution C until density reaches 40% (w/w) sucrose equivalent (measure density with a hand-refractometer)
8. Prepare linear sucrose gradient (16 ml; 30–40% (w/w)) from solutions D and E. Add 12 ml overlay of solution F. **Carefully** underlay gradient with 8 ml of vacuole homogenate.
9. Centrifuge in swing-out rotor for 4 h at 100 000 *g*. Tonoplast membranes band at the 15/30% interface. Remove, dilute 1:4 with gradient buffer, and centrifuge at 100 000 *g* for 45 min.

5. Isolation of endoplasmic reticulum and Golgi apparatus

The marker enzyme profiles for ER and Golgi apparatus (GApp) membranes in isopycnic sucrose gradients often overlap so much that the respective peak fractions show a considerable degree of cross-contamination. This is generally more drastic when homogenates and total membrane fractions are prepared under 'low' (0.1 mM) Mg^{2+} conditions (ER present as smooth membranes) than when 'high' (3–5 mM) Mg^{2+} conditions are used (ribosomes retained on the ER). The amount of Mg^{2+} needed to maximize ribosomal retention is critical and may well differ from plant to plant. If too much is present, aggregates of membranes may form and they may sediment to the bottom of the gradient. If there is too little, an artificially high amount of smooth ER is produced.

Whilst the 'density shift' produced in this manner can lead to differences in equilibrium density of over 15% sucrose for ER membranes (see *Figure 1*), it is hardly shown by GApp membranes (24, 25). As such it is therefore a useful pointer to an ER localization when the distribution of a particular enzyme/substrate shows a similar behaviour in response to Mg^{2+} concentration. Mg^{2+} shifting forms the basis for a purification of ER membranes (see *Protocol 5* and *Figure 2a* and *b*), but the concentrations of Mg^{2+} chelators or other agents required to remove ribosomes from a rough ER fraction (10–20 mM EDTA, 10–15 mM pyrophosphate, 100 mM Na_2CO_3 pH 11.5, (26)), are considerably higher than those necessary to produce smooth ER directly in homogenates. Of these three substances one (Na_2CO_3) gives rise to sheets of membrane rather than vesicles. The recoveries of smooth ER produced in this way are low, but the fractions appear to be very pure, since the Golgi marker inosine 5-diphosphatase (IDPase) is no longer measurable.

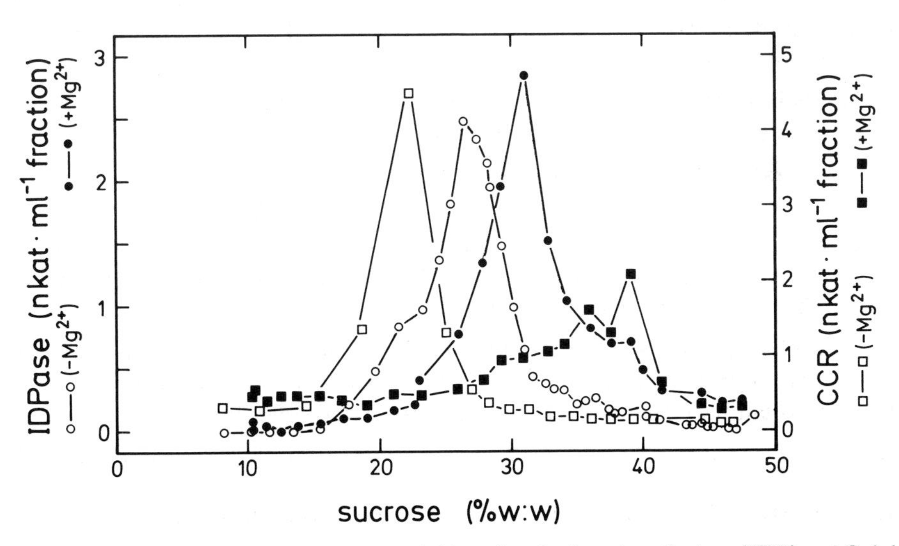

Figure 1. Profiles for marker enzyme activities of endoplasmic reticulum (CCR) and Golgi apparatus (IDPase) on isopycnic linear density gradients of total membrane fractions from developing pea cotyledons prepared under either low (−, 0.1 mM); or high (+; 3 mM) $MgCl_2$ conditions. Unpublished data of I. Hohl.

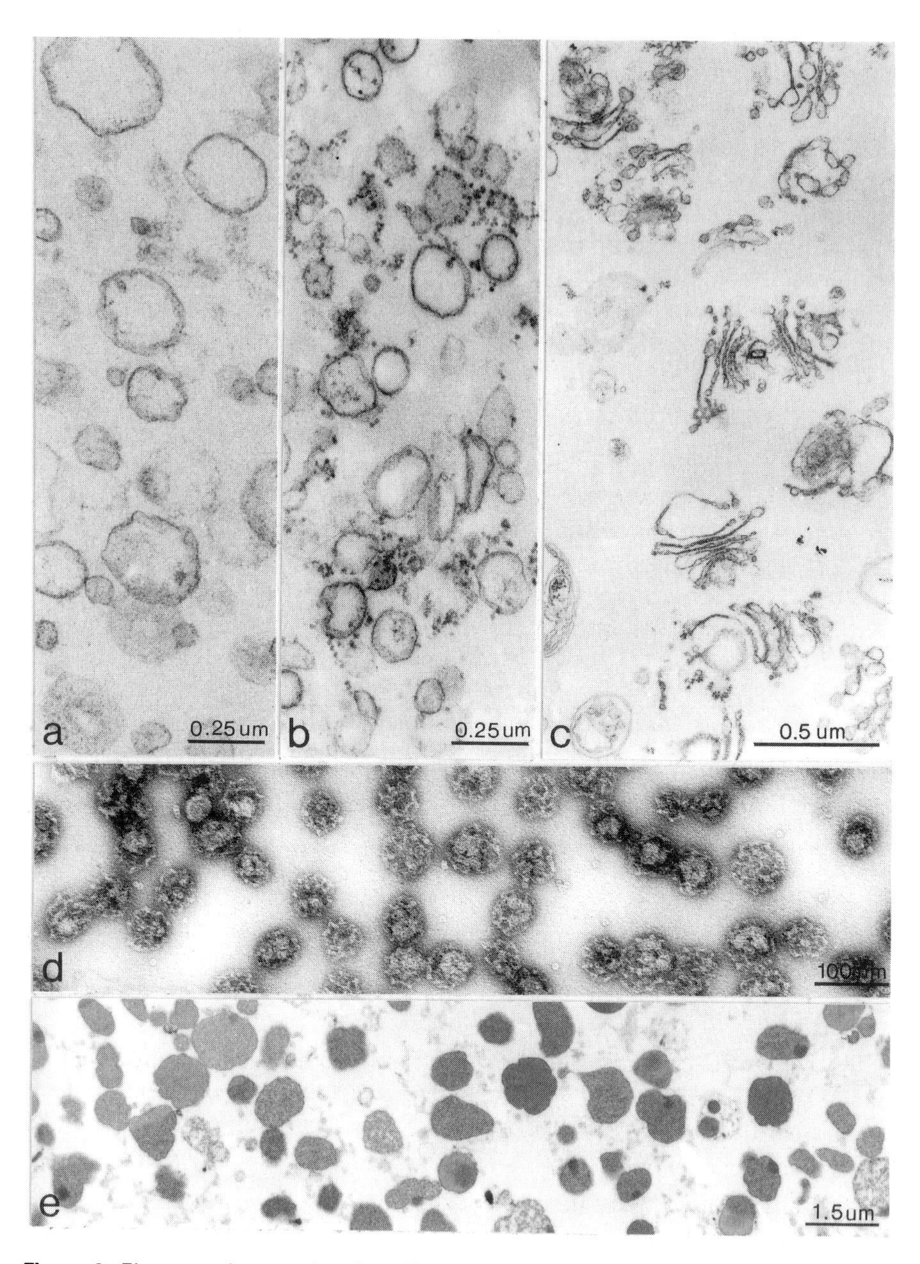

Figure 2. Electron micrographs of purified endomembrane fractions. (a and b) Smooth (a) and rough (b) endoplasmic reticulum fractions isolated from *Chlamydomonas reinhardtii*. (c) Intact dictyosomes isolated from etiolated pea stem segments. (d) Clathrin-coated vesicles isolated from developing pea cotyledons. (e) Peroxisomes isolated from spinach leaves.

Protocol 5. Isolation of endoplasmic reticulum

Reagents:

- DTT, EDTA, Hepes, sucrose, and inorganic chemicals (see *Protocol 1*)
- potassium pyrophosphate $K_4P_2O_7 \cdot 10\ H_2O$

Solutions:

- basic buffer (40 mM Hepes–NaOH pH 7.5 containing 10 mM KCl)
- homogenization medium (0.4 M sucrose, 1 mM DTT, 0.1 mM EDTA, 3 mM $MgCl_2$ in basic buffer)
- sucrose solutions for step gradient A (40%, 35%, and 15% (w/w) sucrose in basic buffer containing 0.1 mM EDTA and 3 mM $MgCl_2$)
- sucrose solutions for step gradient B (50%, 25%, and 15% (w/w) sucrose in basic buffer containing 3 mM EDTA and 0.1 mM $MgCl_2$)

Procedure:

All manipulations are to be carried out at 1–4 °C.

1. Homogenize (either by hand chopping with a razor blade in a Petri dish, or with a commercial herb grinder) 10 g plant tissue in 10 ml prechilled homogenization medium.
2. Filter homogenate through one layer of Miracloth. Centrifuge at 6000 *g* for 20 min to remove mitochondria and other large organelles.
3. Carefully layer supernatant on to 8 ml cushion of 50% (w/w) sucrose in basic buffer. Fill up tube (36 ml) with homogenization medium.
4. Centrifuge at 100 000 *g* for 1 h (swing-out rotor)
5. Total membranes collect at interface to sucrose cushion. Remove with Pasteur pipette. Dilute with homogenization medium minus sucrose to 6 ml (final sucrose concentration must be $<15\%$ (w/w)).
6. Prepare step gradient A in 36 ml tubes (from bottom to top: 8 ml 40%, 10 ml 35%, and 9 ml 15% sucrose solutions) and overlay with resuspended total membranes.
7. Centrifuge at 100 000 *g* for 2 h (swing-out rotor).
8. Rough ER membranes collect at 35/40% sucrose interface. Carefully remove with Pasteur pipette.
9. Dilute with basic buffer containing 10 mM EDTA (or 15 mM pyrophosphate) and leave on ice for 30 min. Centrifuge at 150 000 *g* for 1 h (fixed angle rotor). Resuspend pellet in basic buffer containing 3 mM EDTA.

Protocol 5. *Continued*

10. Prepare step gradient B in 17 ml tubes (from bottom to top: 2 ml 50%, 5 ml 25%, 5 ml 15% sucrose solutions) and overlay resuspended membranes.
11. Centrifuge at 100 000 *g* for at least 2 h (swing-out rotor).
12. Smooth ER membranes collect at 15/25% sucrose interface.

The GApp of a plant cell consists of numerous, discrete dictyosomes, which, in turn, are composed of stacks of cisternae. In order to isolate the GApp, advantage can be taken of the fact that, whilst they have the same isopycnic density in sucrose (24), dictyosomes and individual Golgi cisternae differ considerably in terms of their mass. Thus it is often customary to employ some kind of rate zonal centrifugation which allows the (heavier) dictyosomes to be separated from other (lighter) elements in a total or microsomal membrane preparation. Preventing the unstacking of cisternae is therefore of paramount importance in such an isolation procedure. It entails the use of low-shear forces during homogenization and also involves the inclusion of protective agents such as dextran (0.5–2% (w/v) (27)) or glutaraldehyde (0.3–0.6% (v/v) (28)) in the homogenizing medium (see *Figure 2c*). Since the latter is a fixative the optimum concentration for preserving structural integrity as well as for preserving marker enzyme activity must be determined in advance on total membrane preparations. *Protocol 6* for GApp isolation is based on that given in Ray *et al.* (25). The original method used glutaraldehyde, but dextran can be substituted with equal success. It appears to function equally well with etiolated and green tissue, although the conditions for the initial rate zonal centrifugation may vary from tissue to tissue.

Protocol 6. The isolation of intact dictyosomes

Reagents:

- antiproteases (see *Table 2*)
- DTT, EDTA, Hepes, sucrose, and inorganic chemicals (see *Protocol 1*)
- dextran T 250 (DEX-250, Sigma D-7265)

Solutions:

- basic buffer (40 mM Hepes–NaOH pH 7.0 containing 10 mM KCl, 0.1 mM EDTA, 3 mM $MgCl_2$)
- homogenization medium (0.4 M sucrose, 1 mM DTT, 1% (w/v) DEX-250, antiproteases as needed in basic buffer)

- sucrose solutions for density gradients (13, 25, 30, and 45% (w/w) sucrose in basic buffer)

Procedure:

All manipulations are to be carried out at 1–4 °C.

1. Homogenize 10 g tissue (hand chop for 10 min with a razor blade in a Petri dish on ice) in 10 ml homogenization medium.
2. Pass homogenate through one layer of Miracloth. Rinse debris with 1 ml homogenization medium.
3. Centrifuge filtered homogenate at 3000 *g* for 10 min.
4. Layer 6 ml supernatant on to linear (13–30% (w/w) sucrose, 30 ml) sucrose gradient. Centrifuge at 22 500 *g* for 20 min in swing-out rotor.
5. Fractionate into 1.5 ml fractions. Determine density (using a hand refractometer) and run IDPase assays.
6. Pool fractions between 21 and 25% sucrose (corresponding to second peak of IDPase activity) and layer 6 ml on to linear (25–45% (w/w) sucrose, 30 ml) sucrose gradient. Centrifuge at 100 000 *g* for 2 h in swing-out rotor.
7. Fractionate; dictyosomes collect between 30 and 34% sucrose.

6. Isolation of transport vesicles (secretory vesicles and clathrin-coated vesicles)

The ER, GApp, and plasma membrane are in communication with one another through a number of transport vesicles. The secretory pathway has three such vesicles: transition vesicles (TV), responsible for transport between the ER and the GApp (29); non-clathrin-coated vesicles (NCCV), which accomplish transport between cisternae in the Golgi stack (30); and secretory vesicles (SV), which are released from the GApp and eventually fuse with the plasma membrane (29). Also exiting from the *trans* face of the GApp are clathrin-coated vesicles (CCV), which in animal cells transport hydrolases to the lysosomes (31) and in plants carry proteins to storage vacuoles (32). The endocytotic pathway involves both coated (CCV) and smooth vesicles, and has, as intermediate organelles *en route* to the lysosome/vacuole, so-called endosomes (33). In plants this compartment is represented by multivesicular bodies and the partially coated reticulum (34).

The above named vesicular intermediates, as well as organelles of the endosomal/lysosomal type, have been successfully isolated from animal sources on numerous occasions (e.g. see references 35 (TV), 36 (NCCV), 37 (SV), 38 (CCV), and 39 (endosomes)). TV (40) and CCV (41) have also been purified from yeasts. Only TV, SV, and CCV have so far been isolated from plant tissues. Vesicles resembling NCCV at the periphery of *cis* and *median* cisternae are occasionally visible in thin sections of the plant GApp. If the procedure is to be employed which led to their isolation from animal cells (induction of vesicles budding *in vitro* through incubation of Golgi cisternae with GTPγS, NEM, and cytosolic factors (36)), large amounts of GApp membranes are required which would seem to make their isolation from plant tissue technically unfeasible. On the other hand, the *in vitro* induction of TV and their release, by incubating with cytosolic factors, UTP, and ATP has been successfully performed on higher plant ER fractions (42), and the authors were able to separate the TV from the donor ER membranes by free-flow electrophoresis. Whether the latter procedure can be substituted by more easily available centrifugal methods remains to be seen.

Because of their accumulation in very large amounts at the growing tips of germinating pollen tubes (see 43 for a dramatic depiction of this), this cell type has been chosen as the starting material for SV isolation from plant sources. Two methods for SV isolation from pollen tubes have been published. In one (44) a homogenate is passed through a series of filters of decreasing pore size, culminating in a 0.45 μm filter. The 0.45 μm filtrate is enriched in SV. Alternatively, and resulting in much higher yields, one can use a centrifugation procedure (45, see *Protocol 7*). Electron microscopy of these SV fractions reveals a homogeneous population with little recognizable contamination. Glucan synthase activity has been detected in pollen tube SV (46) but further enzymic characterization has not been carried out.

Protocol 7. The isolation of secretory vesicles from pollen tubes

Reagents:

- antiproteases (see *Table 2*)
- DTT, EDTA, Hepes, KCl, and sucrose (see *Protocol 1*)
- inorganic chemicals H_3BO_3 and $Ca(NO_3)_2$)

Solutions:

- germination medium (10% (w/v) sucrose, 0.01% (w/v) H_3BO_3, 3 mM $Ca(NO_3)_2$ in distilled H_2O
- basic buffer (100 mM Hepes–NaOH pH 7.2 containing 1 mM EDTA, 10 mM KCl, and 1 mM DTT)
- homogenizing medium (0.3 M sucrose plus antiproteases as required in basic buffer)
- sucrose solutions for density gradients (0.5, 0.7, 0.9, 1, 1.1, 1.5, and 2 M sucrose each in basic buffer).

Procedure:

All manipulations are to be carried out at 4 °C except where stated.

1. Germinate 1 g pollen (*Lilium*, *Nicotiana*, or *Petunia* can be used) in 10 ml germination medium at 30 °C for 2–3 h. Gently centrifuge at 500 *g* and take up pellet in prechilled 5 ml homogenizing medium.
2. Homogenize in commercial herb grinder (three bursts of 10 sec each).
3. Filter through one layer of Miracloth. Centrifuge filtrate for 5 min at 3000 *g*. Discard pellet.
4. Layer supernatant on to a sucrose step gradient (from botton to top: 7 ml each of 2, 1.5, 1, and 0.5 M sucrose solutions). Centrifuge at 100 000 *g* for 90 min in a swing-out rotor.
5. Remove material at 0.5/1 M interface; dilute 2-fold with basic buffer.
6. Layer on to a sucrose step gradient (from bottom to top: 7 ml each of 1.1, 0.9, 0.7, and 0.5 M sucrose solutions). Centrifuge at 100 000 *g* for 90 min in a swing-out rotor.
7. SV collect at the 0.7/0.9 M interface.

Other studies have shown that both SV and GApp cisternae possess IDPase activity, which is localized primarily at the *trans* face of the dictyosome (see reference 24, chapter 5 and references therein), but SV apparently lack the Golgi marker glucan synthase I (47). This together with the fact that SV are smaller and lighter than GApp cisternae, allows effective separations of the SV through rate zonal centrifugation (25, 47). Both renografin and sucrose have been used for this purpose and the yield fo SV can be increased by treating the tissue prior to homogenization with cytochalasin B (25) which is known to prevent SV transport to the plasma membrane (48). However, the fractions of SV so obtained are by no means free of contamination (in particular tonoplast and ER are present). Nevertheless such fractions may well be of use for kinetic studies on cell wall biosynthesis.

CCV have been isolated from both green (49) and etiolated higher plant tissues (50) as well as from suspension-cultured cells (e.g. 51). Because of their size (<100 nm diameter) they are essentially post-microsomal elements in plant homogenates. Such fractions (material pelleting between 40 000 and 125 000 *g*) are grossly contaminated with ribosomes, and, in the case of photosynthetically active tissue, ribulose bisphosphate carboxylase (RubisCo) as well. The former can be removed by treating with RNase (50) or by chelating with EDTA (52). The RubisCo separates in linear density gradient centrifugations due to its smaller size and lower density (49). Whether RNase or EDTA is used depends on the tissue involved. With zucchini hypocotyls EDTA alone is sufficient to eliminate ribosome contamination, but with developing pea cotyledons (a tissue extremely rich in CCV) in RNase treatment is still necessary. RNase digestions can, however, be carried out at 4 °C thereby minimizing the danger of proteolysis (*Protocol 8*).

CCV can be recognized in post-microsomal fractions by negative staining (see *Figure 2d*), and on the basis of the clathrin heavy chain polypeptide at 190 kDa in SDS–PAGE. Correct identification of the light chains of clathrin can only be made when suitable precautions against proteolysis are taken (52, 53).

There have been reports (50) that plant CCV have a lower equilibrium density in sucrose than their animal counterparts (38). This is certainly due to partial degradation of the coat proteins. When adequately protected against proteolysis equilibrium densities similar to those for animal CCV (1.24 g/cm^3; 50% (w/w) sucrose) are observed for CCV from plant sources (51, 52). However, exposure to such concentrations for periods longer than 5–6 h leads to dissociation of the coat proteins (54). As a consequence the attainment of isopycnic conditions in less than this time can only be achieved by employing vertical or fixed-angle rotors operating at very high speeds (> 150 000 *g*). Alternatively, solutions of Ficoll in D_2O can be employed instead of sucrose (49).

Protocol 8. Isolation of clathrin-coated vesicles

Reagents:

- antiproteases (aprotinin, E-64, leupeptin, pepstatin, and *o*-phenanthroline, see *Table 2*)
- DTT, EDTA, sucrose, and inorganic chemicals (see *Protocol 1*)
- ethylene glycol *bis* (β-aminoethylether) *N,N,N′,N′*-tetraacetic acid (EGTA, Sigma E-4378)
- fatty-acid-free BSA (FA-BSA, Sigma A-6003)
- 2-(*N*-morpholino) ethane sulphonic acid (Mes, Biomol 06010)
- pancreatic ribonuclease A (RNase, Serva 34388)

Solutions:

- basic buffer (0.1 M Mes–NaOH pH 6.4 containing 1 mM EGTA, 1 mM DTT, and 0.5 mM $MgCl_2$)
- homogenizing medium (3 mM EDTA, 1 mM *o*-phenanthroline, 2% (w/w) FA-BSA, 2 μg/ml aprotinin, 2 μg/ml E-64, 2 mM leupeptin, 0.7 μM pepstatin)
- RNase solution (3 mg/ml RNase in 65 mM KH_2PO_4–NaOH buffer pH 6.5 containing 10 mM $MgCl_2$)

- density gradient solutions (10% (w/v), 40% (w/v), 25% (w/w), 30% (w/w), and 55% (w/w) sucrose in basic buffer)

Procedure:

All manipulations are to be carried out at 4 °C.

1. Homogenize (Waring blender, 3 × 15 sec) 250 g tissue in 500 ml homogenizing medium (add two or three drops of anti-foam reagent).
2. Filter brei through four layers of gauze. Centrifuge filtrate at 40 000 *g* for 30 min. Discard pellets.
3. Centrifuge supernatant at 125 000 *g* for 75 min. Discard supernatant.
4. Resuspend pellet in 16–20 ml homogenizing medium (use a tight-fitting glass Potter–Elvejhem homogenizer). Alternatively, when necessary, resuspend pellet in 3 ml RNase solution; incubate for 20 min at 4 °C; centrifuge at 9000 *g* for 20 min; discard pellet, use supernatant.
5. Layer over step gradient (8 ml 40% (w/v), 8 ml 10% (w/v) sucrose solutions) in 38 ml centrifuge tube. Centrifuge at 70 000 *g* for 45 min (swing-out rotor). Remove material at 10/40% interface and in the 10% layer. Dilute 3-fold with basic buffer and centrifuge at 125 000 *g* for 75 min (fixed-angle rotor).
6. Resuspend pellet in 3 ml homogenizing medium. Add 1.5 ml to each of two linear gradients (25–55% (w/w); 14 ml) in 17 ml centrifuge tubes; centrifuge at 170 000 *g* for 3 h (vertical rotor; braking time 30 min).
7. Combine fractions corresponding to 43–53% sucrose. Dilute 3-fold with basic buffer and centrifuge at 125 000 *g* for 75 min (fixed-angle rotor).
8. Resuspend pellet in 5 ml homogenizing medium. Layer over linear gradient (5–30% (w/w); 30 ml) in 38 ml tube. Centrifuge at 100 000 *g* for 90 min. CCV collect in the 10–20% sucrose fractions.

7. Isolation of microbodies

Because of their importance in plant metabolism, microbodies have been intensively studied for two decades (for reviews see 24, 55). Many aspects of their structure, biochemistry, and origin, and the synthesis and import of their enzymic content are now well established. Depending on their principal function: glycolate breakdown in leaf photorespiration or *β*-oxidation in fat–sugar conversion in germinating seeds, we speak of either peroxisomal or glyoxysomal type microbodies. In both cases, however, catalase is a major enzyme responsible for the breakdown of H_2O_2 which is produced in these organelles. It is therefore used as a specific marker enzyme for microbodies in subcellular fractionation.

Their size (up to 1.5 μm in diameter) and high protein content make microbodies very dense organelles. Earlier separations involved isopycnic centrifiguation in linear sucrose gradients, but in such gradients the activity profile for catalase often overlaps with that of mitochondrial markers. It has therefore become customary to use other gradient media, especially Percoll (see *Protocol 9*) for adequate separation of microbodies from other dense organelles (56, 57). Excellent, homogeneous preparation scan be obtained in this way (see *Figure 2e*).

Protocol 9. Isolation of peroxisomes from spinach leaves

Reagents:

- EDTA, Hepes, NaOH, sucrose (see *Protocol 1*), and FA-BSA (see *Protocol 8*)
- cysteine (Merck 2838)
- mannitol (Sigma M-4125)
- (3-[N-morpholino]propane sulphonic acid) (Mops, Sigma M 1254)
- Percoll (Sigma P-1644)
- polyvinylpyrrolidone-40 (PVP-40, Sigma P-6755)
- propane 1,2-diol (Sigma P-1009)

Solutions:

- homogenization medium (30 mM Mops–NaOH pH 7.5 containing 0.35 M mannitol, 4 mM cysteine, 1 mM EDTA, 0.2% (w/v) FA-BSA, and 0.6% (w/v) PVP-40)
- wash medium A (20 mM Mops–NaOH pH 7.2 containing 0.3 M mannitol, 1 mM EDTA, and 0.2% (w/v) FA-BSA)
- wash medium B (1 mM Hepes–NaOH pH 7.5 containing 0.25 M sucrose)
- density gradient solutions (21% 27%, 45%, and 60% (v/v) Percoll in 10 mM Mops–NaOH pH 7.2, 0.25 M sucrose, and 0.2% (w/v) FA-BSA; the 21 and 27% solutions contain, in addition, 100 mM propane 1,2-diol)

Procedure:

All manipulations are to be carried out at 4 °C.

1. Derib 100 g leaves; wash in deionized H_2O, and cut into 2–3 cm strips.
2. Homogenize in 100 ml prechilled medium with mortar and pestle. Filter homogenate through eight layers of gauze. Centrifuge filtrate at 5000 *g* for 1 min. Discard pellet.

3. Centrifuge supernatant at 20 000 *g* for 5 min. Carefully resuspend the pellet (using a Potter–Elvejhem homogenizer) in 1.5 ml wash medium A and layer on to Percoll step gradient (from bottom to top: 6 ml 60%, 14 ml 45%, 8 ml 27%, 8 ml 21% Percoll solutions) in 38 ml centrifuge tubes. Centrifuge at 12 000 *g* for 12 min (swing-out rotor, no braking). Collect fractions between 45 and 60% Percoll, pool, dilute at least 6-fold with wash medium B, and centrifuge at 12 000 *g* for 12 min.
4. Resuspend pellet in 2 ml wash medium B and centrifuge at 3000 *g* for 10 min. Repeat. Peroxisomes are in the final pellet.

8. Assays for marker enzymes

The gradual enrichment of an organelle in progressive subcellular fractions is monitored by assaying for the activity of its marker enzymes (see *Table 1*). Usually this is done directly on aliquots of gradient fractions or on resuspended pellets (protein concentration varying between 0.1 and 0.3 mg/ml). Many of the assays are colorimetric (e.g. those involving phosphate release, namely those for IDPase, cs-PPase, cs-vi-ATPase, and as-ni-ATPase) but some involve radioactive nuclides (e.g. assays for glucan synthases). Spectrophotometers (with recording facilities) and scintillation counters are therefore necessary. The assay mixtures have a total volume of 1 ml or less and can be easily performed in Eppendorf®, or similar, plastic tubes. Enzyme activities should be given in katal. *Protocol 10* gives details of the assay conditions for the most well-known membrane-bound marker enzymes. For extra information the reader is referred to the excellent article by Widell and Larsson (58).

Protocol 10. Marker enzymes

Reagents:

- digitonin, KCl, KH_2PO_4, $MgCl_2$, sucrose (see *Protocol 1*), Tris (see *Protocol 2*), EGTA, and Mes (see *Protocol 8*)
- 1-amino-2-naphthol-4-sulphonic acid (ANS, Serva 13080)
- antimycin A (Serva 13690).
- ascorbic acid (Sigma A-0278)
- *Bis-Tris*-propane (Sigma B-6755)
- cellobiose (Sigma C-7252)
- α-cellulose (Sigma C-8002)
- cytochrome *c* (Biomol 03466)

Protocol 10. *Continued*

- inosine-5′diphosphate (Sigma I-4375)
- Malachite Green–HCl (Sigma M-9636)
- (Mg)-adenosine 5′-triphosphate (Mg-ATP, Sigma A-0770
- β-nicotinamide adenine dinucleotide (NADH, Biomol 16132)
- oligomycin (Sigma O-4876)
- sodium dodecyl sulphate (SDS, Serva 20760)
- spermine (Sigma S-3256)
- Triton X-100 (Sigma F-0733)
- trichloroacetic acid (TCA, Sigma T 4885)
- uridine 5-diphosphoglucose (UDPG (unlabelled), Sigma U-4625)
- [^{14}C]UDPG (NEN NEC-403)
- inorganic chemicals ($(NH_4)_6Mo_7O_{24}\cdot 2H_2O$, $CaCl_2$, $MgSO_4$, KCN, $K_4P_2O_7\cdot 10\ H_2O$, NaN_3, $NaHSO_3$, sodium citrate ($C_6H_5Na_3O_7\cdot 2H_2O$), $Na_2S_2O_4$, $Na_2MoO_4\cdot 2\ H_2O$, $NaNO_3$, Na_3VO_4, Na_2SO_3, and H_2SO_4)

A. *'CAT' (59) catalase (EC 1.11.1.6)*

Solutions:

- buffer (50 mM KH_2PO_4–KOH pH 7.5 containing 0.3 M sucrose)
- 10 mM H_2O_2

Procedure:

1. Pipette together 980 µl and 10–20 µl fraction in chamber of an O_2 electrode at 20 °C.
2. Add 10 µl H_2O_2 and mix. Measure O_2 release.
3. Calibrate O_2 electrode with
   (a) O_2 saturated H_2O (0.56 µmol O_2/ml)
   (b) H_2O with sodium dithionite (add half a spatula tip full to H_2O in chamber).

B. *'CCR' (60); antimycin A insensitive NADH:cytochrome* c *reductase (EC 1.6.99.3)*

Solutions:

- 'CCR mix' (50 mM Tris–acetate pH 7.5 containing 80 µM oxidized cytochrome *c*, 1 µM KCN, 5 mM antimycin A (from a 5 mM stock solution in ethanol))
- 50 mM NADH (in distilled H_2O)

Procedure:

1. Add 20 μl fraction to 950 ml CCR mix in microcuvette at room temperature.
2. Monitor non-specific reduction at 550 nm for 2 min.
3. Add 30 μl NADH. Monitor ΔE_{550} for a further 2 min or until linearity is reached.
4. Determine difference in rates for NADH-specific and non-sepcific reductions.

C. *'cs-PPase' (61, 62): inorganic pyrophosphatase (EC 3.6.1.1)*

Solutions:

- 'PPase mix' (30 mM Tris–Mes pH 8.0 containing 0.1 mM $MgSO_4$, 0.2 mM ammonium molybdate, 2 mM oligomycin, 50 mM KCl, 0.04% (v/v) Triton X-100, 2 mM potassium pyrophosphate (added fresh immediately prior to reaction)
- 7% (w/v) SDS in distilled H_2O
- 5% (w/v) molybdate (stir 25 g ammonium molybdate in 55.5 ml 96% H_2SO_4 and make up to 500 ml with distilled H_2O)
- ANS reagent (add 0.25 g ANS to 90 ml of 15% (w/v) $NaHSO_3$ solution and stir at 37 °C for 30 min in the dark; add 3 ml 10% (w/v) Na_2SO_3 during the stirring in 0.5 ml aliquots; allow to cool, filter, and make up to 100 ml with 15% (w/v) $NaHSO_3$; store in the dark)

Procedure:

1. Pipette together 100 μl fraction, 250 μl PPase mix, 150 μl 30 mM Tris–Mes pH 8.0 into Eppendorf tube. Incubate at 37 °C for 30 min.
2. Terminate reaction by adding 500 μl SDS.
3. Add 250 μl molybdate, then 250 μl ANS reagent. Allow to stand at room temperature for 15 min.
4. Measure extinction at 660 nm against reagent control.
5. Determine released phosphate from calibration curve (previously prepared with KH_2PO_4, concentration range 20–100 nmol).

D. *'as-ni-ATPase' and 'cs-vi-ATPase' (63, 64): anion-stimulated, nitrate-inhibitable H^+-ATPase (EC 3.6.1.3); cation-stimulated vanadate-inhibitable H^+-ATPase (EC 3.6.1.35)*

Solutions:

- 'ATPase mixes' for as-ni-ATPase (3 mM Mg-ATP, 5 mM $MgSO_4$, 55 mM KCl, 1 mM sodium molybdate, 1 mM NaN_3, 35 mM Tris–HCl pH 8.0, +/– 50 mM $NaNO_3$)

Protocol 10. *Continued*

- 'ATPase mixes' for cs-vi-ATPase (3 mM Mg-ATP, 5 mM $MgSO_4$, 55 mM KCl, 1 mM sodium molybdate, 1 mM NaN_3, 35 mM Tris–Mes pH 6.5 +/– 70 μM sodium orthovanadate)
- Malachite Green reagent (4.2 g ammonium molybdate dissolved in 100 mM 4 M HCl; add 135 mg Malachite Green–HCl and 300 ml distilled H_2O; stir for 20 min at room temperature; filter; store at 4 °C)
- citrate buffer (34 g sodium citrate dissolved in 90 ml 1 mM sodium orthovanadate, bring the pH to 7.0 with solid *bis-tris* propane; make up to 100 ml).

Procedure:

1. Add 10 μl fraction (2–5 μg protein) to 90 μl ATP mix (in an Eppendorf tube). Incubate for 5–20 min at 37 °C.
2. Terminate reaction by adding 800 μl Malachite Green reagent.
3. Wait 6 min, add 100 μl citrate buffer, wait further 10 min.
4. Measure extinction at 660 nm. Prepare calibration curves with 0–20 nmol inorganic phosphate in the presence of ATP mix. Nitrate and vanadate-inhibitable ATPase activities are determined by substracting the differences between assays with and without 50 mM $NaNO_3$ or 70 μM sodium orthovanadate.

E. 'GS I' (65): glucan synthase I (EC 2.4.1.12)

Solutions:

- 'GS I buffer' (110 mM Mes–NaOH pH 7.0 containing 1.1 mM EGTA, 0.55 mM $MgCl_2$)
- 'GS I mix' (110 mM Mes–NaOH pH 7.0 containing 1.1 mM EGTA, 120 mM $MgCl_2$, 1.6 mM unlabelled UDPG
- [^{14}C]UDPG in distilled H_2O (12 GBq/mmol)
- saturated α-cellulose in absolute ethanol

Procedure:

1. Dilute gradient fraction 3:1 with homogenizing buffer and centrifuge at 100 000 *g* for 1 h. Resuspend membrane pellet in GS I buffer.
2. Pipette together 200 μl resuspended membranes (approximately 10 μg protein), 50 μl GS I mix, and 10 μl [^{14}C]UDPG (740 Bq) in Eppendorf tube. Incubate at 30 °C for 1 h.
3. Terminate reaction by adding 260 μl ethanol–cellulose suspension. Place in boiling water bath for 1 min. Cool and store at −20 °C for at least 2 h.

4. Centrifuge at 9000 *g* for 10 min. Remove supernatant by aspiration.
5. Resuspend pellet in 1 ml 70% ethanol. Centrifuge at 9000 *g* for 10 min. Remove supernatant by aspiration. Repeat twice.
6. Cut off tip of Eppendorf tube containing pellet and transfer to scintillation vial. Add scintillant cocktail; shake well and determine radioactivity.

F. *'GS II' (66): glucan synthase II (EC 2.4.1.34)*

Solutions:

- 'GS II solution I' (110 mM Mes–NaOH pH 7.0 containing 1.1 mM EGTA, 0.55 mM $MgCl_2$)
- 'GS II solution II' (as for GS II solution I with the addition of 5 mM $CaCl_2$)
- 'GS II buffer' (GS II solutions I and II mixed in 2:3 (v/v); the resulting solution has 10 μM free Ca^{2+})
- UDPG solution I (165 mM Mes–NaOH pH 7.0 containing 1.65 mM EGTA, 0.83 mM $MgCl_2$, 0.04% (w/v) digitonin, 33 mM cellobiose, and 1.65 mM unlabelled UDPG
- UDPG solution II (as for solution I with the addition of 7.5 mM $CaCl_2$)
- 'GS II mix' (UDPG solutions I and II mixed 2:3 (v/v); the resulting solution has 10 μM free Ca^{2+})
- spermine solution (make 50 mM stock solution in distilled H_2O; dilute 2 ml to 20 ml with 0.04% (w/v) digitonin solution; take 17.6 ml and make up to 20 ml with same digitonin solution)
- [^{14}C]UDPG and ethanol–cellulose suspension as for *Protocol 10*E

Procedure:

1. Dilute gradient fraction 3:1 with homogenizing buffer and centrifuge at 100 000 *g* for 1 h. Resuspend membrane pellet in GS II buffer.
2. Pipette together 50 μl resuspended membranes (approximately 10 μg protein), 30 μl GS II mix, 20 μl spermine, 10 μl [^{14}C]UDPG (740 Bq) in Eppendorf tube. Incubate at 30 °C for 15 min.
3. Terminate reaction and work up product as for GS I.

G. *'IDPase' (67, 68): latent inosine 5-diphosphatase (EC 3.6.1.6)*

Solutions:

- 'IDP mix' (50 mM Tris–HCl pH 7.2 containing 10 mM KCl, 4 mM IDP, +/− 0.2% (v/v) Triton X-100)
- 10% (w/v) TCA

Protocol 10. *Continued*

- colour reagent (3 M H_2SO_4, 2.5% (w/v) ammonium molybdate·$4H_2O$, 10% (w/v) ascorbic acid, and H_2O in proportions 1:1:1:6 by volume)

Procedure:

1. Add 50 µl fraction to 50 µl IDP mix in Eppendorf tube. Incubate at 37 °C for 30 min.
2. Terminate reaction by adding 500 µl prechilled TCA. Allow to stand at 4 °C for at least 2 h.
3. Centrifuge at 9000 *g* for 10 min.
4. Add 200 µl supernatant to 1.8 ml colour reagent. Incubate at 37 °C for 1 h.
5. Measure extinction at 820 nm against reagent control.
6. Determine released phosphate from calibration curve (previously prepared with KH_2PO_4, concentration range 0–250 nmol). Substract −Triton values from +Triton values for measure of detergent stimulation (latency).

References

1. Hamano, Y. T. (1984). *Plant Cell Physiol.*, **5**, 1469.
2. Boehringer Mannheim GmbH (1987). *Biochemica Information*. Mannheim.
3. Galliard, T. (1974). *Methods Enzymol.*, **30**, 520.
4. Van Driessche, E., Beeckmans, E., Dejaegere, R., and Kanarek, L. (1984). *Anal. Biochem.*, **141**, 184.
5. Scherer, G. and Morré, D. J. (1978). *Plant Physiol.*, **62**, 933.
6. Drucker, M., Hinz, G., and Robinson, D. G. (1993). *J. Exp. Bot.*, **44**, Suppl. 283.
7. Andreae, M., Blankenstein, P., Zhang, Y.-H., and Robinson, D. G. (1988). *Eur. J. Cell Biol.*, **47**, 181.
8. Sandelius, A. S. and Morré, D. J. (1990). In *The plant plasma membrane* (ed. C. Larsson and I. M. Møller), pp. 44–75. Springer Verlag, Berlin.
9. Birnie, G. D. and Rickwood, D. (1979). *Centrifugal separations in molecular and cell biology*. Butterworth, London.
10. Ford, T. C., and Graham, G. M. (1991). *An introduction to centrifugation*. Bios, Oxford.
11. Canut, H., Brightman, A., Boudet, A. M., and Morré, D. G. (1988). *Plant Physiol.*, **86**, 631.
12. Larsson, C., Kjellbom, P., Widell, S., and Lundborg, T. (1984). *FEBS Lett.*, **171**, 271.
13. Larsson, C., Widell, S., and Sommarin, M. (1988). *FEBS Lett.*, **229**, 259.
14. Steck, T. L. (1974). *Methods Membrane Biol.*, **2**, 245.

15. Larsson, C., Widell, S., and Kjellbom, P. (1987). *Methods Enzymol.*, **148**, 558.
16. Sze, H. (1985). *Annu. Rev. Plant Physiol.*, **36**, 175.
17. Chanson, A. (1990). *Plant Sci.*, **71**, 199.
18. Mäder, M. and Chrispeels, M. J. (1984). *Planta*, **160**, 330.
19. Sandelius, A. S., Penel, C., Auderset, G., Brightman, A., Millard, M., and Morré, D. J. (1986). *Plant Physiol.*, **81**, 177.
20. Scherer, G. and vom Dorp, B. (1988). In *Cell-free analysis of membrane traffic* (ed. D. J. Morré, K. E. Howell, and G. M. W. Evans), pp. 269–79. Alan R. Liss, New York.
21. Leigh, R. A. and Branton, D. (1976). *Plant Physiol.*, **58**, 656.
22. Leonard, R. T. (1987). *Methods Enzymol.*, **148**, 82.
23. Martinoia, E., Heck, U., and Wiemken, A. (1981). *Nature*, **289**, 292.
24. Robinson, D. G. (1985). *Plant membranes*. John Wiley, New York.
25. Ray, P. M., Eisinger, W. R., and Robinson, D. G. (1976). *Ber. Dtsch Bot. Ges.*, **89**, 121.
26. Fujiki, Y., Hubbard, A. L., Fowler, S., and Lazarow, P. B. (1982). *J. Cell Biol.*, **93**, 97.
27. Morré, D. J. (1971). *Methods Enzymol.* **22**, 130.
28. Morré, D. J., Mollenhauer, H. H., and Chambers, J. E. (1965). *Exp. Cell. Res.*, **38**, 672.
29. Morré, D. J., Kartenbeck, J., and Franke, W. W. (1979). *Biochim. Biophys. Acta*, **559**, 71.
30. Orci, L., Malhotra, V., Amherdt, M., Serafini, T., and Rothman, J. E. (1989). *Cell*, **56**, 357.
31. Marquard, T., Braulke, T., Hasilik, A., and von Figura, K. (1987). Eur. J. Biochem., **168**, 37.
32. Hoh, B., Schauermann, G., and Robinson, D. G. (1991). *J. Plant Physiol.*, **138**, 309.
33. van Deurs, B., Peterson, O. W., Olsner, S., and Sandvig, K., (1989). *Int. Rev. Cytol.*, **17**, 131.
34. Robinson, D. G. and Hillmer, S. (1990). *Physiol. Plant.*, **79**, 96.
35. Schweizer, V., Matter, K., Ketcham, C. M., and Hauri, H. P. (1990). *J. Cell Biol.*, **113**, 45.
36. Malhotra, V., Serafini, T., Orci, L., Shepherd, J. C., and Rothman, J. E. (1989). *Cell*, **58**, 329.
37. Tartakoff, A. M. (1987). *The secretory and endocytic paths*. John Wiley, New York.
38. Pearse, B. M. F. (1983). *Methods Enzymol.*, **98**, 320.
39. Storrie, B., Pool, R. R., Sachdeva, M., Maurey, K. M., and Oliver, C. (1984). *J. Cell Biol.*, **98**, 108.
40. Groesch, M. E., Rushola, H., Bacon, R., Rossi, G., and Ferro-Novick, S. (1990). *J. Cell Biol.*, **111**, 45.
41. Mueller, S. C. and Branton, D. (1984). *J. Cell Biol.*, **98**, 341.
42. Morré, D. J., Nowack, D. D., Paulik, M., Brightman, A. O., Thornborough, K., Yirn, J., and Auderset, G. (1989). *Protoplasma*, **153**, 1.
43. Lancelle, S. A. and Hepler, P. K. (1992). *Protoplasma*, **167**, 215.
44. Van der Woude, W. J., Morré, D. J., and Bracker, C. E. (1971). *J. Cell Sci.*, **8**, 331.

45. Engels, F. M. (1973). *Acta Bot. Neerl.*, **22**, 6.
46. Helsper, J. P. F. G., Veerkamp, J. H., and Saasen, M. M. A. (1977). *Planta*, **133**, 303.
47. Taiz, L., Murry, M., and Robinson, D. G. (1983). *Planta*, 158, 534.
48. Mollenhauer, H. H. and Morré, D. J. (1976). *Protoplasma*, **87**, 39.
49. Depta, H., Freundt, H., Hartmann, D., and Robinson, D. G. (1987). *Protoplasma*, **136**, 154.
50. Depta, H., Holstein, S. E. H., Robinson, D. G., Lützelschwab, M., and Michalke, W. (1991). *Planta*, **183**, 434.
51. Mersey, B. G., Griffing, L. R., Rennie, P. J., and Fowke, L. C. (1985). *Planta*, **163**, 317.
52. Demmer, A., Hinz, G., Schauermann, G., Holstein, S. E. H., and Robinson, D. G. (1992). *J. Exp. Bot.*, **44**, 23.
53. Lin, H.-B., Harley, S. M., Butler, J. M., and Beevers, L. (1992). *J. Cell Sci.*, **103**, 1127.
54. Nandi, R. K., Trace, G., van Jaarsveld, P. P., Lippoldt, R. E., and Edelhoch, H. (1982). *Proc. Natl. Acad. Sci USA*, **79**, 5881.
55. Tolbert, N. E. (1980). *Biochem. Plants*, **1**, 359.
56. Mettler, I. and Beevers, H. (1980). *Plant Physiol.*, **66**, 555.
57. Yu, C. and Huang, A. H. C. (1986). *Arch. Biochem. Biophys.*, **245**, 125.
58. Widell, S. and Larsson, C. (1990). In *The plant plasma membrane* (ed. C. Larsson and I. M. Møller), pp. 16–42. Springer Verlag, Berlin.
59. Stegink, S. J., Vaughn, K. C., Kunce, C. M., and Trelease, R. N. (1987). *Physiol. Plant.*, **69**, 211.
60. Sauer, A. and Robinson, D. G. (1985). *Planta*, **164**, 287.
61. Walker, R. R. and Leigh, R. A. (1981). *Planta*, **153**, 150.
62. Fiske, C. H. and Subarow, Y. (1957). *J. Biol. Chem.*, **66**, 375.
63. Gallagher, S. R. and Leonard, R. T. (1982). *Plant Physiol.*, **70**, 1335.
64. Lanzetta, P. A., Alvarez, L. J., Reinach, P. S., and Candia, O. A. (1979). *Anal. Biochem.*, **100**, 95.
65. Ray, P. M. (1979). *Plant organelles*, ed E. Reid. Methological Surveys (B), Biochemistry, **9**, pp. 135–46. Ellis Horwood, Chichester.
66. Kauss, H. L. and Jeblick, W. (1986). *Plant Sci.*, **43**, 103.
67. Nagahashi, J. and Nagahashi, S. L. (1982). *Protoplasma*, **114**, 174.
68. Chen, R. S., Toribara, T. Y., and Warner, H. (1956). *Anal. Chem.*, **28**, 1756.

12

Protein transport into intact chloroplasts and isolated thylakoids

C. ROBINSON

1. Introduction

The biogenesis of the chloroplast is a complex process involving a great deal of protein traffic. Although the chloroplast itself synthesizes a number of proteins, most of the resident proteins (about 80%) are imported from the cytosol (reviewed in reference 1). These proteins must be both specifically targeted into the organelle and accurately 'sorted' such that imported proteins are correctly localized within the chloroplast. With the exception of outer envelope membrane proteins, which are imported by a distinct mechanism (2, 3), all imported proteins analysed to date are initially synthesized as larger precursors containing amino-terminal presequences. The presequences have been shown to contain information specifying targeting into the chloroplast and, in some cases, localization within the organelle.

The latter process is of particular interest in view of the structural complexity of the chloroplast, which consists of three distinct membranes (the outer and inner envelope membranes and the thylakoid membrane) enclosing three soluble phases (the intermembrane space, the stroma, and the thylakoid lumen). Cytosolically synthesized proteins are targeted into all six compartments, and intensive efforts have been made to elucidate the mechanisms by which the translocation across membranes and intra-organellar sorting events are carried out. These studies have depended totally on the development of efficient *in vitro* assays for protein translocation across the envelope and thylakoid membranes. The aim of this article is to describe in detail protocols for the import of *in vitro*-synthesized proteins into both intact chloroplasts and isolated thylakoids.

2. Chloroplast isolation

2.1 Choice of starting material

The choice of plant species is critical because only a few have been found to be suitable for the isolation of chloroplasts capable of efficient protein import. The most widely used are dwarf varieties of pea (*Pisum sativum*), for example Feltham First, which we obtain from Sharpes Seeds. Spinach chloroplasts can also be used, although the plants take rather longer to grow. In general, monocots such as wheat and barley are a poor source of import-competent chloroplasts, probably because intact chloroplasts are difficult to isolate in reasonable quantities from these species. Monocot proteins, however, appear to be imported efficiently and faithfully by dicot chloroplasts (4).

2.2 Growth of pea seedlings

We routinely grow pea seedlings in compost (for example, Levington's multipurpose) for about 7–10 days at 18–22 °C. Two factors are critical:

(a) The light intensity should be relatively low (40–50 $\mu E/m^2/s$), otherwise starch accumulates in the chloroplasts, leading to lysis during organelle isolation.

(b) Only young tissue (2–3 days after leaf emergence) should be used, because chloroplasts from mature leaves have lost the capacity to import proteins.

2.3 Isolation of intact chloroplasts

When isolating chloroplasts, it is important to work through the protocol rapidly after the initial tissue homogenization step: isolated chloroplasts do not retain import-competence indefinitely. In addition, the organelles must be kept cold at all times. The grinding medium should be placed in a −20 °C

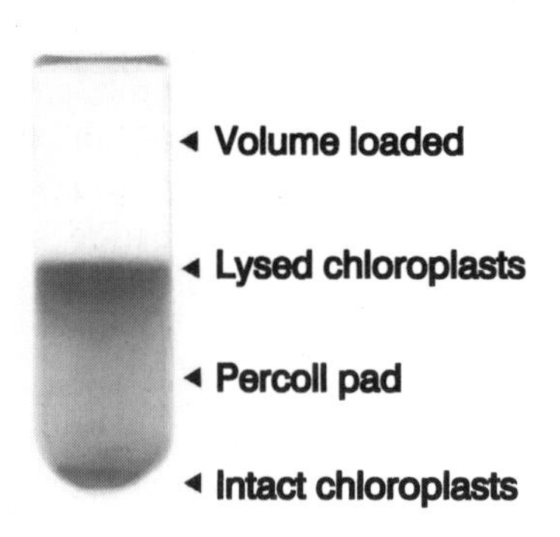

Figure 1. Purification of intact chloroplasts by centrifugation through a Percoll pad. A mixture of broken and intact chloroplasts was centrifuged through a Percoll pad. The resulting pellet contains intact organelles whereas broken chloroplasts fail to penetrate through the Percoll.

freezer until an ice slurry forms; the grinding vessel should also be stored in a freezer before use. In our laboratory, we use a Polytron homogenizer (Northern Media Supplies) but many other types of blender can give satisfactory results. Finally, it should be emphasized that chloroplasts must be handled carefully; being large organelles, they are very susceptible to shearing forces, so pipetting should be avoided where possible. When pipetting small volumes, we widen the apertures of disposable tips by cutting off the ends.

Protocol 1. Isolation of intact chloroplasts

- grinding medium (0.35 M sucrose, 25 mM Hepes–NaOH, pH 7.6, 2 mM EDTA)
- sorbitol medium (50 mM Hepes–KOH, pH 8.4, 0.33 M sorbitol)
- 40% Percoll in sorbitol buffer
- 80% aqueous acetone

1. Harvest leaves from pea seedlings and mix with semi-frozen grinding medium at a ratio of 20 g leaves per 100 ml medium.
2. Homogenize the leaves with two 3 sec bursts of the Polytron at 75% full speed.
3. Strain the homogenate gently through eight layers of muslin to remove debris.
4. Pour the suspension into 50 ml or 100 ml centrifuge tubes and centrifuge at 4000 *g* for 1 min. Discard the supernatant in one motion (the pellets are quite firm at this stage) and wipe the insides of tubes.
5. Resuspend the pellet gently in a small volume (4–8 ml) of sorbitol medium using a cotton swab or small paint brush, and layer the suspension on to an equal volume of 40% Percoll (Pharmacia) in sorbitol buffer. Centrifuge at 2500 *g* for 7 min (with the brake off). Intact chloroplasts are pelleted whereas lysed organelles fail to penetrate through the Percoll pad (illustrated in *Figure 1*).
6. Wash the pellet in 5 ml sorbitol medium and resuspend the pellet in 1 ml sorbitol medium. Check the intactness of the organelles under phase-contrast microscopy; intact organelles appear bright green, often with a surrounding halo, whereas broken chloroplasts appear darker and more opaque. The majority of the organelles (up to 95%) should be intact, although 50% intactness should give reasonable results.
7. Determine the chlorophyll concentration of the suspension as follows: remove a small aliquot (e.g. 20 μl) into 80% aqueous acetone and

Protocol 1. Continued

measure the absorbance at 645 and 663 nm. The chlorophyll concentration is equal to the dilution factor multiplied by [(20.2 × A_{645}) + (8.02 × A_{663})]. Adjust the concentration to 1 mg/ml using sorbitol medium. The chloroplasts are now ready for use in chloroplast import assays, and should be used as rapidly as possible.

3. Synthesis *in vitro* of nuclear-encoded chloroplast proteins

3.1 *In vitro* transcription of cloned DNA

A number of methods are available for the generation *in vitro* of synthetic mRNA, following the isolation of a full-length cDNA clone (5). The transcription protocol detailed below, which can be used with either SP6 or T7 RNA polymerase, is useful because the transcription products can be used directly to programme a cell-free translation system, without first phenol- or salt-extracting the RNA.

Protocol 2. *In vitro* transcription of cDNA

1. Prepare clean, preferably CsCl-purified DNA in an appropriate transcription vector at a concentration of 1 μg/μl in water.
2. Mix:
   - 2 μl DNA
   - 15 μl mix (40 mM Tris–Cl, pH 7.5, 6 mM $MgCl_2$, 2 mM spermidine, 10 mM DTT, 0.5 mM ATP, CTP, and UTP, 50 μM GTP, 100 μg/ml bovine serum albumin)
   - 20 units RNasin (Promega)
   - 0.25 units monomethyl cap ($m^7G(5')ppp(5')G$)
   - 1 μl of SP6 RNA polymerase (BRL, 15 units/ml) or T7 RNA polymerase (BRL, 50 units/ml)
3. Incubate for 30 min at 37 °C, add 1 μl of 10 mM GTP, and continue incubation for a further 30 min. The products of the transcription reaction can be stored at −80 °C until required.

3.2 *In vitro* translation

In this laboratory, we use the wheatgerm lysate system, prepared as described by Anderson *et al.* (6) for the *in vitro* synthesis of chloroplast protein precursors. The reticulocyte lysate system can also be used, but several groups have found that high levels of this lysate can lyse chloroplasts during import assays. Both systems are commercially available.

We generally find that optimum levels of translation are obtained with about 1 μl of transcription products per 12.5 μl translation reaction, although it is advisable to titrate the amount to be added. As a general rule we would expect 1 μl of ^{35}S-labelled translation product to give rise to a fairly intense band after SDS–PAGE, fluorography, and an overnight exposure to X-ray film.

4. Import of proteins into isolated chloroplasts

4.1 General comments

Reconstitution of chloroplast protein import is usually fairly straightforward, providing that intact chloroplasts are isolated as described above, and precursor proteins containing sufficient label can be prepared. There are no real 'tricks of the trade' to guard against. Every precursor protein tested so far, to my knowledge, has been successfully imported into isolated chloroplasts, although with varying efficiencies. Import takes place post-translationally and therefore the precursors are incubated with chloroplasts after translation is complete. Ideally, freshly prepared chloroplasts should be incubated with fresh translation products, but we have found that most translation products can still be imported after freezing at −80 °C. It is, however, important to note that import-competence is rapidly lost if the precursors are subjected to several rounds of freeze–thawing. The basic import assay conditions, modified from original protocols (7), are given below. In this assay, the chloroplasts are illuminated in order to generate, by photophosphorylation, the stromal ATP necessary to drive protein import. If conditions are being considered which may prevent ATP generation by this method, it is possible to drive import using added ATP. In this case, the chloroplasts should be preincubated at 25 °C for 15 min with 5 mM MgATP in order to allow the stromal ATP concentration the reach the optimal level.

Protocol 3. Import of proteins into isolated chloroplasts

A. *The basic import assay*

- sorbitol medium (as in *Protocol 1*)
- 200 mM methionine in sorbitol medium
- translation mixture (from Section 3)
- SDS–PAGE sample buffer
- proteinase K (Sigma; 1 mg/ml in sorbitol medium)
- 2 mM phenyl methyl sulphonyl fluoride (PMSF) in sorbitol medium (Sigma)

Protocol 3. *Continued*

1. Prepare intact chloroplasts and *in vitro*-synthesized proteins as described in sections 2 and 3.
2. Mix:
   - 55 μl chloroplasts
   - 10 μl 200 mM methionine (in sorbitol medium)
   - 4–8 μl translation mixture (depending on efficiency of translation)
3. Incubate at 25 °C for 20–40 min in an illuminated water bath at an intensity of 300 $\mu E/m^2/s$. The tubes should be gently shaken every 5 min or so to prevent the chloroplasts from settling out.
4. After incubation, dilute one sample with 5 ml sorbitol medium and pellet the chloroplasts by centrifugation at 4000 *g* for 2 min. Then resuspend the chloroplasts in a small volume of sorbitol medium, mix with one volume of SDS–PAGE sample buffer, and boil for 5 min. This sample contains imported proteins plus any precursor molecules which are bound to the chloroplast surface.
5. After incubation, add proteinase K to a final concentration of 50 μg/ml to a second sample (or more, if the protein of interest is particularly resistant to digestion) from a 1 mg/ml stock in sorbitol medium. Incubate on ice for 5 min, then dilute with 5 ml sorbitol medium, pellet as in step 4, and resuspend in the same volume of sorbitol medium but containing 2 mM PMSF (which partially suppresses the proteinase K). Mix with one volume of SDS sample buffer that is **already boiling** (this is essential in order to inactivate the proteinase K rapidly). This sample should contain protease-protected, and therefore imported, protein, which in the vast majority of cases will be of the mature size.

Note: unlabelled methionine is included in the import incubation mixture to prevent the high specific activity, labelled methionine in the translation mix from being incorporated into protein by the chloroplast protein synthesis machinery. If a different labelled amino acid is used in the translation system, a corresponding amount of unlabelled amino acid should be present in the import incubation.

B. *Procedures to localize imported proteins*

The basic assay detailed above is useful for determining whether or not a given precursor protein is imported and processed to the mature size, but additional tests to determine the location of the imported protein are often required. In the case of envelope proteins, this is a difficult task, because these membranes account for such a small proportion of total chloroplast membrane. Rather than attempting to purify the membranes, it is more practical merely to enrich the envelope fraction by removing the majority of thylakoid membranes (8). Preparation of stromal and thylakoid fractions is, however, quite straightforward:

- sorbitol medium
- 10 mM Hepes–KOH, pH 7.5, 5 mM $MgCl_2$

1. After the import incubation, dilute the suspension with 5 ml sorbitol medium and pellet the chloroplasts as in A above.
2. Lyse the organelles by resuspending the pellet in 10 mM Hepes–KOH, pH 7.5, 5 mM $MgCl_2$, and incubate on ice for 5 min.
3. Centrifuge in a microfuge for 5 min to generate a stromal supernatant and a thylakoid pellet. If required, it is possible to determine whether the imported protein is located in the thylakoid lumen by treating the thylakoids with protease; this is carried out as essentially as described for intact chloroplasts in section A above. Examples of this type of fractionation procedure to localize imported thylakoid lumen proteins are given in reference 9.

5. Import of proteins into isolated thylakoids

5.1 The import pathway for thylakoid lumen proteins

In higher plants and green algae such as *Chlamydomonas reinhardtii*, most of the proteins in the thylakoid lumen are synthesized in the cytosol and targeted into the thylakoid lumen by a two-phase pathway. Examples of such proteins include plastocyanin and the 33, 23, and 16 kDa proteins of the oxygen-evolving complex (OEC). The proteins are initially synthesized as larger precursors and then transported into the stroma and processed to intermediate forms by the stromal processing peptidase. The intermediate forms are then transported across the thylakoid membrane and processed to the mature size by a thylakoidal processing peptidase (10, 11). Using the intact chloroplast import assay described above, the later stages of this pathway are difficult to analyse in any detail, simply because these events occur very rapidly. However, techniques have recently been developed to reconstitute protein import by isolated thylakoids, and the assay conditions are given below.

5.2 A light-driven assay for thylakoidal protein import

The translocation of the 33, 23, and 16 kDa OEC proteins into isolated thylakoids was reconstituted by mixing precursor proteins with isolated pea thylakoids and incubating the mixture in the light (12, 13). Further analyses showed that the transport of all three proteins across the thylakoid membrane requires the trans-thylakoidal proton gradient (9). Otherwise, however, the proteins do exhibit some differences in requirements for import into thylakoids: import of the 23 and 16 kDa proteins requires only the presence of precursor protein and thylakoids, whereas import of the 33 kDa protein also requires the presence of concentrated stromal extract (12). We have recently found that this reflects the involvement of a soluble translocation factor in the transport mechanism (14). It is not yet clear whether this factor is required for the import of other lumenal proteins, but clearly, when attempting to achieve import of an untested protein, it would be logical to conduct assays in both the presence and absence of stroma. The basic assay conditions are given below, but two points should be emphasized:

(a) To my knowledge, *in vitro* import has, to date, been achieved only with pea thylakoids. We have not actually tested thylakoids from other species, but colleagues from the laboratory of R. G. Herrmann (Munich) have informed me that spinach thylakoids import poorly, if at all.

(b) Not all thylakoid lumen proteins require a proton gradient for transport across the thylakoid membrane. Plastocyanin and the CFo 2 subunit of the ATPase do not require an energized membrane, and the precise energy requirements for the translocation of these proteins are not known. Nevertheless, we have found that these proteins can be imported in the assay described below, although import of plastocyanin is rather inefficient (R. B. Klosgen, R. G. Herrmann, and C. Robinson, unpublished results).

Protocol 4. Import of proteins into isolated thylakoids

- 10 mM Hepes–KOH, pH 8.0
- 100 mM $MgCl_2$
- 2 mg/ml thermolysin or 500 μg/ml proteinase K (Sigma)
- SDS–PAGE sample buffer
- 10 mM EDTA or 2 mM PMSF

1. Prepare a pellet of intact pea chloroplasts as described in *Protocol 1*, steps 1–6.
2. Lyse the chloroplasts by resuspending the pellet in 10 mM Hepes–KOH, pH 8.0 to give a chlorophyll concentration of 1 mg/ml. Leave on ice for 5 min to ensure complete lysis.
3. Centrifuge in a microfuge at ~10 000 *g* for 5 min to generate a stromal supernatant and a thylakoid pellet. Keep the stromal extract on ice until required.
4. Wash the thylakoids twice in 10 mM Hepes–KOH, pH 8.0, and then

resuspend the pellet in either the same buffer or the stromal extract to 1 mg/ml chlorophyll.

5. Import incubations contain:
- 30 μl thylakoid suspension
- 5 μl 100 mM $MgCl_2$
- 3–8 μl translation products, prepared as in Section 3
- make up the volume to 50 μl with 10 mM Hepes–KOH, pH 8.0, and incubate at 25 °C for 30 min under illumination (300 $\mu E/m^2/s$)

6. After incubation, divide each sample into two equal portions. To one, add protease (200 μg/ml thermolysin or 50 μg/ml proteinase K); and incubate on ice for 30 min; leave the other sample on ice. After the incubation dilute each sample with 1 ml 10 mM Hepes–KOH, pH 8.0, and pellet the thylakoids in a microfuge (10 min spin). Resuspend the pellet in a small volume (e.g. 50 μl) of Hepes buffer containing protease inhibitor (10 mM EDTA for thermolysin or fresh 2 mM PMSF for proteinase K) and mix with SDS–PAGE sample buffer that is **already boiling** (to ensure rapid inactivation of the protease).

When the samples are analysed by SDS–PAGE and fluorography, the sample that is not protease-treated should contain up to three labelled polypeptides: the full precursor, the intermediate form which is generated by the stromal processing peptidase (if stromal extract was added), and the mature-size protein. Only the mature-size protein should be evident in protease-treated samples. *Figure 2* illustrates the results of a typical import assay using the precursor of the 23 kDa OEC protein as a substrate.

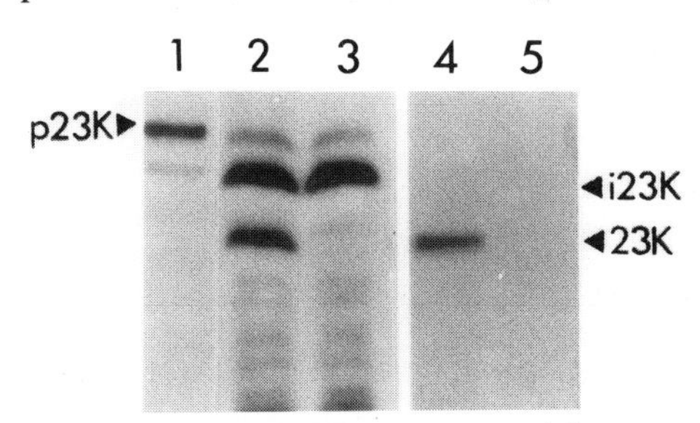

Figure 2. Light-driven import of the 23 kDa oxygen-evolving complex protein by isolated thylakoids. The precursor of the 23 kDa OEC protein (p23K; lane 1) was synthesized by transcription and translation of a cDNA clone and incubated with a suspension of pea thylakoids in stromal extract in the light (lanes 2 and 4) or in the dark (lanes 3 and 5). After incubation, samples were analysed directly (lanes 2 and 3) or after treating the thylakoids with protease (lanes 4 and 5). The figure shows that, in both light and dark incubations, the precursor protein is converted to the intermediate form (i23K) by stromal processing activity. However, mature-size protein (23K) is produced only in the light; this polypeptide species is resistant to protease digestion, showing that import into the thylakoid lumen has taken place.

It should be pointed out that relatively few thylakoidal proteins have been tested using this import assay. When assaying for the import of an untested thylakoid protein, it would therefore be wise to test first for import of a protein which is known to be efficiently imported, in order to confirm that the assay system is working. The best substrates to date are the 23 and 16 kDa OEC proteins (12, 13).

References

1. Smeekens, S., Weisbeek, P., and Robinson, C. (1990). *Trends Biochem. Sci.*, **15**, 73.
2. Li, H., Moore, T., and Keegstra, K. (1991). *Plant Cell*, **3**, 709.
3. Salomon, M., Fischer, K., Flugge, U.-I., and Soll, J. (1990). *Proc. Natl Acad. Sci. USA*, **87**, 5778.
4. Kirwin, P. M., Meadows, J. W., Shackleton, J. B., Musgrove, J. E., Elderfield, P. D., Mould, R., Hay, N. A., and Robinson, C. (1989). *EMBO J*, **8**, 2251.
5. Melton, D. A., Krieg, P., Rabagliciti, M. R., Maniatis, T., Zinn, K., and Green, M. R. (1984). *Nucleic Acids Res.*, **12**, 7035.
6. Anderson, C. W., Straus, J. W., and Dudock, B. S. (1983). *Methods Enzymol.*, **101**, 635.
7. Grossman, A. R., Bartlett, S. G., Schmidt, G. W., Mullett, J. E., and Chua, N.-H. (1982). *J. Biol. Chem.*, **257**, 1558.
8. Flugge, U. I., Fischer, K., Gross, A., Sebald, W., Lottspeich, F., and Eckershorn, C. (1989). *EMBO J.*, **8**, 39.
9. Mould, R. M. and Robinson, C. (1991). *J. Biol. Chem.*, **266**, 12189.
10. Hageman, J., Robinson, C., Smeekens, S. and Weisbeek, P. (1986). *Nature*, **324**, 567.
11. James, H. E., Bartling, D., Musgrove, J. E., Kirwin, P. M., Herrmann, R. G., and Robinson, C. (1989). *J. Biol. Chem.*, **264**, 19573.
12. Mould, R. M., Shackleton, J. B., and Robinson, C. (1991). *J. Biol. Chem.*, **266**, 17286.
13. Klosgen, R. B., Brock, I. W., Herrmann, R. G., and Robinson, C. (1992). *Plant Mol. Biol.*, **18**, 1031.
14. Hulford, A., Hazell, L., Mould, R. M., and Robinson, C. (1994). *J. Cell Biol.*, (in press).

13

Ion-selective microelectrodes

A. J. MILLER

1. Introduction

1.1 Uses of ion-selective microelectrodes

The word **microelectrode** has come to mean a glass micropipette which is pulled into a fine tip at one end and filled with an aqueous salt solution. The junction between the salt solution inside the microelectrode and the input to the electrometer amplifier is provided by a half-cell. There are different types of half-cell, but usually the metal contact is AgCl-coated silver wire and the salt solution is 0.1 M KCl. The simplest microelectrodes measure voltage and when inserted into cells measure the membrane potential, in mV, between the inside and outside of the cell. An **ion-selective microelectrode** contains an ion-selective membrane in the tip of the glass micropipette and is responsive to both the membrane potential and the activity (not concentration) of the ion sensed by the selective membrane.

Ion-selective microelectrodes are used to measure ion gradients across membranes. These measurements can be made outside and inside cells. For example, ion fluxes at the surface of roots can be measured by using either ion-selective microelectrodes (1) or an ion-selective vibrating probe. Intracellular measurements have been used to give important information on the compartmentation of nutrients, dynamics of cellular ion activities (e.g. in intracellular signalling), and transport mechanisms, particularly the energy gradients for ion transport. To make intracellular measurements it is necessary simultaneously to measure the membrane potential either by insertion of a second electrode or, for small cells, by combining the ion-selective and voltage-measuring electrodes into a **double-barrelled microelectrode**. The main criticism of intracellular measurements made with microelectrodes is that they report the ion activity at a single point within the cell. This will result in incomplete information if there are significant ion gradients within the cytoplasm of a single cell as occur in some situations. Overall, the chief advantages of using ion-selective microelectrodes are that

- they offer a non-destructive method of measuring ions within cells
- they do not change the activity of the ion being measured

- they permit simultaneous measurement of the electrical and chemical gradients across membranes
- they are relatively cheap, when compared with other methods for measuring intracellular ions, and once purchased, the same equipment can be used to measure a range of different ions

1.2 Theory

The theoretical background has already been described by many authors (see reference 2 and references therein) and will only be outlined here. The properties of an ion-selective microelectrode are defined by several characteristics:

- detection limit
- selectivity
- slope
- response time

The ideal relationship between electrode output (mV) and the activity (a_i) of the ion of interest (i) is log-linear and is described mathematically by the Nernst equation. Calibration of the electrode against a range of standard solutions should ideally, yield a slope (s) of 59 mV (at 25 °C) per decade change in the activity of a monovalent ion. In practice, however, the situation is more complicated than this because no ion-selective electrode has ideal selectivity for one particular ion and under most conditions there is more than one ion present in the sample solution. Hence contributions to the overall electro-motive force (EMF) made by each interfering ion, j, must be taken into account. In this situation, the Nicolsky–Eisenman equation, a modified Nernst equation, describes the EMF:

$$\text{EMF} = \text{E} + \text{s}\cdot\log\,[\text{a}_i + \text{K}_{ij}^{\text{pot}}(\text{a}_j)^{zi/zj}]$$

where K_{ij}^{pot} is the so-called selectivity coefficient of the electrode for the ion i with respect to ion j. This term expresses, on a molar basis, the relative contribution of ions i and j to the measured potential.

The parameters s and K_{ij}^{pot} are the two main characteristics defining any type of ion-selective electrode. The slope should be a near ideal Nernstian response when an electrode is calibrated against ion activity, but s is temperature sensitive (see section 3.4). The selectivity coefficient measures the preference of the sensor for the detected ion i relative to the interfering ion, j. It can be determined by the separate solution method, the fixed interference method, or the fixed primary ion method. For ideally selective membranes, or for samples containing no other ions with the same net charge as the ion in question, K_{ij}^{pot} must be zero. A selectivity coefficient of below 1 indicates a preference for the measuring ion *i* relative to the interfering ion *j*, and vice versa for a selectivity coefficient greater than 1. The K_{ij}^{pot} values should not be considered to be constant parameters that characterize membrane selectivity under all conditions; the values are dependent both on the method used for determination and on the conditions under which the calibrations are made. The fixed interference method is most commonly used to calculate the selectivity coefficient, and it is the method recommended by the International Union of Pure and Applied Chemistry (2). Whichever type of method is used, the one used should always be stated.

Another important parameter of an ion-selective microelectrode is the detection limit, which is the lowest ion activity that can be detected with confidence and is defined by the intercept of the two asymptotes of the Nicolsky response curve (see *Figure 1*). In practice, the detection limit seems to depend on the tip geometry and composition of the microelectrode's ion-selective membrane. Finer or smaller diameter tips have higher detection limits, while composition affects detection in ways that can only be determined experimentally. The presence of interfering ions alters the detection limit (e.g. reference 3). The electrodes provide no useful information below their detection limits and for maximum benefit should be used in the linear portion of their calibration curves. The response time of ion-selective electrodes can be important when measuring changes in ion activities. This microelectrode parameter is dependent on many factors, including tip geometry, membrane composition, and resistance. Response time can be measured during the calibration as the time taken for the voltage to adjust when ion activity at the tip is changed.

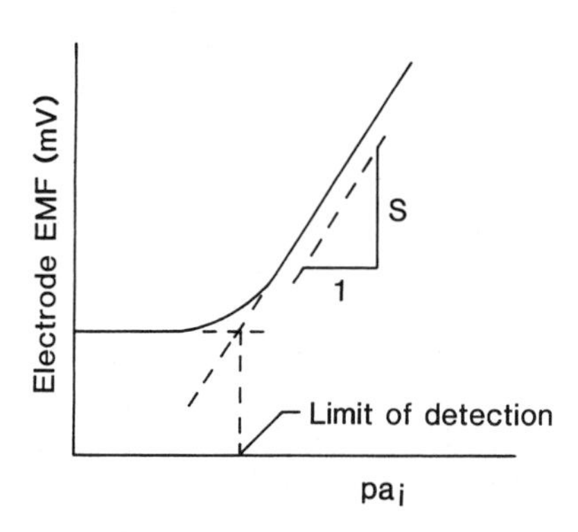

Figure 1. A schematic representation showing an ideal ion-selective microelectrode calibration curve. The slope, s, is the change in EMF per decade change in activity of a monovalent anion, *i*, which is equivalent to 59.2 mV at 25 °C; the limit of detection is defined as described in the text and is also indicated. pa_i is the $-\log_{10}$ of the activity of ion *i*.

2. Types of ion-selective microelectrode

There are three major types of ion-selective electrode, all of which can be minaturized for use in plant cells. These are solid state, glass, and liquid (or fluid) membrane electrodes. Solid-state microelectrodes have been used to measure pH or Cl^- inside plant cells (4). Recessed-tip glass microelectrodes have been made using pH-selective glass (5). These two types of micro-electrode have largely been superceded for intracellular measurements by liquid-membrane electrodes so only the latter will be described here. Liquid membrane sensors are commercially available for a wide range of ions (see *Table 1*).

To make an ion-selective microelectrode, the tip of the electrode is filled with an ion-sensing chemical cocktail which gives a voltage output of different values when placed in solutions containing different activities of the ion. Therefore when the electrode is inserted in the cell, the voltage measured gives a direct indication of the ion activity inside the cell. This situation is complicated by the voltage across the cell membrane; the ion-selective electrode will sense this in addition to voltage due to the activity of the ion of

Table 1. Some examples of sensors for liquid membrane ion-selective microelectrodes and some of their properties

Ion	Sensor molecule(s)	Detection limit	Major interfering ions in plant cells
Ca^{2+}	ETH 129	10 nM	H^+, K^+, Mg^{2+}
	ETH 1001	40 nM	H^+, K^+, Mg^{2+}
Cl^-	Mn(III)TPPC[c]	1–5 mM	acetate, HCO_3^-, SCN^-, NO_3^- above pH 7.6
H^+	tridodecylamine	above pH 9	K^+
	ETH 1907	pH 9	K^+
K^+	Valinomycin	100 μM	Ca^{2+}, HN_4^+
Mg^{2+}	ETH 5214	200 μM	Ca^{2+}, K^+
Na^+	ETH 227[b]	3 mM	Ca^{2+}, H^+, K^+
	ETH 157[b]	2 mM	H^+, K^+
NH_4^+	nonactin[b]	2 μM	K^+
NO_3^-	MTDDA·NO_3[c]	0.5 mM	Cl^-, NO_2^-, SCN^-

[a] Mn(III)TPPC = 5,10,15,20-tetraphenyl-21H,23H-porphin manganese (III) chloride (6).
[b] Data from Fluka Chemicals.
[c] MTDDA·NO_3 = methyl tridodecylammonium nitrate (3).
The detection limits quoted are from calibration in solutions approximating to cytoplasmic composition; values will also depend on the tip diameter but the values above are for tips less than 1 μm in diameter. This means that lower detection limits are possible for extracellular measurements where larger tip diameters can be used. All the above sensor molecules (or the precursor for synthesis for MTDDA·NO_3) can be obtained from Fluka Chemicals.

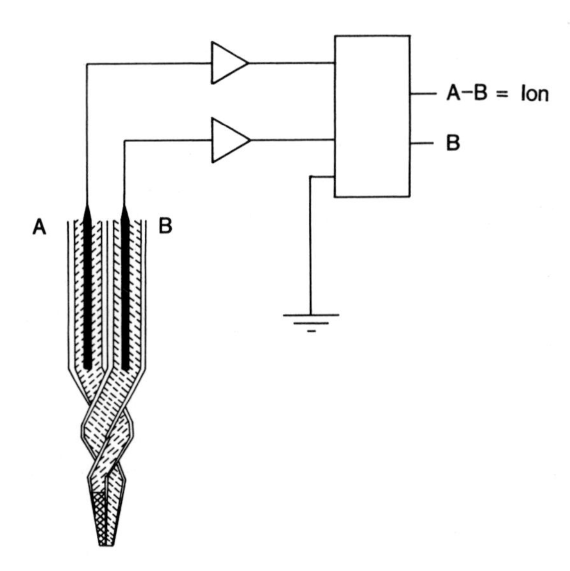

Figure 2. Diagram of a liquid membrane double-barrelled ion-selective microelectrode suitable for intracellular recording. The microelectrode is made by twisting together double-barrelled glass. (A) Ion-sensing barrel with ion-selective cocktail in the tip (cross-hatched area), (B) Cell voltage recording barrel. The output of B is subtracted from the output of A and converted to ion activity using a calibration curve such as that in *Figure 1*.

interest (see Introduction). To obtain the output for the ion alone, the cell voltage must be subtracted. This is done by using either two single electrodes or a double-barrelled electrode in which the ion-sensing electrode is combined with a cell voltage-measuring electrode (see *Figure 2*). Both output voltages are measured against a reference ground electrode in the external solution. The ion activity is determined from the calibration curve after subtracting the membrane potential.

3. Making ion-selective microelectrodes

The preparation of ion-selective microelectrodes can be divided into four main stages:

(a) pulling of glass micropipettes
(b) silanization of the inside surface of the ion-selective electrode or barrel
(c) backfilling
(d) calibration

The preparation of a nitrate-selective cocktail for backfilling microelectrodes is described in *Protocol 1*. A detailed generalized method which is suitable for all of the different types of ion-selective microelectrode is described in *Protocol 2*. The background to each stage is described here.

3.1 Pulling

Microelectrodes should be prepared to give dimensions suitable for impaling the target cell type. Double-barrelled microelectrodes can be prepared by twisting two single pieces of filamented borosilicate glass or using glass which is already fused together. Filamented glass has a glass fibre attached to the inner wall; this fibre assists backfilling by providing a hydraulic conduit along which the solution can flow by capillarity. This twisting is done using an electrode puller which both heats the glass and pulls it in a way predetermined by the operator. The heating is paused for the two barrels to be twisted around one another then the heating and pulling continues. There are various different types of microelectrode puller (7). The most important feature is reproducibility, which ensures that when an optimum microelectrode shape for a particular cell type has been prepared, it can be exactly duplicated many times.

Microelectrodes are usually made from borosilicate glass, although the harder aluminosilicate glass is also sometimes used. Multi-barrelled glass of varying dimensions can be purchased from suppliers (e.g. Hilgenberg). This type of glass seems to be the best for ion-selective microelectrode work. An alternative type of double-barrelled glass called 'theta' glass can be used; this has a single thin glass wall between the two pre-formed barrels. Adjacent ion-selective barrels may mutually interfere because the thin glass walls at the electrode tip have an impedance that may be as low as the impedances of the liquid ion-exchangers so that the measured potential depends on the potential across the glass as well as the potential across the liquid ion-exchanger. This problem is more acute when 'theta' glass is used because the final glass partition in the tip is much thinner. Both barrels of glass should have an internal filament to assist with backfilling. Identification of the different barrels can be done by marking the different barrels or cutting to different lengths (see *Protocol 2*). Wear safety glasses at all times when pulling and breaking glass.

Before preparing the ion-selective microelectrode it is important to determine that the glass microelectrodes filled with 0.1 M KCl can be used to impale cells and measure stable resting membrane potentials sensitive to metabolic inhibitors (in the usual range for the cell type, in the bathing solution used). An estimate of the tip geometry of the microelectrode is provided by measuring its electrical resistance when filled with KCl, larger tips having lower resistances. For tips of 2–0.1 μm diameter the electrical resistances of ion-selective microelectrodes are usually in the GΩ range, while microelectrodes filled with 0.1 M KCl have 10^3-fold smaller resistances in the MΩ range. The dimensions of the microelectrodes are usually a compromise between obtaining a stable membrane potential and a good calibration response (detection limit).

3.2 Silanizing

The inside of the glass micropipettes must be given a hydrophobic coating to allow the formation of a high resistance seal between the glass and the hydrophobic ion-selective membrane. The barrel designated to be ion-selective is silanized by placing a few drops of a solution of 1% (w/v) silanizing agent in chloroform on its blunt open end. There is a range of different silanizing agents which can be used at this concentration but dimethyl-dichlorosilane or tributylchlorosilane are most common. Care must be taken to ensure that the reagent does not enter the membrane potential-measuring barrel. **Beware**: silanizing agents are corrosive and toxic, so protective glasses and gloves must be worn and glass must be treated in a fume hood. The microelectrode is then placed under a heating lamp giving a temperature of 140 °C at the micropipette surface. The silanizing solution quickly vaporizes giving the ion-selective barrel a hydrophobic coating. After silanization there should be no liquid residue remaining in the microelectrode tip before backfilling.

3.3 Backfilling

There are two steps to backfilling, the first (a) uses a cocktail to form the ion-selective membrane in the microelectrode tip and the second (b) usually a minimum of 48 h later, uses an aqueous salt solution to provide contact between this membrane and the Ag/AgCl metal electrode (in the base of the microelectrode holder). Both steps are made much simpler by using filamented glass to make the microelectrodes and can be achieved using a syringe and fine needle (30 G).

(a) The electrodes are backfilled with the sensor cocktail containing several different types of component:

- an ion-selective molecule, sensor or exchanger
- membrane solvent or plasticizer
- additives e.g. lipophilic cation/anion
- a membrane matrix to solidify the ion-selective membrane—this is essential for measurements in cells possessing turgor

For many ions, the membrane cocktail can be purchased already mixed and it is advisable to start by using the commercial mixture. However, the individual cocktail components can be bought from chemical suppliers and preparing the cocktail oneself is cheaper. For commercially available liquid membrane cocktails the membrane matrix is not normally included. A matrix is needed if microelectrodes are to be used in plant cells, because turgor will displace a liquid membrane from the electrode tip, thereby changing or eliminating the sensitivity to the measuring ion (8–10). The matrix used is normally a high molecular weight poly(vinyl chloride) (PVC) polymer, but can also include nitrocellulose for additional strength.

Of all the components, the ion-selective sensor is the main factor determining the electrode's properties (e.g. slope, selectivity, limit of detection); however, the membrane solvent can, by an unknown mechanism, also alter the properties such as the lifetime, stability, and selectivity of the electrode. Additionally, membrane additives, such as lipophilic ions, can be used to improve the performance of microelectrodes. The roles played by each component are described in detail by Ammann (2). The final optimum cocktail composition must be found by trial and error (testing the performance of electrodes made using slightly altered composition). Good electrodes should have a low detection limit, a near ideal slope, and a small selectivity coefficient for physiologically important interfering ions.

Protocol 1. Preparation of nitrate-selective cocktail for backfilling ion-selective barrel

- methyl tridodecyl ammonium nitrate (MTDDA·NO_3) 6% (w/w) (3)
- 2-nitrophenyl octyl ether 65% (w/w) (Fluka 73732)
- methyltriphenyl phosphonium bromide 1% (w/w) (lipophilic cation) (Sigma M7883)
- nitrocellulose 5% (w/w) (Whatman 7184002)
- PVC 23% (w/w) high molecular weight polymer (Fluka 81392)
- tetrahydrofuran (THF, Sigma T5267)

1. Weigh the MTDDA·NO_3 (3 mg), nitrocellulose (2.5 mg), PVC (11. 5 mg), and lipophilic cation (0.5 mg) into a 1 ml glass screw-topped vial using a balance accurate to 0.1 mg.
2. Add nitrophenyl octyl ether (32.5 mg) to the vial using a microcapillary on the balance pan.
3. Dissolve the cocktail in approximately 4 vol. of THF. This should be dispensed using a glass syringe and metal needle with no plastic components[a]. The cocktail takes at least 30 minutes to dissolve completely.
4. This cocktail can then be stored at 4 °C for several weeks; it is ready for use in *Protocol 2* and is enough to make about 70 nitrate-selective microelectrodes.

[a] THF will dissolve some types of plastic disposable syringe.

(b) Backfilling with aqueous salt solution can also be done using a fine needle and syringe or a plastic disposable pipette tip can be gently heated and pulled into a fine capillary suitable for inserting inside the glass barrel. It is important to try to displace all the air from the interface between the ion-selective membrane and the backfilling solution. Sometimes a hair or cat's whisker can help to do this.

3.4 Calibration

Ion-selective microelectrodes can be calibrated using concentration or activity. In fact, the electrodes respond to changes in activity (see Section 1.2) and it is the activity that is the important parameter for all biochemical reactions. Therefore calibrating with ion activity gives a microelectrode output which can be used directly without any assumptions of the intracellular activity coefficient for the ion. For these reasons the calibration of microelectrodes generally uses solutions which resemble the intracellular environment in terms of interfering ions and ionic strength. Calibration of pH microelectrodes is easy because standard pH buffers can be used and simply checked with a pH meter. For other types of ion-selective microelectrode the calibration solutions may need to contain a pH buffer and a background salt solution to give an ionic strength approximately equivalent to that inside the cell. Care must be taken in the choice of these additional ions: they must not give significant interference over the range of measurements. In other words, the microelectrodes must have very small selectivity coefficients for these background ions. The calibration solutions are usually chosen to be approximately 0.14 M ionic strength. There are few examples of whole sap analysis to suggest what an appropriate figure might be, but for giant algal cells this value would seem reasonable (11). The use of computer programs to calculate ion activity and the availability of a wide range of ion-selective macroelectrodes make it easier to prepare calibration solutions for all types of ion-selective microelectrode. Furthermore, calibration solution recipes have been published for some ion-selective microelectrodes (Ca^{2+} (12), Mg^{2+} (13), and NO_3^- (3)). Be aware that some calibration solutions use concentration not activity, and also that the term 'free' ion usually means the concentration (rather than the activity) of unbound ion, particularly for Ca^{2+} and Mg^{2+}. The calibration of calcium-selective microelectrodes for intracellular measurements requires the use of calcium buffering agents such as EGTA because of the very low concentrations being measured (12).

The ion-selective microelectrode can be calibrated in the microscope chamber (see later) where intracellular measurements will be made or in a U-shaped glass funnel alongside the microscope. Finally, note that the slope of the calibration curve is temperature sensitive and so both calibrations and intracellular measurements should be done at the same temperature. If the temperature of the calibration solutions is 4 °C and the cell is at 20 °C, the slope of the electrode calibration for a monovalent ion will be 55 mV per decade change in activity, not the 58 mV expected at 20 °C.

Protocol 2. Preparation of ion-selective microelectrodes

- dimethyldichlorosilane (Fluka 40136), 1% (v/v) silanizing agent in chloroform
- THF (Sigma T5267)
- ion-selective cocktail (Fluka) with PVC added or cocktail from *Protocol 1*
- PVC (high molecular weight polymer, Fluka 81392)
- nitrocellulose (Whatman 7184002)

1. Double-barrelled glass microelectrodes are pulled and twisted using an electrode puller.
2. Break back one barrel to give a short barrel, this will be the membrane potential-recording barrel. The barrel can be broken back using square-ended pliers or a razor blade on the edge of a metal plate (wear safety glasses).
3. Place microelectrodes under a heating lamp for at least 30 min to dry before silanizing.
4. Use a disposable syringe and 25 G needle to introduce silanizing agent into the blunt end of the longer barrel which will become the ion-selective barrel. **Warning**: this must be done in a fume hood, under a heating lamp, because the silanizing vapour is toxic and corrosive. Keep the double-barrelled micropipette under the lamp for a further 30 min.
5. If the commercial ion-selective cocktail for backfilling does not contain PVC or nitrocellulose, add one or both by first dissolving these matrix components in excess THF (4 vol.) and then mixing with commercial cocktail. The quantity of PVC can range from 10 to 30% (w/w) and that of nitrocellulose will be 5% (w/w) of cocktail when THF has evaporated (see *Protocol 1*).
6. Backfill ion-selective barrel of microelectrode with cocktail dissolved in THF/PVC mixture. Use a glass syringe (1 ml) and an all-metal 30 G needle (Scientific Laboratory Supplies Ltd.)
7. Store the microelectrodes (tip downwards) in a silica-gel dried environment for at least 48 h. During this time most of the THF evaporates leaving a solvent-case membrane plug in the tip of the designated ion-selective barrel.
8. When the THF has evaporated, the ion-selective barrel can be

backfilled with an aqueous salt solution containing the ion to be measured.

9. Most types of ion-selective microelectrode require 'conditioning' for a minimum of 30 min in an aqueous solution of the ion to be measured. This process involves immersing the tip in the solution containing a high concentration (e.g. 100 mM) of the ion to be measured.
10. Calibration of microelectrodes can be performed in the chamber built to take the plant tissue, or using a U-shaped funnel.

General practical points:

- handle microelectrodes with forceps
- when dispensing THF pour a few millilitres from the stock bottle into a clean glass beaker, after first rinsing the beaker with a little freshly dispensed THF; cover the beaker with Parafilm, then dispense further THF using a glass syringe and needle by piercing the film cover with the needle (this helps to reduce solvent vapour and prevents contamination of THF)
- if more than one type of cocktail is used, employ a different syringe for each type of cocktail; it is best to dedicate a syringe for one particular cocktail only, and thus avoid any contamination by other ion sensors
- calibrate starting with the highest concentration, and calibrate only in the range in which you expect to be working. There is no point in exposing electrodes unnecessarily to low ion concentrations as most types of ion-selective membrane respond badly to long exposures at very low concentrations.

3.5 Fault finding and some possible problems

The best approach is to try to solve problems by a process of elimination. Firstly, establish whether a problem occurs in the circuitry or is specific to the ion-selective microelectrodes. The circuitry can be tested by putting a broken-tipped KCl-filled microelectrode in place of the ion-selective microelectrode. The broken-tipped microelectrode should give a stable zero output. It may be necessary to recoat the Ag/AgCl contact in the half-cell or there may be a wiring problem. Noisy recordings can be caused by poor earthing or by air bubbles in the backfilling solutions. If the circuitry has no problems then the ion-selective microelectrode must be the cause. When the ion-selective microelectrode does not respond to the calibration solutions then the membrane can be checked by deliberately breaking the tip to expose a larger area of ion-selective membrane. Breaking the tip can displace the ion-selective membrane from the tip so it is important to measure the resistance to check it is still in the GΩ range. If the broken tip gives a good response to changes in ion activity then the problem is independent of the composition of the membrane. When the microelectrode tip diameter becomes too fine the output from the ion-selective electrode will no longer respond to changes in ion activity.

3.6 Storage of ion-selective microelectrodes

For long-term storage, the ion-selective microelectrodes should be stored without backfilling, in a silica-gel dried sealed container in the dark. This can be done in a screw-capped glass jar containing dry silica gel, with the microelectrodes attached to the inner wall using Plasticine or Blu-tack. Ion-selective microelectrodes stored for several years in this way can still give a reasonable performance when backfilled.

4. Intracellular measurements

The equipment needed for intracellular measurements is similar to that for normal electrophysiological measurements in plant cells and general details have been described in reviews on electrophysiological methods (7). The equipment needed is listed below:

- electrometer with high input impedance
- micromanipulator
- microscope with stage-mounted perfusion chamber
- Faraday cage
- PC microcomputer with analog/digital (A/D) interface
- chart recorder

Only a few special features of equipment for ion-selective microelectrode work will be described here—most of these are needed because ion-selective microelectrodes have high impedances. The microscope must be the fixed-stage type with a fibre optic light source. The lamp is located outside the Faraday cage to avoid interference which would be caused by including any mains voltage cable inside the Faraday cage. The chamber to hold tissue is mounted on the microscope stage and designed so that the tissue can be continuously perfused with the experimental nutrient solution throughout the measurement to avoid a concentration gradient developing at the surface of the tissue. A high impedance electrometer amplifier (e.g. World Precision Instruments, model FD 223) is required because the ion-selective micro-electrodes have very high resistances. The amplifier must have a high input impedance at least 1000 times higher than the ion-selective electrode e.g. 10^{15} Ω. Furthermore the input leakage current from the electrometer must be low so that no significant offset voltage (>1 mV) is produced across the ion-selective electrode. Other useful facilities include GΩ range resistance tester and a difference-voltage output so that a direct output equivalent to cell ion activity (A–B in *Figure 2*) can be obtained. The electrometer output can be passed to a chart recorder and also via an A/D converter to a microcomputer. Data handling and processing are made much easier using software such as that developed by I. R. Jennings at the Biology Department, University of York, UK (3).

Several criteria for acceptable measurements can be defined. Firstly, after impalement the ion-selective microelectrode should be recalibrated and should give a very similar response to that shown before the cell impalement, particularly at activities similar to those measured *in vivo*. Sometimes the recalibration shows a displacement up or down the *y* axis (mV output). More often the detection limit of the ion-selective microelectrode has changed but provided the measurement was on the linear response range of the electrode calibration curve this is not usually a reason to disregard the result. Sometimes the performance of the ion selective microelectrode can even improve with the detection limit actually becoming lower. For this reason, it may be best to impale a cell quickly with a new tip before calibrating prior to measuring the activity in the cell. A comparison between the electrical resistance of the ion-selective microelectrode before and after an impalement provides a good indicator of whether the tip will recalibrate. If the resistance decreases below 1 GΩ, the ion-selective membrane has probably been displaced during impalement and the electrode will not recalibrate. Throughout the recording the state of the cell can be assessed by monitoring the membrane potential (which should remain stable unless deliberately perturbed) or processes like cytoplasmic streaming.

When measuring changes in intracellular ion concentrations, artefacts can be caused by the differential response times of the two barrels; the ion-selective barrel usually has a longer response time than the membrane potential-sensing barrel. This can be corrected for when the response time of the electrode is known (5). The electrode response time can limit detection of rapid changes in ion activity.

In plant cells, identifying the internal cell compartment (cytoplasm or vacuole) in which the microelectrode tip is located can be a problem for some ions and it may be necessary to grow the plant under conditions in which two populations of measurements can be identified. Alternatively, a triple-barrelled microelectrode can be used where one barrel is pH- or Ca^{2+}-selective. Large gradients of protons and Ca^{2+} ions are known to exist across the tonoplast, with the cytoplasm maintained at relatively constant values (pH 7.2, 100 nM Ca^{2+}) so compartment identification is possible. Another approach is to use tissues where the two major cell compartments can be identified under the microscope, e.g. root hairs, or cell cultures which have no large vacuole. However, identifying which compartment the electrode is in can still be problematic, particularly if the electrode indents the tonoplast but does not penetrate it.

Leaking of salts from the tip of the membrane potential-sensing barrel has been reported (14); this may be a particular problem in small cells. Diffusion of ions from the membrane potential-sensing barrel could give high local gradients of ions at the tip of a double-barrelled microelectrode. It may be important to try measurements where different types of backfilling solution are used in the reference barrel. Large leaks should affect membrane potential and monitoring this should indicate possible problems.

A further possible problem can arise when using ion-selective microelectrodes with inhibitors. Some inhibitor chemicals are highly lipophilic and will readily dissolve in the ion-selective membrane. These chemicals can poison the membrane but this will be demonstrated during the recalibration of the ion-selective microelectrode.

One last point concerning statistical analysis of data concerns the calculation of means. These should be calculated using the data that are distributed normally, that is using the log activity or output voltages not the actual activities (15). Therefore when mean activity value is used it can only be expressed with 95% confidence limits, whereas −log [activity] can be given standard errors or standard deviations.

References

1. Henriksen, G. H., Bloom, A. J., and Spanswick, R. M. (1990). *Plant Physiol.*, **93**, 271.
2. Ammann, D. (1986). *Ion-selective microelectrodes, principles, design and application*. Springer-Verlag, Berlin.
3. Miller, A. J. and Zhen, R.-G., (1991). *Planta*, **184**, 47.
4. Coster, H. G. L. (1966). *Aust. J. Biol. Sci.*, **19**, 545.
5. Sanders, D. and Slayman, C. L. (1982). *J. Gen. Physiol.*, **80**, 377.
6. Kondo, Y., Bührer, T., Seiler, K., Frömter, E., and Simon, W. (1989). *Pflügers Arch.*, **414**, 663.
7. Blatt, M. R. (1991). *Methods Plant Biochem.*, **6**, 281.
8. Felle, H. and Bertl, A. (1986). *J. Exp. Bot.*, **37**, 1416.
9. Sanders, D. and Miller, A. J. (1986). In *Molecular and cellular aspects of calcium in plant development* (ed. A. J. Trewavas), pp. 149–63. Plenum Press, New York.
10. Reid, R. J. and Smith, F. A. (1988). *J. Exp. Bot.*, **39**, 1421.
11. Okihara, K. and Kiyosawa, K. (1988). *Plant Cell Physiol.*, **29**, 21.
12. Tsien, R. Y. and Rink, T. J. (1981). *J. Neurosci. Method*, **4**, 73.
13. Blatter, L. A. and McGuigan, J. A. S. (1988). *Magnesium*, **7**, 154.
14. Blatt, M. R. and Slayman, C. L. (1983). *J. Memb. Biol.*, **72**, 223.
15. Fry, C. H., Hall, S. K., Blatter, L. A., and McGuigan, J. A. S. (1990). *Experimental Physiol.*, **75**, 187.

14

Microsampling and measurements of solutes in single cells

D. TOMOS, P. HINDE, P. RICHARDSON, J. PRITCHARD and W. FRICKE

1. Introduction

Plants are high-pressure hydraulic machines. Cell turgor pressure (which is often in the 0.5–1 MPa range) is central to growth, support of non-woody tissues, penetration of soil by roots, and such mechanical processes as stomatal action. This mechanically useful pressure is derived from chemical energy by the active transport of solutes across selectively permeable membranes. Considerable strides are being made towards understanding the molecular basis of such transport processes. A relatively neglected area is the cell-to-cell variation in solutes (1). Quantification of these in individual cells has relied largely on X-ray fluorescence analysis of sectioned material and on the use of microelectrodes. In this chapter we give procedures for obtaining liquid microsamples from individual cells in intact tissues and the quantitative analysis of solutes in them. Currently the techniques are being applied routinely only to vacuolar samples. Their application to cytoplasmic material is in principle the same but obtaining such samples is technically far more challenging.

In addition to individual solutes, we also describe a method for measuring osmotic pressure which provides an independent measure of the *total* solute concentration of the cell-sap. The presence of additional solutes can be identified by comparing this value with the sum of the solutes measured.

There are advantages to be gained from obtaining and analysing microsamples rather than studying material *in situ*. Once isolated, they are readily amenable to a wider range of direct physicochemical analysis. In many cases a series of sequential analyses can be performed on the same single-cell sample. (This can include such physical parameters as turgor pressure if a pressure probe is available.) This also allows for replication and statistical appraisal of data. In some cases (e.g. sugars) it is difficult to envisage other methods of measuring individual cell concentrations. Moreover, the sample can be 'spiked' with material to provide an internal standard and as a control for interfering factors.

The protocols have evolved from the use of the pressure probe (2) to measure single-cell hydrostatic pressure. Most, however, are independent of water relations studies and do not require a pressure probe. They describe the isolation of a volume of cell-sap in the picolitre range from the cell (Section 3), the measurement of its osmotic pressure (Section 4), the measurement of its elemental composition by X-ray fluorescence (Section 5), and the measurement of a range of metabolizable solutes (organic and inorganic) by enzyme-linked microscope fluorimetry (Section 6). We have used the technique for both roots and shoots. Cells with volumes down to 20–30 pl and smaller are generally amenable to study. Surface cells are easiest, but a technique is given for analysing cells to a depth of up to approximately 1 mm.

2. Capital equipment

The following capital equipment is needed for the analyses:

- capillary puller (e.g. Harvard Apparatus)
- capillary microforge (e.g. de Fonbrune/Alcatel)
- a picolitre (our workshops) or nanolitre (Clifton Technical Physics) osmometer modified according to (3)
- micromanipulator (e.g. Prior, Narashige)
- energy dispersive X-ray microanalytical equipment (e.g. Link Analytical QX2000 fitted to Hitachi S520 scanning electron microscope
- microscope photometer (e.g. Leitz Fluovert and MPV system)
- sampling pressure probe (conventional pressure probe (2) modified as described in reference 3 or available from our workshops)

3. Extraction and manipulation of single-cell samples

3.1 Outline of technique

Samples are obtained by inserting a fine microcapillary filled with water-saturated paraffin oil or low-viscosity silicone oil directly into the target cell. The microcapillary is mounted on a micromanipulator and the entire process is observed under a stereo microscope (magnification zoom up to ×200). The pressure of the cell contents forces cell-sap into the capillary. Following removal of the capillary from the cell, the microsample is ejected under liquid paraffin for analysis. For *highly vacuolate cells*, such as root cortex or leaf epidermis, it is *assumed* that the sample obtained is vacuolar. This is concluded largely on the basis that the relatively large volume obtained generally precludes significant cytoplasmic contamination. Such contamination may be assessed by assaying cytoplasmic enzymes (e.g. malate dehydrogenase—see *Protocol 6* for a basis) or other constituents, in the microsample. For cells with a larger proportion of cytoplasm, such as leaf mesophyll, this can be significant.

3.2 Apparatus

3.2.1 Preparation of microcapillaries

Two types of microcapillary are used. The first is a drawn-out capillary used for obtaining samples from cells and for manipulations where a precise volume is not needed. The second is a constriction pipette that is used for measuring identical aliquots of samples, standards, and reagents. The exact volume generally need not be known as samples are measured by comparison with standards using the same pipettes.

Depending on the circumstances (listed below) each type of capillary can be used either silanized (to render them hydrophobic; *Protocol 2*) or non-silanized.

Microcapillaries are manufactured from 1 mm glass capillary tubing (e.g. Clark Electromedical Instruments) using a commercial capillary puller. The exact dimensions can be adjusted by altering settings on the puller. (The most useful tip shape will depend on the application and nature of the cell. For example, 'stocky' inflexible tips are needed for tough cell walls such as those of *Cladophora*.) After pulling, the tips have an aperture of about 0.1 μm. (Such tips are used directly in electrophysiological work, see Chapter 13.) To allow entry of sap into the tip for sampling, however, the extreme tip of the microcapillary must be removed by gently brushing it against the surface of a Perspex block, or similar solid object (under the microscope). Alternatively it may be dipped in a rapidly stirred beaker of abrasive slurry (e.g. 10% (v/v) suspension of 0.05 μm alumina). (Some laboratories bevel the tips mechanically on a turntable, while others use a jet of abrasive slurry.) Tips with apertures in the range of 1–3 μm are used.

3.2.2 Preparation of constriction pipettes

These are prepared from the microcapillaries in a commercial microforge. Place the capillary into the microforge and bring the barrel of the tip up to the heating wire, switch on the current through the micro-element for 2–3 s while watching through the microscope, and compensate for any movement of the element (due to thermal expansion) using the microforge positional controls. The constriction must leave the bore easily visible since flow of solution will be restricted if it is too narrow. Constriction pipettes of different volumes can be made by varying the diameter of the capillary barrel and/or by placing the capillary so that the microforge heating wire is further from the capillary tip. Constriction pipettes can be made to handle droplets as small as 1 pl. (Often the constriction is only used as a datum point, smaller volumes being gauged by eye. The errors involved are small.)

The pipettes can either be calibrated absolutely using a radioactive solute of high specific activity or, more conveniently, approximate volumes can be measured by expelling a droplet of water from the constriction pipette into a volume of oil and calculating the volume from the diameter of the spherical droplet measured against a microscope eyepiece graticule. In practice, however, they are used for both samples and standards in such a way as to eliminate the need for knowledge of absolute volumes.

A more complete description of microcapillary preparation is given in Lowry and Passonneau (4).

3.3 Extraction of samples

3.3.1 Sample collection in air from surface cells

Samples are obtained by inserting an oil-filled microcapillary into the cell (*Protocol 1*). The cell turgor forces the cell-sap into the capillary tip. It is essential that the sample is then removed from the cell as rapidly as possible ($<<1$ sec) since the loss of cell turgor will draw pure water into the cell by osmosis (2). This dilutes the cell contents and sample.

Protocol 1. Extraction of microsamples

- microcapillary (see section 3.2.1)
- water-saturated liquid paraffin (BP grade): shake liquid paraffin with distilled water, spin in a bench centrifuge to remove undissolved water, and filter through filter paper to remove dust
- silicone oil (AS4 Wacker Chemie)
- micromanipulator (e.g. Prior or Narashige type) mounted on a hinged base to allow a pitching movement
- stereo microscope (e.g. Leitz Wild M8) with zoom up to approximately ×200)
- pneumatic control consisting of plastic tubing (1 mm diameter), 50 ml syringe barrel, and solenoid valve for tubing operated by a foot switch (*Figure 1*)

1. Fill a microcapillary with water-saturated liquid paraffin (or silicone oil) using a 10 ml plastic syringe barrel fitted via a short plastic tube to a needle from a precision pipette (e.g. CSGE type 100/250/500).
2. Attach the capillary to the fine plastic tubing (*Figure 1*) and mount onto the micromanipulator. Operation of the valve isolates the tubing from the atmosphere. Ensure that the microcapillary tip is filled with oil.

3. With the valve open and while observing under the microscope, bring the capillary up to the tissue and introduce the tip into a cell. On penetration of the cell wall, pressure in the cell will rapidly force sap into the capillary.
4. Immediately (see Section 3.3.1) remove the tip from the tissue. The sample is now present in the barrel of the capillary and ready to be assayed.

3.3.2 Sampling under solution from surface cells

If tissue must be sampled under the bathing medium (e.g. root cells) the sampling capillary must be silanized (*Protocol 2*) to prevent uptake of bathing solution into the tip (and so contamination of the sample) by capillary action when the tip is removed from the tissue. Otherwise the technique is as in *Protocol 1*. If the sap/oil boundary is seen to move relative to the capillary upon and following removal from the cell, the sample must be discarded.

Protocol 2. Silanization of microcapillaries and constriction pipettes

- dimethyl-dichlorosilane (Sigma) (the vapour of this compound is irritating to the eyes, so use it in a fume cabinet at all times)
- 200 °C oven (in a fume cabinet)
- 50 μl syringe or automatic pipette (e.g. Gilson)
- capillary holder made of an aluminium block 30 mm × 30 mm × 50 mm drilled part way through with holes 1.1 mm diameter

1. Place microcapillaries, tip uppermost, in the holder in a glass beaker covered with aluminium foil and heat in an oven at 200 °C for 2 h.
2. Inject 50 μl dimethyl-dichlorosilane through the foil and leave in the oven for a further 30 min.
3. Capillaries can be stored indefinitely.

3.3.3 Sampling with the pressure probe

Sap can be extracted using a conventional pressure probe (2) by lowering the pressure in the probe to atmospheric pressure. However, simply dropping the pressure with the probe motor may result in dilution of the cell-sap as noted above. To minimize this, the probe must be modified by introducing a solenoid valve that facilitates almost instantaneous reduction of the probe pressure to atmospheric pressure (3). Using this 'sampling pressure probe' turgor can first be measured in the normal manner; the valve is then opened and sap is forced into the capillary. The tip is removed from the tissue as above.

3.3.4 Extraction from cells up to 1 mm below tissue surface with a sampling pressure probe

Simply pushing a microcapillary through tissue will result in contamination of the sample by the contents of the overlying cells. To avoid this the probe tip is pressurized to a slightly higher degree than the cells to be sampled. The capillary tip is introduced into the tissue to the required depth, which can be measured using an LVDT (linear voltage displacement transducer) (4) or by using a speck of dirt on the capillary as a reference point against an eyepiece graticule. Then the probe sampling valve is opened and the tip rapidly (i.e. $<<1$ sec) removed from the tissue. With practice the capillary will contain sap from a cell at the correct depth.

These cells below the surface (e.g. leaf mesophyll or root cortex) present the greatest difficulties. The cell assayed can rarely be seen: its position can only be inferred from its depth and the anatomy of the tissue. (It is not generally possible to measure the turgor pressure and also extract samples from the same cell for solute analysis for these sub-surface cells.) Additionally it is not possible to determine if the tip is cleanly in a cell before sampling so variability may be greater than when sampling surface cells. At present, however, this is the only universal method for extracting sap from the deeper cells of an intact tissue. (We currently usefully enter some mesophyll and bundle sheath cells of cereals *via* the stomatal aperture. Such individual approaches clearly depend on tissue architecture.)

3.4 Sample storage and manipulation

The sample in the capillary can be expelled immediately into water-saturated liquid paraffin (oil) on an osmometer stage where it can be held for a number of hours without water loss. Subsequently samples can be stored in a fridge at 4 °C in silanized microcapillaries for longer periods of time. (It is possible to store several samples in the same tip, each separated by oil. We feel, however, that the disadvantages of this outweigh the advantages.) Store the capillaries vertically (tip upwards) at 4 °C in a humid container. Under these circumstances osmotic pressure increases by about 1% a day for a 50 pl sample. Smaller samples will have a greater loss of water. Changes in concentration can be accounted for if osmotic pressures are measured immediately on obtaining the sample and again, after storage, immediately before analysis.

4. Measurement of osmotic pressure

4.1 Outline of method

Osmotic pressure is determined by measuring the melting point of the frozen solution (the freezing point of an ideal 1 molal solution is depressed by 1.855 °C). Due to supercooling such microdroplets may need to be reduced to temperatures approaching −40 °C in order to induce freezing. Additional technical details are given in Malone and Tomos (5).

Protocol 3. Use of picolitre osmometer

- abrasive powder (0.05 μm alumina)
- water-saturated liquid paraffin (*Protocol 1*)
- freezing point depression osmometer and stereo microscope
- osmotic pressure standards (e.g. Wescor)
- microcapillaries (Section 3.2.1)
- glass coverslips

A. *Preparation of glass coverslips (5)*

1. Abrade a microscope coverslip by rubbing it over a slurry of abrasive powder on a glass plate. This provides nucleation sites for ice crystals to help avoid supercooling of sample droplets.
2. Wash the slurry off the glass with distilled water and degrease in acetone.
3. Blacken the unabraded side of the glass with a permanent marker. When dry, scratch a grid pattern in the ink with a scalpel blade. This aids visualizing the droplets later.
4. Cut the coverslip to size (approximately 5 mm × 5 mm) with a blade, polish the abraded side with lens tissue, and blow off adhering dust with an air brush.

B. *Preparation and operation of the osmometer*

1. Turn on the water supply then switch on the power and direct the dry-air supply[a] (to limit condensation on cold surfaces) through the osmometer.
2. Put a thin film of heat transfer paste (e.g. RadioSpares) over the osmometer cold stage.
3. Place the patterned coverslip (abraded side uppermost) on to the heat sink paste and press down gently with a pair of forceps.
4. Put a droplet of water-saturated liquid paraffin (approximately 5 μl) in the centre of the glass slide.

Protocol 3. *Continued*

5. Using the micro-droplet manipulation system (*Figure 1*) mounted on the micromanipulator draw distilled water into a capillary (distilled water and standards can be held in holes drilled in a Perspex block within working distance of the osmometer stage).
6. Place a ring of water droplets (each roughly 100 pl in volume) 1–2 mm in diameter on to the glass, under the oil. This will minimize abstraction of water from the samples into the oil.
7. Place sample and standard droplets inside the ring of water droplets along the white lines of the scratches on the underside of the glass. Place a fresh glass coverslip over the osmometer aperture and then direct the air dry supply over the surface of this upper coverslip to avoid condensation.
8. Turn the osmometer dial below the lowest expected osmotic pressure and put the osmometer into the freeze cycle.
9. When the temperature has equilibrated, slowly raise the temperature using the osmometer dial while observing the frozen droplets. (Switch off the dry-air supply to the coverslip as the melting point is approached, since condensation is relatively slow at temperatures near zero and the air supply interferes with accurate measurement.) For each droplet, record the reading on the thermistor/thermocouple output at which the last ice crystal melts. With practice, considerable control can be exercised over the ice crystals by small alterations in temperature, which improves the accuracy of the measurements.
10. Read the osmotic pressure for each sample from the standard curve prepared from the melting points of the standard droplets (an accuracy of ± 5 milli-osmolal can be achieved).
11. Samples can be kept on the osmometer stage at 4 °C for a number of hours. The precise length of time largely depends on how many times the air supply is directed at the surface of the cold oil. Every time the covering glass coverslip is removed for access, the air must be on to prevent excessive condensation on to the oil surface and this will necessarily speed the drying of the samples. For long experiments we suggest having on this air supply for no longer than 3–4 min per hour—depending on the ambient relative humidity.

[a] With both our own machines and our modified Clifton osmometer we use two air jets. One is directed over the surface of the oil under which the osmotic pressure is measured. This prevents condensation forming on the oil and diluting the samples. This jet is used as sparingly as possible, only being turned on when the cold oil is exposed to the atmosphere. The second jet is directed at a glass coverslip that covers the oil reservoir. This is to clear condensation that obstructs microscopic observation of the melting micro-droplets.

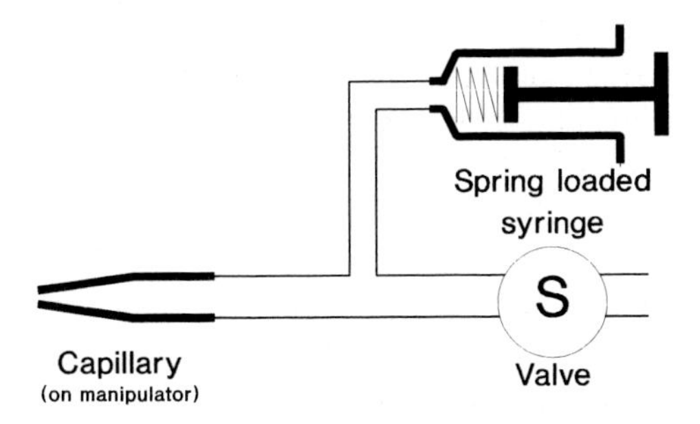

Figure 1. Sample manipulation apparatus. Samples are drawn up by closing the solenoid (S) and creating a vacuum in the syringe. At the required time the solenoid is opened and the entry of solution into the capillary stops. For expelling samples the syringe is pressurized. (A usual alternative is to replace the solenoid by the use of clenched teeth.)

5. X-ray microanalysis of dried microdroplets

5.1 Outline of method

Elements emit X-rays, the characteristics of which are element specific, when bombarded with energetic electrons. These can be used to quantify the amount of an element present in a sample. Standard methods of X-ray microanalysis of biological samples usually involve energy dispersive analysis (6) of frozen, freeze-substituted, or freeze-dried material *in situ*. These techniques are generally difficult to calibrate accurately. This can be overcome by analysis of microdroplets first extracted from the cell.

The method used involves drying droplets of fixed volume on to a thin-film support on an electron microscope grid alongside standards. The key feature, however, is the inclusion of an internal standard (e.g. RbF) in each sample droplet. The integral of the Rb L_α/L_β peak is used to normalize the data.

Sample handling is again performed under water-saturated liquid paraffin. Since the procedure of pipetting can take some hours, water loss is reduced by cooling the oil to approximately 4 °C using the osmometer stage as the support. The grid is held under a 5 mm depth of liquid paraffin in a well, constructed from aluminium. Condensation at this temperature is limited by minimal use of a stream of dry air passing over its surface.

Identical volumes of sample and internal standard (RbF in mannitol) are mixed in turn. (This 1:1 ratio can be maintained with a different pipette if the original one becomes blocked or broken.) At this stage, samples can also be 'spiked' with elements to test for interference between elements.

It is important to avoid the formation of large crystals in the final dried droplet. The drying step, which is achieved by evaporation in a stream of warm air, is therefore as rapid as possible and all precautions are taken to avoid crystallization before this step.

The Rb concentration in the standard should be similar to that of the major element expected in the sample. Mannitol provides a matrix in which the elements of interest are interspersed. In addition to the original intention of mimicking and swamping the background caused by carbon-containing components of the sample mannitol also minimizes the loss of chloride from the sample and thoroughly mixes the sample as it 'boils' under the electron beam.

5.2 Choice of sample support

We use copper folding grid (100/200 mesh) as the membrane support since it is inexpensive and, being non-magnetic, is easy to handle with stainless steel watchmaker's forceps. If signals from a copper grid interfere with sample peaks (for example a Cu peak can cause difficulties in the case of Na analysis) a range of other materials is available. A variety of membrane materials is used in electron microscopy. We routinely use a Pioloform membrane, but there is no reason why others should not work provided that they are tough enough to cope with the rigours of pipetting and sample drying and are insoluble in the solvents used to remove the liquid paraffin. In addition they must offer a surface that is sufficiently hydrophilic to allow sample deposition, yet sufficiently hydrophobic to prevent spreading of the droplet. Hyatt and Marshall (7) have successfully used a nylon film (Dupont Elvamide 8061) that fulfils the above conditions, but they do not recommend the use of Celloidin, Parlodion, or Formvar. For those not wishing to prepare their own grids some companies (e.g. Agar Scientific Ltd) offer a custom service and can prepare grids to order.

The Pioloform solution we use to prepare the grids (1 g in 100 ml solvent) is twice as concentrated as that normally used for EM work. This significantly improves both the strength and ease of preparation without any noticeable adverse effects on the background radiation.

5.3 Manipulation of samples

Protocol 4. Manipulation of samples for X-ray microanalysis

A. *Pipetting of the samples and standards*

- water-saturated liquid paraffin held in an aluminium container shallow enough to allow easy pipetting but deep enough to prevent sample evaporation (2 mm is typical)

- Pioloform-coated (from 1% solution) folding copper grids (200/100 mesh; Agar Scientific Ltd)
- samples and standards contained within constriction pipettes (*Protocol 1*) or in an oil reservoir on the osmometer stage
- rubidium fluoride (AR grade) (200 mM) prepared in saturated mannitol (AR)
- cooling stage able to maintain samples in liquid paraffin at approximately 4 °C (e.g. osmometer stage or plate cooled by water bath)
- silanized constriction pipettes (*Protocol 2*) of approximately equal volumes sufficiently small to allow the sub-sampling of extracted vacuolar samples
- standards prepared to contain roughly the concentrations of elements expected (0–120%); mixed or individual standards of each element may be used (take care to avoid precipitates)

1. Place the grid under the liquid paraffin with the membrane uppermost.
2. Deposit a sample from the storage constriction pipette (or directly from the reservoir on the adjacent osmometer) on to the Pioloform surface on the 200 mesh side of the grid. (One sample per square in a pattern that can be identified later in the scanning electron microscope.) Prepare as many replicates as required. The 100 mesh side can be useful for holding samples for sub-sampling.
3. Add an identical volume of the Rb/mannitol solution into the droplets using the same constriction pipette.
4. Repeat on fresh squares with the standard calibration solutions. (Some five replicates of five points are used).
5. Wash and dry the grid. (*Protocol 4*B).

B. *Drying of samples*

- folding grid with the pipetted samples in liquid paraffin
- water-saturated hexane (HiPerSolv grade) at 4 °C (two × 10 ml vials)
- water-saturated 2-methyl butane (isopentane) at 4 °C (one × 10 ml vial)
- hair dryer mounted on a stand
- watchmaker's forceps

Be aware of the fire/explosive hazard associated with isopentane and hexane and the use of a hair dryer. Organic solvents are also toxic: avoid breathing the vapours. Take care to avoid dust being drawn on to the samples by the forced draught of the fume cupboard.

1. Turn on the hair dryer and allow it to run for a minute to remove accumulated dust and warm up.

Protocol 4. *Continued*

2. With a watchmaker's forceps use the 100 mesh grid as a handle to submerge the sample rapidly in the chilled hexane. Rinse off the liquid paraffin by agitating the grid in the hexane for 4 sec.
3. Repeat the rinse in the second hexane bath.
4. Rinse in the isopentane. Flick the grid once quickly to remove excess isopentane (yet not allowing all isopentane to evaporate) and position the grid in the hair dryer airstream for 5 sec to evaporate (air temperature of 55 °C). (The samples can be stored indefinitely on the grids in gelatin capsules stored over silica gel. Limit shaking, however, as deformation of the film at this stage is a major cause of loss of samples.)
5. Mount the grid directly on to a carved carbon block (Section 5.4) with double sided adhesive tape, using the 100 mesh portion of the grid, and earth with colloidal graphite paste (or a sliver of aluminium foil). Do not breathe directly on to the samples to avoid rehydration and subsequent recrystallization of the samples.
6. The block can be stored upright in a stoppered tube containing a drying agent (e.g. silica gel or phosphorus pentoxide) until analysed.

5.4 X-ray analysis

In our system the sample grid is maintained at an angle of approximately 45 ° to both detector and electron beam, on a carved carbon block that allows a gap of approximately 3 mm between the grid and the carbon surface. The block is mounted on a conventional aluminium sample stub.

The dried droplet is scanned with an electron raster just large enough to cover the entire droplet. Droplets are analysed under an accelerating voltage of 14 kV for approximately 1 min. Counting rates are of the order of 1000 c.p.sec. These settings allow for the analysis of the most common elements encountered in extracted plant sap. (Higher voltages may be necessary for elements much heavier than calcium.) Some elements, particularly chlorine, are susceptible to sublimation under high beam currents. We find that experiments incorporating internal standards indicate that this loss is not a problem. There is some evidence to suggest that the presence of organic substances may stabilize the samples. Chlorine loss has also been attributed to the formation of volatile HCl and alkalinization of acidic samples by addition of LiOH may ameliorate this. Carbon coating of grids is not usually necessary, though it does aid photography of the finished grid.

5.5 Processing of results

We use two approaches to determine concentrations from the X-ray spectra.

Both utilize the internal Rb standard, which is a considerable improvement over the use of peak/background ratios for quantification. The simpler procedure is to use an atomic ratio method (7) by which the sample signals are quantified against the Rb peak of the droplet (of known Rb concentration). The relevant constant of proportionality is re-calculated each analysis session from a droplet in which the concentrations of both Rb and the elements of interest are known. The second procedure is to add an identical volume of Rb internal standard to both sample and standard droplets. A standard curve can then be prepared for each element of interest against the Rb peak (which can be an arbitrary value). The unknown is then normalized against its own Rb peak and read off the standard curve.

Care must be taken with overlapping peaks. For example, an allowance must be made for measurements of calcium if potassium is present in the sample, since the potassium K_β peak overlies the calcium K_α peak. We employ a measuring window large enough to encompass both peaks and subtract a percentage of the potassium peak integral to account for this. (For our window settings this is exactly 10%. Check this for your own machine and settings.)

6. Measurement of metabolites

6.1 Outline of method

Metabolites are analysed by means of enzymatic assays linked to the oxidation or reduction of NAD(P)(H). The change in NAD(P)H fluorescence measured is proportional to the amount of metabolite initially present in the reaction mixture. Calibration is achieved by adding identical volumes of standard solutions and of the sample to be analysed to replicate droplets of reaction mixture.

The following protocols describe modifications of commonly used enzymatic assays (4, 8) which make them suitable for analysis of samples of small volume (≥10 pl). All assays are performed under liquid paraffin to minimize water loss. The change in fluorescence is recorded at various times until the reaction is completed, i.e. until there is no further metabolite-dependent change in fluorescence. Note that the samples may contain metabolites and enzymes that interfere with some assays. Design controls accordingly.

6.2 Measurement of fluorescence

The mercury exciting lamp should be switched on at least 10 min before measurements are made and its stability should be checked by continuous measurement of a blank. (We find lamp stability to be the major source of difficulties with this technique. Ensure that the supplier replaces any lamps that are not stable for the guaranteed period—they are not cheap!) The measurement diaphragm should appear slightly larger than the excitation diaphragm which, in turn, should appear slightly larger than the droplet. Incident light intensity in the room should be minimized and use of air conditioning (or a room with stabilized temperature) is strongly recommended.

6.3 Individual assays

Of the large number of possible assays based on NAD(P)H fluorescence (4) three assays are described. That for nitrate (*Protocol 5*) is based on the catalysis by nitrate reductase (EC 1.6.6.2) (NR) of the reduction of nitrate to nitrite (NADPH → NADP). The assay for malate (*Protocol 6*) involves the reduction of malate to oxaloacetate catalysed by malate dehydrogenase (EC 1.1.1.37) (MDH)(NAD → NADH). Glutamate oxaloacetate transaminase (EC 2.6.1.1) (GOT) is used to pull over the equilibrium of the MDH reaction in the direction of NADH formation.

An assay for glucose, fructose, and sucrose (*Protocol 7*) is based on the oxidation of glucose-6-phosphate to 6-phosphogluconolactone with concurrent formation of fluorescent NADPH catalysed by glucose-6-phosphate dehydrogenase (EC 1.1.1.49) (Glc6PDH). Hexokinase (EC 2.7.1.1) (HK), phosphoglucose isomerase (EC 5.3.1.9) (PGI), and invertase (EC 3.2.1.26) are added in sequence to measure glucose, fructose, and sucrose. Note that *two* glucose phosphate molecules are derived from each sucrose.

Bovine serum albumin (Factor V) (BSA) is added to each assay to stabilize the enzymes in the micro-droplets. It is advisable to run a control blank omitting enzymes to detect components that interfere with the assays.

Protocol 5. Nitrate assay

Apparatus:

- a fluorescence microscope (e.g. Leitz Fluovert) equipped with a photometer (e.g. Leitz MPV Compact photometer and computer)
- a filter block (e.g. Letiz filter block A) with the following optical properties: excitation filter: 340–380 nm; splitting dichromatic mirror: 400 nm; suppression filter: 430 nm
- constriction pipettes of a range of volumes (Section 3.2.2)
- micro-pipetting system (*Protocol 1*)
- wells made by gluing aluminium rings (made by cutting 2 cm diameter aluminium tubing into 1–3 mm sections) to glass microscope slides with Araldite (Ciba-Geigy Plastics, Cambridge)—these may be used repeatedly; between assays they should be washed in 1% Decon 90, followed by rinses in distilled water and propan-1-ol and dried at 80 °C; immediately before use clean with a lens tissue and air brush

Chemicals:

- reagents used are of analytical grade; all enzymes can be obtained from Boehringer and water is purified by reverse osmosis and deionization
- water-saturated liquid paraffin (*Protocol 1*)
- NADPH (tetrasodium salt, 98% Boehringer)
- TEA (triethanolamine hydrochloride)
- KOH
- BSA
- Nitrate reductase (NR)
- FAD (added as a cofactor of NR)

Solutions:

- buffer: 480 mM TEA, adjusted to pH 7.6 with KOH (this solution is stable for up to 4 weeks at 4 °C)
- assay cocktail: a 100 μl stock is made by mixing 20 μl buffer, 15 μl 12 mM NADPH[a], 10 μl 40 μM FAD, 10 μl 1% (w/v) BSA, and 45 μl water (make this solution fresh daily and keep at 4 °C)

Method:

1. Fill the well with water-saturated liquid paraffin (*Protocol 1*).
2. Using a 'Hamilton'-type microsyringe, expel drops (approximately 1–2 μl) of assay solutions and standards near the edge of the well. (These provide reservoirs from which appropriate volumes may later be taken with a constriction pipette.)
3. Pipette the required number of droplets of assay cocktail with a pipette of approximately 4 nl volume. (The number required is determined by the number of samples and standards and their replicates.)
4. Add standards with a silanized constriction pipette (approximately 12 pl). Standards may contain up to 300 mM nitrate.
5. Transfer the sample from a storage microcapillary or from the osmometer stage to the well.
6. Add aliquots of the sample to the assay cocktail droplets using the same constriction pipette as used for the standards.
7. Record the fluorescence of each droplet.
8. Add identical volumes (approximately 500 pl) of nitrate reductase (0.44 U/ml) (the initiating factor) to each droplet.
9. Incubate for 40–80 min at room temperature. Since the rate of reaction varies with enzyme concentration, and since pipette volumes are

Protocol 5. *Continued*

inherently variable, it is advisable to measure the decline in fluorescence at intervals of 20 min in order to establish when the nitrate-dependent reaction is complete. (There may be a nitrate-independent oxidation of NADPH. This can be quantified readily from the time course of reaction.)

[a] This is stable for 2 weeks when made up in 1% (w/v) $NaHCO_3$.

Protocol 6. Malate assay

- apparatus as *Protocol 5*
- water-saturated liquid paraffin (*Protocol 1*)
- glycylglycine (Boehringer)
- glutamic acid (Sigma)
- NAD (free acid, grade 2; Boehringer)
- BSA
- enzymes: MDH and GOT (both in $(NH_4)_2SO_4$; Boehringer)

Solutions:

- buffer: 60 mM glycylglycine + 160 mM glutamic acid, adjusted to pH 10.0 with KOH (this is stable for 1 month at 4 °C)
- assay cocktail: for 470 µl solution mix 240 µl buffer, 50 µl 1% (w/v) BSA, 100 µl MDH (400 U/ml in 0.1% BSA), 80 µl GOT (58 U/ml in 0.1% BSA) (make this solution fresh daily and keep at 4 °C)

Method:

1–7. As *Protocol 5*. (Standards up to 50 mM).

8. Start the reaction by adding identical volumes (approximately 500 pl) of 45 mM NAD (initiating factor) to each droplet.

9. Record the fluorescence for 5–20 min until completion.

Protocol 7. Glucose/fructose/sucrose assay

- apparatus as *Protocol 5*
- ATP
- BSA

- enzymes: hexokinase/glucose-6-phosphate dehydrogenase (HK/Glc6PDH), invertase, and phosphoglucose isomerase (PGI) (Boehringer)
- $MgSO_4$
- Mes 2[*N*-Morpholino]ethanesulphonic acid
- NADP (disodium salt, 98%)
- TEA (triethanolamine hydrochloride)

Solutions:

- buffer I: 480 mM TEA adjusted to pH 7.6 with KOH (stable for up to 4 weeks at 4 °C)
- buffer II: 20 mM Mes adjusted to pH 5.5 with KOH (stable for up to 4 weeks at 4 °C)
- assay cocktail A (glucose assay): for 400 μl of solution mix 40 μl buffer I, 65 μl $MgSO_4$ (70 mM), 40 μl BSA (1% (w/v)), 10 μl HK/Glc6PDH (280/140 U/ml), and 245 μl water
- assay cocktail B (assay for fructose + glucose): for 200 μl of solution mix 197.5 μl assay cocktail A and 2.5 μl PGI (700 U/ml)
- assay cocktail C (assay for fructose + glucose + 2 × sucrose): for 200 μl of solution mix 50 μl buffer II, 33 μl $MgSO_4$ (70 mM), 20 μl BSA (1% (w/v)), 5 μl HK/Glc6PDH (280/140 U/ml), 2.5 μl PGI (700 U/ml), invertase (120 U in 89.5 μl water)
- assay initiation solution: 50 mM NADP, 50 mM ATP made up in buffer I

Method:

1–2. As *Protocol 5*.

3. Pipette on to the slide several identical volumes (approximately 4 nl) of assay cocktails A, B, and C.

4. Add a series of identical volumes of standard (approximately 12 pl and concentration up to 100 mM hexose[a] equivalents) to droplets of solutions A, B, and C, using a silanized pipette. Using the same pipette, add samples to other droplets of solutions A, B, and C.

5. Record the fluorescence for each droplet. (Note that suppliers occasionally add glucose to stabilize invertase. This is accounted for by the zero value of the standard curve.)

6. After 5 min, start the reaction by addition of identical volumes of assay initiation solution (approximately 500 pl) to each droplet in turn.

7. Record the fluorescence of each droplet after 10–15 min, by which time it will have reached completion.

[a] If higher concentrations need to be measured a higher concentration of NADP may be required.

Acknowledgements

We are grateful to Drs Mike Malone, John Williams, and Hans-Werner Koyro for their key roles in establishing these protocols, to Dr Roger A. Leigh for his support throughout, and to Professor Dr Bill Outlaw for his inspiration. The development of protocols and associated work have been supported by several Link Research Group AFRC grants to A. D. T. and R. A. L.

References

1. Tomos, A. D. and Wyn Jones, R. G. (1988). In *Solute transport in plant cells and tissues* (ed. D. A. Baker and J. L. Hall), pp. 220–50. Longman Scientific, London.
2. Hüsken, D., Steudle, E., and Zimmermann, U. (1978). *Plant Physiol.*, **61**, 158.
3. Malone, M., Leigh, R. A., and Tomos, A. D. (1989). *Plant Cell Environ.*, **12**, 919.
4. Lowry, O. H. and Passonneau, J. V. (1972). *A flexible system of enzymatic analysis*. Academic Press, New York.
5. Malone, M. and Tomos, A. D. (1992). *J. Exp. Bot.*, **43**, 1325.
6. Morgan, A. J. (ed.) (1985). *X-ray microanalysis in electron microscopy for biologists*, Microscopy handbook 5. Oxford University Press/Royal Microscopical Society.
7. Hyatt, A. D. and Marshall, A. T. (1985). *Micron Microsc. Acta*, **16**, 39.
8. Boehringer (1980). *Methods of enzymatic food analysis*. Boehringer Mannheim GmbH.

A1

List of suppliers

Agar Scientific Ltd, 66A Cambridge Rd, Stansted, Essex CM24 8DA, UK.
American Cyanamid Co., Bound Brook, NJ, USA.
Amersham International plc, Amersham Place, Little Chalfont, Amersham, Bucks HP7 9NA, UK.
Amersham Life Sciences, Amersham Place, Little Chalfont, Amersham, Bucks HP7 9NA, UK.
BDH, Merck Ltd, Merck House, Poole, Dorset BH15 1TD, UK.
Bethesda Research Labs/Gibco. 8400 Helgerman Court, PO Box 6009, Gaithersburg, MD20884–9980, USA.
Bio-Cell Research Laboratories, Cardiff Business Technology Centre, Senghenydd Rd, Cardiff CF2 4AY, UK.
Bio-Rad Laboratories Ltd, Bio-Rad House, Maylands Ave., Hemel Hempstead, Herts HP2 7TD, UK.
Bio-Rad Laboratories, 2000 Alfred Nobel Drive, Hercules, CA 94547, USA.
Bio-Rad, 3300 Regatta BLVD, Richmond, CA 94804, USA.
Bio-Rad Microscience Division, Bio-Rad House, Maylands Ave, Hemel Hempstead, Herts HP2 7TD, UK.
Biosuppliers, PO Box 835, University of Melbourne, Parkville, 3052 Vic Australia.
Boehringer Mannheim UK Ltd, Bell Lane, Lewes, East Sussex BN7 1LG, UK.
Calbiochem Corporation, PO Box 12087, San Diego, CA 92112–4180.
Calbiochem Corporation (Novabiochem (UK) Ltd, 3 Heathcoat Building, Highfields Science Park, University Boulevard, Nottingham NG7 2QJ, UK.
Ciba-Geigy Plastics, Duxford, Cambridge, UK.
Citifluor Ltd, Connaught Building, City University, Northampton Square, London EC1V 0HB, UK.
Clark Electromedical Instruments, PO Box 8, Pangbourne, Reading RG8 7HU, UK.
Clontech Laboratories Inc., 4030 Fabian Way, Palo Alto, CA 94303–9605, USA.
Dako Ltd. 16 Manor Courtyard, Hughenden Ave, High Wycombe, Bucks HP13 5RE, UK
Dupont (UK) Ltd, Wedgwood Way, Stevenage, Herts SG1 4QN, UK.
Dupont, Barley Mill Plaza, Wilmington, DE 19898, USA.

Eastman-Kodak Ltd, Acornfield Rd, Knowsley Industrial Park North, Liverpool L33 7UF, UK.
Emitech Inc, 3845 FM 1960-West, Suite 345, Houston, TX 77068, USA.
Emitech Ltd, 11 Enterprise Centre, Newton Rd, Ashford, Kent TN24 0PD, UK.
Fluka Chemicals Ltd, The Old Brickyard, New Rd, Gillingham, Dorset SP8 4BR, UK.
Genencor Inc., Baron Steuben Place, Corning, NY 14831, USA.
Gurr (Merck BDH). Merck House, Poole, Dorset BH15 1TD, UK.
Hilgenberg GmBH. D–3509, Malsfeld, Germany.
ICN Biomedicals Ltd, Lincoln Rd, Cressex Industrial Estate, High Wycombe, Bucks HP12 3XJ, UK.
ICN/Flow Labs (ICN Biomedicals Ltd), Eagle House, Peregrine Business Park, Gomm Rd, High Wycombe, Bucks HP13 7DL, UK.
Instrumedics Inc., 521 Warwick Ave, Teaneck, NJ, 07666, USA.
Kanematsu-Gosha Incorp., 3333 South Hope Street, Suite 2800, Los Angeles, CA 90071, USA.
London Resin Company,PO Box 34, Basingstoke, Hants RG21 2NW, UK.
Merck Ltd (BDH) Merck House, Poole, Dorset BH15 1TD, UK.
Merck, D-6100, Darmstadt, Germany.
Molecular Probes Inc., PO Box 22010, 4849 Pitchford Ave, Eugene, OR 97402–0414 USA.
Molecular Probes Inc. (Cambridge Bioscience) 25 Signet Court, Newmarket Rd, Cambridge CB5 8LA, UK.
Narashige Europe Ltd, Unit 7, Willow Business Park, Willow Way, London SE26 4QP, UK.
Narashige USA, Inc., One Plaza Rd, Greenvale, New York 115448, NY, USA.
Nikon Japan, Fuji Building 2–3, Marunouchi 3–Chome, Chiyoda-Ku, Tokyo 100, Japan.
Nikon UK Ltd, Haybrook, Halesfield 9, Telford, Shropshire TF7 4EW, UK.
Omega Optical Inc., PO Box 573, Bratlleboro, VT 05302–0573, USA.
Pelco International, PO Box 492477, Redding, CA 96049–2477, USA.
Pharmacia Inc., 800 Centennial Ave, Piscataway, NJ 08854, USA.
Pharmacia, Davy Ave, Milton Keynes, Bucks MK5 8PH, UK.
Pharmacia, S751 04 Uppsala 1, Sweden.
Pierce, PO Box 117, Rockford, IL 61105, USA.
Polysciences Inc., 400 Valley Rd, Warrington, PA 18976–2590, USA.
Polysciences Ltd, 24 Low Farm Place, Moulton Park, Northampton NN3 1HY, UK.
Prior Scientific Instruments Ltd. Unit 4, Wilbraham Rd, Fulbourn, Cambridge CB1 5ET, UK.
Promega Corp., 2800 Woods Hollow Rd, Madison, WI 53711–5399, USA.

Rone-Poulenc Laboratory Products, Liverpool Rd, Eccles, Manchester M30 7RT, UK.
Savant Inc., Hicksville, NY, USA.
Schleicher and Schuell, Postfach 4, D-3354, Dassell, Germany.
Scientific Laboratories Supplies Ltd., Unit 27, Nottingham South Industrial Estate, Wilford, Nottingham, NG11 7EP, UK.
Sera Labs Ltd, Crawley Down, Sussex RH10 4FF, UK.
Serva Biochemicals, 50 A & S Drive, Paramus, NJ 07652, USA.
Sharpes Seeds, Sleaford, Lincs, UK.
Sigma Chemical Co., Fancy Rd, Poole, Dorset BH17 7NH, UK.
Sigma Chemical Co., PO Box 14508, St Louis, Missouri, USA.
SPI Supplies, Toronto, Canada.
Stratagene Inc., 11011 North Torrey Pines Rd, La Jolla, 92037 CA, USA.
Surgipath Medical Industries Ind., PO Box 769, Ziegler Drive, Graylake, IL 60030, USA.
TAAB Laboratories Equipment Ltd, 3 Minerva House, Calleva, Reading RG7 4RW, UK.
Tissue Tek./Miles Inc., Elkhart, IN 46515, USA.
Vector Laboratories, Inc., 30 Ingold Rd, Burlingame, CA 94010, USA.
Vector Laboratories (UK). 16 Wulfric Square, Bretton, Peterborough PE3 8RF, UK.
Vit-Labs/Vitri Friedrich-Ebert Str 33–35, D-6104, Seeheim-Jugenheim, Germany.
Whatman Labsales Ltd., St Leonard's Rd, Maidstone, Kent ME16 0LS, UK.
Whatman Scientific Ltd, St Leonard's Rd, 20/20 Maidstone, Kent ME16 0LS, UK.
Wilson Sieves, D. Wilson, 2 Long Acre, Common Lane, Hucknall, Nottingham NG15 6QD, UK.
World Precision Instruments, 375 Quinnipac Ave, Newhaven, CT 06513, USA.
Yakult Honsha Co. (Yakult Pharmaceutical Co. Ltd) 1–1–19 Higashi Shinbashi Minato-Ku, Tokyo, 105, Japan.
Zeiss UK, PO Box 78, Woodfield Rd, Welwyn Garden City, Herts AL7 1LU, UK.

Index

Index

Index